T0336013

GALILEO'S MEDICEAN MOONS:
THEIR IMPACT ON 400 YEARS OF DISCOVERY

IAU SYMPOSIUM No. 269

COVER ILLUSTRATION: *Galileo and the Medicean Moons*
composed by Elisa Segato

INTERNATIONAL ASTRONOMICAL UNION

UNION ASTRONOMIQUE INTERNATIONALE

GALILEO'S MEDICEAN MOONS: THEIR IMPACT ON 400 YEARS OF DISCOVERY

PROCEEDINGS OF THE 269th SYMPOSIUM OF THE INTERNATIONAL ASTRONOMICAL UNION HELD IN PADOVA, ITALY JANUARY 6–9, 2010

Edited by

CESARE BARBIERI

Department of Astronomy, University of Padua - Padova, Italy

SUPRIYA CHAKRABARTI

Center for Space Physics, Boston University - Boston, USA

MARCELLO CORADINI

European Space Agency - Paris, France

and

MONICA LAZZARIN

Department of Astronomy, University of Padua - Padova, Italy

CAMBRIDGE
UNIVERSITY PRESS

University Printing House, Cambridge CB2 8BS, United Kingdom

One Liberty Plaza, 20th Floor, New York, NY 10006, USA

477 Williamstown Road, Port Melbourne, VIC 3207, Australia

314-321, 3rd Floor, Plot 3, Splendor Forum, Jasola District Centre, New Delhi - 110025, India

79 Anson Road, #06-04/06, Singapore 079906

Cambridge University Press is part of the University of Cambridge.

It furthers the University's mission by disseminating knowledge in the pursuit of education, learning and research at the highest international levels of excellence.

www.cambridge.org
Information on this title: www.cambridge.org/9780521195560

First published 2010

A catalogue record for this publication is available from the British Library

ISBN 978-0-521-19556-0 Hardback

Table of Contents

Papers

Posters

Preface

Magnifico Rettore, Sindaco della Città di Padova, Autorità, dear Colleagues and students.

It is my greatest pleasure to welcome you in Padova. We have three intense days of work ahead of us, with a large number of oral papers and poster presentations, the surest indication of the interest aroused by the Symposium nr. 269 of the International Astronomical Union. Indeed the first IAU symposium of 2010, but also one of the great events which characterized the International Year of Astronomy. Our great predecessor in this University, Galileo Galilei, has been celebrated in many ways all over the world, and this Symposium will indicate how important his discoveries of 1609-1610 were in the subsequent 400 years, and will be for the several decades of already planned researches from the ground and from space. Many Institutions and persons have contributed to the success of the Symposium, in primis the International Astronomical Union, the European Space Agency ESA, the Committee on Space Research COSPAR, the Italian Space Agency ASI and the Italian National Institute for Astrophysics INAF. It is my pleasure to acknowledge the support received from the University and the Galilean School of Higher Education. As you have seen from the program, the concert will be played by students of our University; Padova has a wonderful musical tradition, with an impressive number of Internationally renowned groups and musicians. We felt however that our young students should be given the opportunity to perform in front of such distinguished audience of scientists coming from all over the world. Although busy with their exams and theses, they accepted with enthusiasm our proposal. Maestro Terrel Stone, one of the world leading liute player, also volunteered to open the concert with two pieces, one quite possibly from Galileo himself and the other from his brother Michelangelo. Padova hosts a scientific and technological park carrying the name of Galileo, http://www.galileopark.it/. This park promotes innovation and transfer of technologies to new, small enterprises of the territory, enterprises often sprung out of novel ideas of our students. The Galileo Park has given five students the opportunity to register at the Symposium. The Mayor of Padova, Mr. Flavio Zanonato, offered the Palazzo San Gaetano facilities. The firm Salmoiraghi and Viganò and the Association of the Friends of the University have also generously contributed. My sincerest thanks to the many persons which have supported the organization of the Symposium, first of all Dr. Monica Lazzarin, who chaired the Local Organizing Committee and coordinated the several students who voluntarily assisted the participants. Dr. Elisa Segato made a very artistic composition for the conference poster and website cover. Silvia Roi and Stefano Salvadori were responsible for the web site www.astro.unipd./galileo. The site contains the whole program and some oral contributions which did not came in time to be included in the Proceedings. This book could not have been produced without the great skill and dedicated effort of Silvia Cervesato.

Cesare Barbieri,
Chairman of the Scientific Organizing Committee
University of Padua, Italy, 6 January, 2010

THE ORGANIZING COMMITTEE

Scientific, Chairs of SOC

C. Barbieri, University of Padua (Italy)
A. Coradini, INAF Rome (Italy)
M. Mendillo, Boston University (USA)
T. Owen, University of Hawaii (USA)

Members of SOC

S. Atreya, University of Michigan (USA)
A. Barucci, Observatoire de Meudon (France)
S. Chakrabarti, Boston University (USA)
M. Coradini, European Space Agency (France)
G. Coyne SJ, Specola Vaticana (Vatican City State)
T. Encrenaz, Observatoire de Paris (France)
B. Foing, ESA/ESTEC (The Netherlands)
M. Grande, University of Wales (United Kingdom)
D. Hall, University of Hawaii (USA)
W. Ip, National Centr. University (China Taipei)
T. V. Johnson, Jet Propulsion Laboratory (USA)
H. U. Keller, MPI-Solar System Res. (Germany)
S. Okano, Tohoku University (Japan)
R. Pappalardo, Jet Propulsion Laboratory (USA)
Nick Schneider, University of Colorado (USA)
R. Williams, Space Telescope Sc. Inst. (USA)
A. Zakharov, Space Res. Inst. (IKI) (Russia)

Local

M. Lazzarin (chair)
E. Segato
L. Vanzan
M. Gazzola
G. Rodeghiero
N. Baccichet
D. Cannone

Acknowledgements

The Symposium is sponsored and supported by the IAU Divisions III (Planetary Systems Sciences), I (Fundamental Astronomy), (IX) Optical and Infrared Telescopes, XI (Space & High Energy Astrophysics), (XII) Union-Wide Activities.

The Local Organizing Committee operated under the auspices of the University of Padua.

Funding by the
International Astronomical Union,
European Space Agency ESA,
Committee for Space Research COSPAR,
Association of the Friends of the University of Padua,
Comune di Padova,
Department of Astronomy of the University of Padua,
Parco Scientifico Tecnico Galileo di Padova,
Accademia Galileiana di Scienze e Lettere ed Arti,
Camera di Commercio di Padova,
and
Salmoiraghi e Viganò,
is gratefully acknowledged.

Participants

Kaare **Aksnes**, University of Oslo, Norway kaare.aksnes@astro.uio.no
Sushil **Atreya**, University of Michigan, USA atreya@umich.edu
Nicola **Bacchichet**, University of Padua, Italy nicola.baccichet@hotmail.it
Cesare **Barbieri**, University of Padua, Italy cesare.barbieri@unipd.it
Maria A. **Barucci**, Observatoire de Paris, France antonella.barucci@obspm.fr
Leopoldo **Benacchio**, INAF OAPD, Italy leopoldo.benacchio@oapd.astro.it
Ivano **Bertini**, University of Padua, Italy ivano.bertini@unipd.it
Francesco **Bertola**, University of Padua, Italy francesco.bertola@unipd.it
Antonio **Bianchini**, University of Padua, Italy antonio.bianchini@unipd.it
Michel **Blanc**, École Polytechnique de Paris, France michel.blanc@polytechnique.edu
Carlo **Blanco**, Catania University, Italy cblanco@oact.inaf.it
Scott **Bolton**, SWRI, USA sbolton@swri.edu
Roger M. **Bonnet**, ISSI, Suisse silvia.wenger@issibern.ch
Wilfried **Bornemann**, EADS Astrium GMBH SAT, Germany wilfried.bornemann@astrium.eads.net
Alexandre **Bortoletto**, Laboratorio Nacional de Astrofisica, Brazil abortoletto@lna.br
Giovanni **Busetto**, University of Padua, Italy giovanni.busetto@unipd.it
Massimo **Calvani**, INAF INAF OAPD, Italy massimo.calvani@oapd.inaf.it
Dario **Cannone**, University of Padua, Italy dario.rigel@hotmail.it
Enrico **Cappellaro**, INAF INAF OAPD, Italy enrico.capellaro@oapd.inaf.it
Gianfranco **Cariolaro**, University of Padua, Italy cariolar@dei.unipd.it
Benedetta L. S. **Carnaghi**, Scuola Galileiana di Studi Superiori, Italy benny winnie green@hotmail.it
John **Casani**, Jet Propulsion Laboratory, USA jrcasani@jpl.nasa.gov
Stefano **Casotto**, University of Padua, Italy stefano.casotto@unipd.it
Sandro **Centro**, University of Padua, Italy centro@pd.infn.it
Supriya **Chakrabarti**, Boston University, USA supc@bu.edu
Sze L. **Cheung**, Ho Koon Astronomical Center, Hong Kong cheung.szeleung@physics.org
Theodore **Clarke**, SWRI, USA tcclarke@earthlink.net
Guy **Consolmagno**, Specola Vaticana, Vatican City gjc@specola.va
Angioletta **Coradini**, INAF Roma, Italy angioletta.coradini@ifsi-roma.inaf.it
Marcello **Coradini**, ESA, France marcello.coradini@esa.int
Enrico M. **Corsini**, University of Padua, Italy enricomaria.corsini@unipd.it
George **Coyne**, S.J., Vatican Observatory, Vatican City gcoyne@as.arizona.edu
Gabriele **Cremonese**, INAF OAPD, Italy gabriele.cremonese@oapd.inaf.it
Elena **Dalla Bontá**, University of Padua, Italy elena.dallabontá@unipd.it
Sandro **D'Odorico**, ESO, Germany sdodoric@eso.org
Therese **Encrenaz**, Paris Observatory, France therese.encrenaz@obspm.fr
Enrico **Flamini**, ASI, Italy enrico.flamini@asi.it
Bernard **Foing**, ESA/ESTEC, The Netherlands Bernard.Foing@esa.int
Sonia **Fornasier**, Université de Paris, France Sonia.Fornasier@obspm.fr
Marcello **Fulchignoni**, Obs. Paris Univ. Denis Diderot Paris 7, France marcello.fulchignoni@obspm.fr
Giuseppe **Galletta**, University of Padua, Italy giuseppe.galletta@unipd.it
Marco **Gazzola**, University of Padua, Italy marco.gazzola.1@studenti.unipd.it
Serena **Gradari**, University of Padua, Italy serena.gradari@unipd.it
Manuel **Grande**, IMAPS Aberystwyth University, UK mng@aber.ac.uk
Cesare **Grava**, University of Padua, Italy cesare.grava@unipd.it
Wing **Ip**, Huen Institute of Astronomy, National Central University, Taiwan wingip@astro.ncu.edu.tw
Torrence **Johnson**, Jet Propulsion Laboratory, Caltech USA Torrence.V.Johnson@jpl.nasa.gov
Horst U. **Keller**, Max Planck Institut für Sonnensystemforschung, Germany keller@linmpi.mpg.de
Margaret **Kivelson**, University of California, USA mkivelson@igpp.ucla.edu
Monica **Lazzarin**, University of Padua, Italy monica.lazzarin@unipd.it
Rosaly **Lopes**, Jet Propulsion Laboratory, USA Rosaly.M.Lopes-Gautier@jpl.nasa.gov
Enrico **Lorenzini**, University of Padua, Italy enrico.lorenzini@unipd.it
Dmitrij **Lupishko**, Institute of Astronomy of Kharkiv V.N. Karazin National University, Ukraine lupishko@astron.kharkov.ua
Sara **Magrin**, University of Padua, Italy sara.magrin@unipd.it
Simone **Marchi**, University of Padua, Italy simone.marchi@unipd.it
Elena **Martellato**, University of Padua, Italy elena.martellato@oapd.inaf.it
Sabino **Matarrese**, University of Padua, Italy matarrese@pd.infn.it
Karen **McBride**, University of California, Los Angeles USA space.mcbride@gmail.com
Michael **Mendillo**, Boston University, USA mendillo@bu.edu
Alessandro **Morbidelli**, Nice Observatory, France Alessandro.MORBIDELLI@obs-nice.fr
Giampiero **Naletto**, University of Padua, Italy naletto@dei.unipd.it
Keith **Noll**, Space Telescope Science Institute, USA noll@stsci.edu
Benoit **Noyelles**, University of Namur, Belgie benoit.noyelles@fundp.ac.be
Shoichi **Okano**, Tohoku University, Japan okano@pparc.gp.tohoku.ac.jp
Alessandro **Omizzolo**, Vatican Observatory, Italy alessandro.omizzolo@oapd.inaf.it
Sergio **Ortolani**, University of Padua, Italy ortolani@pd.astro.it
Tobias **Owen**, University of Hawaii, USA owen@ifa.hawaii.edu
Robert **Pappalardo**, Jet Propulsion Laboratory, California Institute of Technology, USA robert.pappalardo@jpl.nasa.gov
Giulio **Peruzzi**, University of Padua, Italy peruzzi@pd.infn.it
Carl **Pilcher**, NASA Astrobiology Institute, USA carl.b.pilcher@nasa.gov
Giampaolo **Piotto**, University of Padua, Italy giampaolo.piotto@unipd.it
Alessandro **Pizzella**, University of Padua, Italy alessandro.pizzella@unipd.it
Salvatore **Pluchino**, INAF IRA, Italy s.pluchino@ira.inaf.it
Zhanna **Pozhalova**, Research Institute Nikolaev Astronomical Observatory, Ukraine pza1900@mail.ru
Louise **Prockter**, Johns Hopkins University Applied Physics Laboratory, USA Louise.Prockter@jhuapl.edu
Daiana **Ribeiro Bortoletto**, Laboratorio Nacional de Astrofisica, Brasile dbortoletto@lna.br
Alberto **Righini**, University of Florence, Italy alberto.righini@unifi.it
Gabriele **Rodeghiero**, University of Padua, Italy gabriele.rodeghiero@studenti.unipd.it
Daniela **Saadeh**, University of Padua Scuola Galileiana di Studi Superiori, Italy capt.kirk@inwind.it
Piero **Salinari**, INAF ARCETRI, Italy salinari@arcetri.astro.it
Guido **Sangiovanni**, ASI, Italy guido.sangiovanni@comune.vanzago.mi.it
Paul **Schenk**, NASA, USA schenk@lpi.usra.edu
Nick **Schneider**, University of Colorado, USA nick.schneider@lasp.colorado.edu
Elisa **Segato**, University of Padua, Italy elisa.segato@unipd.it
Peter **Smith**, University of Arizona, USA psmith@lpl,arizona.edu
Dava **Sobel**, Discover Magazine, USA ds23@optonline.net
Carlo G. **Sodema**, University of Padua and Accademia Galileiana, Italy galileiana@libero.it
Roberto **Somma**, Thales Alenia Space, Italy roberto.somma@thalesaleniaspace.com
Daniele **Sorini**, University of Padua Scuola Galileiana di Studi Superiori, Italy caparezza89@alice.it

Fabrizio **Tamburini**, University of Padua, Italy fabrizio.tamburini@unipd.it
Msssimo **Tarenghi**, ESO, Chile mtarengh@eso.org
Federico **Tosi**, INAF IFSI, Italy federico.tosi@ifsi-roma.inaf.it
Robert **Williams**, STSCI, USA wms@stsci.edu
Alexander **Zakharov**, Space Research Institute, Russia zakharov@iki.rssi.ru
Lev **Zelenyi**, Space Research Institute, Russia lzelenyi@iki.rssi.ru

CONFERENCE PHOTOGRAPHS
(photos by Matteo Danesin, info@matteodanesin.it)

Figure 1. Registration under the severe eyes of Galileo from his chair.

Figure 2. The Rector of the University of Padua, Prof. Giuseppe Zaccaria (top panel), and the President of IAU, Prof. Roger J. Williams (bottom panel), greet the authorities and the participants in the magnificent Aula Magna of the University of Padua.

Figure 3. The Director of the Galilean School of Excellence, Prof. Carlo Umiltà, and the Chairman of the Scientific Committee, Prof. Cesare Barbieri, greet the participants to the Concert (top panel). A moment of the concert, a piano solo (bottom panel).

Figure 4. A view of the conference dinner held in Palazzo Zacco (Army's Officers Club), and the gorgeous cake.

Figure 5. An intense expression of Maestro Terrell Stone playing a composition by Michelangelo Galilei on his lute.

Figure 6. The beautiful images of the Medicean Moons prepared by P. Schenk, displayed in the Agorà of Palazzo San Gaetano.

Address by the Rector of the University of Padua Prof. Giuseppe Zaccaria

Authorities, Dear Collegues, Dear speakers, Ladies and gentlemen On behalf of the University of Padua, I am very happy to welcome you to this prestigious Symposium of the International Astronomical Union.

The discovery of the Medicean Moons by Galileo Galilei took place in Padova from January 7 to 15, 1610. The discovery added four new worlds to the known solar system and blew the Earth-centered universe. *Sidereus Nuncius* not only had a major influence on the development of astronomy and science, but also upon religious and philosophical theories and social behaviour.

The conference, taking place in the very same place of the discovery and exactly 400 years later, will address the following themes:

- The discovery of the Medicean Moons, their history and influence on science and humanity
- The Medicean Moons, Jupiter's system, the legacy of NASA's Galileo mission, future missions to Jupiter
- Our solar system after Galileo Galilei, the grand vision
- New telescopes, new solar systems, new people out there?

The main aims are:

- to recall the memorable event and examine its influence on science and humanities
- to examine the status of present knowledge of Jupiter, the Medicean Moons, and the full Jovian system, including prospects for advancing our understanding by future space missions and future telescopes
- to expound the contemporary vision of our solar system, of the many extra-solar planetary systems, and the expectations of discovering new intelligent inhabitants beyond our solar system.

Galileo and the natural philosophers of the 17th century envisioned the planets and moons in our solar system to be habitable worlds.

By the mid-20th century, the view had changed: the Plurality of Worlds applies to objects in orbits around stars elsewhere in the Galaxy.

Today, at the beginning of the 21st century, ongoing and planned space missions to our Moon, to Mars, to Jupiter and the Medicean Moons, to Saturn and Enceladus, again reveal a fascination with alien life in our solar system, but this time the focus is on potential host sites for extremely primitive forms of life nurtured by the presence of water. Water and other crucial molecules for sustaining life will undoubtedly be discovered on some of these neighboring worlds by the Extremely Large Telescopes and dedicated orbital missions of the next decades. The prospects for finding Earth-like, habitable, and perhaps even inhabited worlds are real and clearly before us. Galileo's discovery of the Medicean Moons thus continues to point the way towards a deeper understanding of our place in the Universe - but now from the linked perspectives of the physical and life sciences.

The conference is attended by many distinguished scientists from all Europe, Russia, Ukraine, the United States of America, Brazil, Hong Kong, Taiwan.

Among them, two 'laureati honoris causae' of our University, Father George Coyne S.J. (Specola Vaticana) and Dr. Torrence V. Johnson (Jet Propulsion Laboratory, California Insitute of Technology).

The conference therefore not only celebrates a crucial event in the past history of our University, but also underlines its continued primary role in ongoing and future astronomical researches.

I thank you all for you participation in this important scientific event and wish you a very fruitful Symposium and a pleasant story in Padua.

Giuseppe Zaccaria, Rector of the University of Padua
Padova, 6 January 2010

Address by the President of the International Astronomical Union
Prof. Robert Williams

Galileo's observations of Jupiter's satellites are among the most important astronomical observations ever made. With one simple observation he demonstrated that 2000 years of belief was incorrect and the earth was not the center of all objects in the universe. Some ideas die hard, especially when they represent established dogma. Thus, Galileo's observations and conclusions met serious resistance. As Simon & Garfunkel wrote in a popular song from the 1960s: "A man hears what he wants to hear and disregards the rest." We must acknowledge that even we scientists are not immune to our own dogma. We have similar tendencies, but there is an important difference. The difference in science is that questioning and doubt and re-evaluation of conclusions, based on new data, are accepted and expected. They are valued as a part of the process of understanding. Getting the public to appreciate this is one of the most important aspects of science.

We are now 400 years after Galileo's observations of the Medicean satellites, and these satellites are once again the focus of attention for another important question: might life exist in one of them or is life unique to earth. There is as much dogma about this question as there was about the geocentric universe in Galileo's time. The answer to this question would have implications as profound as Galileo's demonstration that the earth was not the center of the universe. We will not be able answer that question this week.

This is the fascination of astronomy. We address important questions. The entire Jovian system has remained a subject of intense study in astronomy. Whether comets strike it or it becomes the archetype of exoplanets, which we call 'hot Jupiters'. The IAU is pleased to sponsor this meeting as a means for understanding Jupiter's satellites better, and on behalf of the IAU I am pleased to add my welcome to you as participants.

Robert Williams, President of the IAU
Padova, 6 January 2010

Papers

Galileo's Medicean Moons: their impact on 400 years of discovery
Proceedings IAU Symposium No. 269, 2010
C. Barbieri, S. Chakrabarti, M. Coradini & M. Lazzarin, eds.

doi:10.1017/S1743921310007192

Galileo's telescopic observations: the marvel and meaning of discovery

George V. Coyne, S. J.
Vatican Observatory
Vatican City
email: gcoyne@as.arizona.edu

Abstract. During the very last year of what he himself described "as the best [eighteen] years of his life" spent at the University of Padua, Galileo first observed the heavens with a telescope. In order to appreciate the marvel and the true significance of those observations we must appreciate both the intellectual climate in Europe and the critical intellectual period through which Galileo himself was passing at the time those observations were made. Through his studies on motion Galileo had come to have serious doubts about the Aristotelian concept of nature. What he sensed was lacking was a true physics. He was very acute, therefore, when he came to sense the significance of his observations of the moon, of the phases of Venus, of the moons of Jupiter and of the Milky Way. The preconceptions of the Aristotelians were crumbling before his eyes. He had remained silent long enough, over a three month period, in his contemplations of the heavens. It was time to organize his thoughts and tell what he had seen and what he thought it meant. It was time to publish! In so doing he would become one of the pioneers of modern science. For the first time in over 2,000 years new significant observational data had been put at the disposition of anyone who cared to think, not in abstract preconceptions but in obedience to what the universe had to say about itself.

Keywords. Aristotle, Copernicus, moons of Jupiter, phases of Venus, Sidereus Nuncius, telescope.

During the very last year of what he himself described "as the best [eighteen] years of his life" spent at the University of Padua, Galileo first observed the heavens with a telescope. In order to appreciate the marvel and the true significance of those observations, we must appreciate both the intellectual climate in Europe and the critical intellectual period through which Galileo himself was passing at the time those observations were made.

Galileo was the first true observational astronomer † but he was also an experimentalist. In fact, it was precisely through his dedication as an experimentalist, and in particular through his studies on motion, that he had come to have serious doubts about the Aristotelian concept of nature. What he sensed was lacking was a true physics. The world models, inherited from the Greeks, were purely geometrical and the geometry was based upon preconceived philosophical notions about the nature of objects in the universe: all objects had a natural place in the universe and consequently they had a natural motion.

† My claim that Galileo was the first true observational astronomer requires some justification. Galileo did not invent the telescope; he improved it for the precise purpose of astronomical observations. Nor, it seems, was he the first to use the telescope to observe the heavens. There is evidence that Thomas Digges of England, using a rudimentary reflecting telescopic invented by his brother Leonard, saw myriads of stars about thirty years before Galileo's first observations. Even if this is true, the observations of Digges did not become known and had no scientific impact. Galileo not only observed; he intuited the great importance of his observations and he communicated them rapidly to the whole cultured world of his day. It is for that reason that I feel justified in naming him the first true observational astronomer.

But there was no experimental justification for these preconceptions. They were simply based upon a philosophical idea of the degree of perfection of various objects.

But, in addition to his attachment to experiment and the sense for physics that derived from it, Galileo also nourished the idea that the true physical explanation of things must be simple in the richest meaning of that word. To be more specific, among several possible geometrical models the nature of the physical world would see to it that the simplest was the truest. Thus, as early as 1597, at the age of thirty-three and only five years after the beginning of his teaching career in this university, he was able to state in a letter to Kepler:

"... already for many years I have come to the same opinion as Copernicus † and from that point of view the causes of many natural effects, which undoubtedly can not be explained by the common hypothesis, have been revealed by me."

One senses in such statements as this by Galileo that, although he did not yet have the physical explanation, he realized that it must be a simple and unifying one. For Galileo, the motion of falling bodies and the motion of the planets had something in common and geometrical explanations were not sufficient. Physics was required.

Before we turn our gaze upon Galileo with his perfected telescope pointed to the heavens, I would like to attempt to recover his state of mind at that moment. This is admittedly a very tendentious thing to do, but I think it is important to attempt to do so for the sake of understanding what we might possibly mean by "discovery". He was nearing the end of a relatively long, tranquil period of teaching and research, during which he had come to question at its roots the orthodox view of the known physical universe. But he had as yet no solid physical basis upon which to construct a replacement view. He sensed a unity in what he experienced in the laboratory and what he saw in the heavens. But his view of the heavens was limited, although there was some expectation that, since with his telescope he had seen from Venice ships at sea at least ten times the distance at which they could be seen by the naked eye, he might go a bit beyond that limit. He was uncertain about many things in the heavens. He had seen an object suddenly appear as bright as Jupiter and then slowly disappear; he had been able to conclude that it must be in the realm of the fixed stars, but he could venture nothing about its nature. Did he have expectations that he would with the telescope find out for certain whether the world was Copernican? I expect not. His expectations were not that specific. He simply knew that the small instrument he had worked hard to perfect, if he had already convinced his patrons of its value for military purposes, was surely of some value for scientific purposes. That in itself, although it may seem trite to us, was a major discovery ‡. In brief, I propose to you a Galileo who was extremely curious, anxious to resolve some fundamental doubts and clever enough to know that the instrument he held in his hands might contribute to settling those states of mind.

Obviously not everything happened in the first hours or even the first nights of observing. The vault of the heavens is vast and varied. It is difficult to reconstruct in any

† Historians debate endlessly as to when Galileo first became personally convinced of the correctness of Copernicanism. Judging from his statement of "already for many years" and from other indications he must have certainly been leaning towards Copernicanism during the first years of his teaching at Pisa, which began in 1589.

‡ It indeed was a major discovery to intuit the importance of the telescope for investigating the universe. In the first note I have remarked that Thomas Digges may have actually been the first to observe with a telescope but it appears that he did so in a rather perfunctory fashion and without an appreciation of its value for science, or at least he did not communicate that science to the world.

detail the progress of Galileo's observations; but from October 1609 through January 1610 there is every indication that he was absorbed in his telescopic observations. At times his emotional state breaks through in his correspondence. He makes a climatic statement in this regard in a letter of 20 January 1610, some weeks after his observations of the Medicean moons of Jupiter, when he states: "I am infinitely grateful to God who has deigned to choose me alone to be the first to observe such marvelous things which have lain hidden for all ages past". For Galileo these must have been the most exhilarating moments of his entire life. The observations will be carefully recorded in the *Sidereus Nuncius* but denuded for the most part, and by necessity, of their emotional content. What must have been, for instance, the state of mind of Galileo when for the first time he viewed the Milky Way in all of its splendor: innumerable stars resolved for the first time, splotches of light and darkness intertwined in an intriguing mosaic? He will actually say little about this of any scientific significance; and rightly so, since his observations had gone far beyond the capacity to understand. He could, nonetheless, be ignorant and still marvel.

But he will be very acute and intuitive when it comes to sensing the significance of his observations of the moon, of the phases of Venus, and, most of all, of the moons of Jupiter. The preconceptions of the Aristotelians were crumbling before his eyes. He had remained silent long enough, over a three month period, in his contemplations of the heavens. It was time to organize his thoughts and tell what he had seen and what he thought it meant. It was time to publish! It happened quickly. The date of publication of the *Sidereus Nuncius* can be put at 1 March 1610, less than two months after his discovery of Jupiter's brightest moons and not more than five months after he had first pointed his telescope to the heavens. With this publication both science and the scientific view of the universe were forever changed, although Galileo would suffer much before this was realized. For the first time in over 2,000 years new significant observational data had been put at the disposition of anyone who cared to think, not in abstract preconceptions but in obedience to what the universe had to say about itself. Modern science was aborning and the birth pangs were already being felt. We know all too well how much Galileo suffered in that birth process. That story has been told quite well even into most recent times †. I prefer to leave it as recorded and turn rather to some thoughts about discovery as referred to Galileo's telescopic observations.

I would like to suggest three components contained in the notion of discovery: newness, an opening to the future and, in the case of astronomical discovery, a blending of theory and observation. Discovery means that something new comes to light and this generally happens suddenly and unexpectedly. While I would not exclude that one can plan and even predict what is to be discovered, this is generally not the case. Galileo's telescopic discoveries surely brought us completely new and unexpected information about the universe. Taken as a whole that information was the first new significant observational data in over 2,000 years and it dramatically overturned the existing view of the universe. It looked to the future. Were there other centers of motion such as seen with Jupiter and its moons? Did other planets like Venus show phases and changes in their apparent sizes? And what to make of those myriads of stars concentrated in that belt which crosses the sky and is intertwined with bright and dark clouds? All of these were questions for the future. Although neither Galileo nor any of his contemporaries had a well developed

† An excellent up-to-date study of the Galileo affair up until the most recent statements of John Paul II is: A. Fantoli, *Galileo: For Copernicanism and for the Church* (Vatican Observatory Publications: Vatican City State, 2003) Third English Edition; distributed by the University of Notre Dame Press.

comprehensive theory of the universe, Galileo clearly intuited that what he saw through his telescope was of profound significance. His discoveries were not limited to looking; they involved thinking. Henceforth no one could reasonably think about the universe in the tradition of Aristotle which had dominated thinking for over two millennia. A new theory was required.

Did Galileo's telescopic discoveries prove the Copernican system? Did Galileo himself think that they had so proven? There is no simple answer to these questions, since there is no simple definition of what one might mean by proof. Let us limit ourselves to asking whether, with all the information available to a contemporary of Galileo's, it was more reasonable to consider the Earth as the center of the known universe or that there was some other center. The observation of at least one other center of motion, the clear evidence that at least some heavenly bodies were "corrupt", measurements of the sun's rotation and the inclination of its axis to the ecliptic and most of all the immensity and density of the number of stars which populated the Milky Way left little doubt that the Earth could no longer be reasonably considered the center of it all. Of course, a more definitive conclusion will be possible in the coming centuries with the measurement of light aberration, of stellar parallaxes and of the rotation of the Foucault pendulum. As to Galileo, his telescopic discoveries, presented in a booklet of fifty pages, the *Sidereus Nuncius*, will become the substance of his Copernican convictions lucidly presented in his *Dialogue on the Two Chief World Systems*, a work which he promised would appear "in a short while" but which actually appeared only twenty-two years later. His own convictions are clear, for instance, from his own statement in the *Dialogue*: " ... if we consider only the immense mass of the sphere of the stars in comparison to the smallness of the Earth's globe, which could be contained in the former many millions of times, and if furthermore we think upon the immense velocity required for that sphere to go around in the course of a night and a day, I cannot convince myself that anyone could be found who would consider it more reasonable and believable that the celestial sphere would be the one that is turning and that the globe would be at rest". But Galileo was also wise enough to know that not everyone could be easily convinced. In a letter to Benedetto Castelli he wrote: " ... to convince the obstinate and those who care about nothing more than the vain applause of the most stupid and silly populace, the witness of the stars themselves would not be enough, even if they came down to the Earth to tell their own story". While he could not bring the stars to Earth, he had, with his telescope, taken the Earth towards the stars and he would spend the rest of his life drawing out the significance of those discoveries.

Reference

Fantoli, A. 2003, *Galileo: For Copernicanism and for the Church*, Vatican Observatory Publications: Vatican City State, Third English Edition; distributed by the University of Notre Dame Press

Galileo's Medicean Moons: their impact on 400 years of discovery
Proceedings IAU Symposium No. 269, 2010
C. Barbieri, S. Chakrabarti, M. Coradini & M. Lazzarin, eds.

doi:10.1017/S1743921310007209

Popular perceptions of Galileo

Dava Sobel

Discover Magazine
USA
email: ds23@optonline.net

Abstract. Among the most persistent popular misperceptions of Galileo is the image of an irreligious scientist who opposed the Catholic Church and was therefore convicted of heresy–was even excommunicated, according to some accounts, and denied Christian burial. In fact, Galileo considered himself a good Catholic. He accepted the Bible as the true word of God on matters pertaining to salvation, but insisted Scripture did not teach astronomy. Emboldened by his discovery of the Medicean Moons, he took a stand on Biblical exegesis that has since become the official Church position.

Keywords. Science and faith, astronomy and the Bible, Index of Prohibited Books, blindness.

Popular perceptions and misperceptions about Galileo abound. I will not try to list all of them here, but rather focus on the one that is the most problematic. Because Galileo helped unfold the true nature of the universe, in apparent contradiction to the Bible, his story has come to symbolize the struggle of science against faith. This is the most enduring myth about Galileo: the idea that he was a thoroughly modern – and irreligious – scientist, who clashed with a Church blind to reason.

I think the International year of Astronomy has finally convinced the world that Galileo did not invent the telescope. And events of this year have helped thousands, if not millions of people see the things he saw, through an instrument that had been little more than a toy before he improved it. His development of the telescope changed astronomy almost

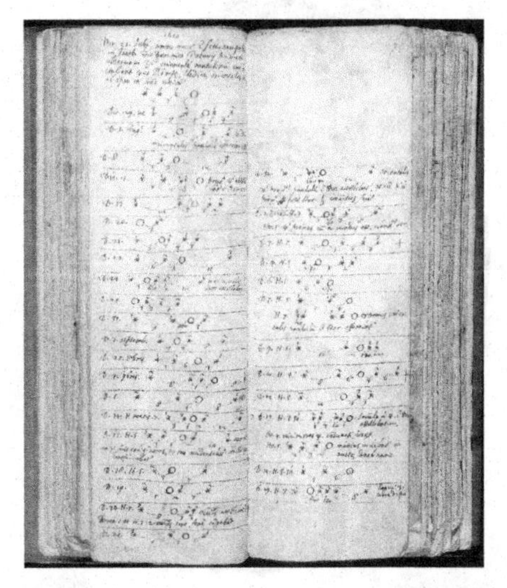

Figure 1. Galileo's record of his continuing observations of the Medicean Moons through summer and fall of 1610. His assessments of the angular distances separating the bodies from one another stand up to modern scrutiny.

overnight from a science of knowing the positions of bodies, to a science that could discover their *composition.*

Since I'm talking about Galileo the man, I must point out that there are two big anniversaries to be recognized this week: the discovery of the Medicean Moons, of course, but also Galileo's death, during the night of January 8, 1642.

He discovered the moons while he was a professor of mathematics, here, in this university-although, as you may know, he had no university degree himself, having completed his mathematical studies with private tutors. By this age, he was also the father of two daughters and a son, all born in this city, with baptismal certificates filed in the local parish registry of San Lorenzo. Galileo had met their mother, Marina Gamba, in Venice, and brought her to Padova, though they lived in separate houses, and he never married her.

During his eighteen years here, which he called the happiest of his life, Galileo had many friends, including the Venetian friar Paolo Sarpi. He and Galileo regularly exchanged ideas about physics and astronomy. Sarpi actually beat Galileo to a correct interpretation of the phenomenon of Earth shine. (Leonardo, the first to figure it out, did not publish. Kepler did publish an explanation in 1604, attributing credit to Maestlin, but Sarpi stated the true cause in a notebook entry from 1588.)

You may recall that the reigning pope of this period, Paul V, took action against the Republic of Venice in 1606. He excommunicated the entire Venetian senate, and placed the Veneto under an interdict. This meant that the clergy were prohibited from offering the mass or administering the sacraments. Galileo's youngest child, the boy, was born that summer, but he was baptized anyway, because Sarpi had convinced the clergy to ignore the Interdict, on the grounds that it was a political act and an abuse of power. At the end of one year, in April 1607, the Interdict was rescinded. Soon after that, assassins made an attempt on Sarpi's life, but he survived it. This was a good thing for us, because it was again Sarpi who, in 1608, told Galileo the interesting reports he had heard about a new instrument that could make faraway objects appear close and large.

Figure 2. Engraving of Galileo at age 60, based on Ottavio Leoni's portrait in chalk, 1624. The downward cast of his right eye suggests a diagnosis for one of the disease conditions affecting his vision.

Although several of Galileo's contemporaries also turned telescopes to the sky, he was the first to conduct systematic observations of other worlds, to draw conclusions about their substance and structure, and to publish his findings. These appeared in an explosive little best-seller called *Sidereus Nuncius*, a title translated as either *The Starry Messenger* or *The Starry Message*, which was printed in Venice on March 13, 1610. And sold out within one week. (A new translation is just out from Bill Shea.)

It is known that Galileo had read and accepted the ideas of Copernicus years before he began his telescopic investigations-at least as early as 1597, when, in a letter to Kepler, he declared himself a timid "Copernican." But it took his discoveries to convince him that the Earth really did revolve around the Sun, as one of the planets. And that he could at last speak out publicly about these convictions.

Like Galileo, Copernicus was a good Catholic, who not only worked for the Church in Poland, but dedicated his famous book, *On the Revolutions of the Heavenly Spheres*, to the pope of his time, Paul III.

What separates Copernicus from Galileo is a lapse of years, during which time an ecumenical council took place at Trent. It began in 1545, two years after Copernicus's death, and concluded in 1563, the year before Galileo was born. Its mission was to contain the Protestant heresy. And one of the outcomes of the Council was a formal and strict position on the interpretation of Holy Scripture. Whereas Martin Luther advocated every man's right to a personal take on Bible reading, the Catholic Church insisted that all the faithful adhere to the official interpretation promulgated by the holy fathers.

Copernicus had been well aware that his theory was open to attack by persons he called "babblers" who would twist and "badly distort some passage of Scripture to their purpose." It was for that reason he dedicated the book to Paul III, whom he credited with understanding that "Astronomy is written for astronomers." While Copernicus's book was in press, he suffered a fatal stroke, and never had to answer to critics. He also remained unaware that the proofreader, who was a Protestant minister, had inserted a cautionary note to readers, stipulating that the heliocentric theory was strictly for the purpose of mathematical calculation, and had no counterpart in reality. "Let no one expect anything certain from astronomy," this statement warned, "lest he... depart from this study a greater fool than when he entered it."

But Galileo's discoveries were not purely mathematical. He was seeing real stars in the heavens that had been invisible before his improvements to the telescope. In the constellation of Orion alone, he estimated several hundred more stars than had been listed in any of the existing star catalogs.

Galileo's observations of the moons of Jupiter were extraordinary, not only for their novelty and the care with which he recorded them, but because they provided evidence for what Copernicus had merely intuited. The Earth was not the center of all celestial motion. There were bodies in orbit around Jupiter. And furthermore, if Jupiter could move through the heavenly ether with all four moons in tow, then what was to stop the earth, which commanded only one moon, from proceeding on its own orbital course around the Sun?

And what was Galileo's response to his own monumental discovery? A prayer of thanksgiving: "I render infinite thanks to God for being so kind as to make me alone the first observer of marvels kept hidden in obscurity for all previous centuries."

For Galileo, there was nothing irreligious about these discoveries or their implications.

In the spring of 1611, Galileo traveled to Rome to defend the truth of his discoveries. Because of the difficulty of producing a good telescope, others had trouble replicating his results, and therefore doubted his claims.

During this 1611 visit, he won the unequivocal endorsement of the Jesuits at the Roman College. Both he and they were committed to figuring out the periods of the Medicean Moons, an extremely arduous project that depended on amassing reams of good data. (The moons seemed to promise a reliable way to determine longitude at sea, which, in Galileo's time, was already a problem of international importance-and one that would not be solved until well into the following century.)

One of the Roman Cardinals taking an interest in Galileo's telescopic findings was the "hammer of the heretics," Roberto Bellarmino, who was attached to the Congregation of the Holy Office of the Inquisition. He specifically requested his fellow Jesuits at the Roman College to evaluate Galileo's work. There were various reasons to question the texture of the Moon's surface, and the nature of the strange appendages that seemed to sprout from Saturn. No astronomer, however, could doubt the reality of the Medicean Moons. Their existence and motion were incontrovertible, but... Did they really offer proof of the Earth's motion? That wasn't nearly so certain.

Interestingly, Cardinal Bellarmino had been the chief spokesman for Rome in the flap with Sarpi over the Interdict.

And Cardinal Bellarmino was well acquainted with the ideas of Copernicus. In fact, scarcely a decade before Galileo's discoveries, Bellarmino had served as a cardinal inquisitor in the trial of a famous heretic who also embraced the heliocentric model.

When the Dominican priest Giordano Bruno took up the Copernican idea, he extended it into an infinite cosmos, where every star was a sun surrounded by its own retinue of planets. This raised the question of life on those other worlds, and salvation for their inhabitants. And you begin to realize that Bruno's ideas were indeed heretical. In fact he was excommunicated twice, by the Catholic Church and the Calvinist. He doubted the divinity of Jesus, the virginity of Mary, and the transubstantiation of the Communion wafer into the body of Christ. But popular myth has reduced all of that to his belief in the Copernican idea-as though his story were merely a prequel to Galileo's. People often ask, if the Inquisition condemned Bruno to burn at the stake, then why didn't it do the same to Galileo?

By 1615, Galileo became aware that the Congregation of the Index was about to ban Copernicus's book, and he traveled to Rome again, to try to forestall any such action, which he feared would be disastrous-both for astronomy (Why ban the book NOW, when evidence in support of it accumulates daily?) and also for the Church (If Catholics are prohibited from teaching and studying these subjects, then the Protestants will unravel the universe, and the Church will be embarrassed.). He went to Rome with two powerful arguments: his theory of the tides, and his conviction that the Bible was a book about how to go to Heaven, not how the Heavens go. He never contested the truth of the Bible, believing it to be the Word of God, but he did not think the Holy Spirit would waste time teaching, by means of Scripture, things that an intelligent person could learn about Nature on his own, using his God-given talents, such as vision, intelligence, manual dexterity, speech. What's more, the Bible said hardly anything about astronomy. And in the few places where it mentioned one or another of the heavenly bodies, it spoke metaphorically, or according to the perceptions of the common man, not as a scientific authority. This was Galileo's big mistake. Because how could he, a layman, dare to interpret the Bible?

While Galileo remained in Rome, the Holy Office of the Inquisition met and voted against the motion of the Earth, which they termed "heretical," "foolish," and "absurd." Before this news became public, Pope Paul V chose Cardinal Bellarmino to meet privately with Galileo, in February of 1616, and tell him to drop the subject. Copernicus was put on the Index of Prohibited Books, supposedly "suspended until corrected." But even after the requisite corrections were issued in 1620, the book stayed on, still prohibited, until the 1835 edition of the Index.

Galileo did as he was told. He stopped talking about cosmology for seven years. He worked on the longitude problem. He studied poetry and wrote literary criticism. By modifying his telescope, he developed a compound microscope.

A popular misconception holds that Galileo went blind by staring at the Sun to observe its spots, but he was too smart to do that. Besides, the pain would have been excruciating. No, he observed the Sun safely, by letting its image fall through the telescope tube onto a sheet of paper, where he could trace the exact outlines of the solar spots. And while doing that, he discovered the Sun's monthly rotation. One more myth about Galileo's blindness suggests that he might not have gone blind, if only the Church had granted permission for him to seek proper medical attention while he was under house arrest. But, Galileo had a lifelong history of eye ailments, and he lived to an age (78) when blindness can befall almost anyone. An ophthalmologist in England, Peter Watson, has been trying for more than a decade to determine the cause of Galileo's blindness, from reports and also from contemporary portraits. This particular image, an engraving based on a chalk study by Ottavio Leoni in 1624, shows the right eye displaced downward, and a swelling of the forehead on that same side, raising the eyebrow and lowering the lid. It's not a definitive diagnosis, but Watson notes these are symptoms of a condition called frontal sinus mucocoele.

While Galileo was busy keeping quiet, a remarkable thing happened. In 1623, Cardinal Maffeo Barberini, Galileo's longtime acquaintance, was elected Pope Urban VIII. Galileo sensed a significant change in the Roman climate. And he returned to Rome to pay his respects and feel out the new pontiff on a possible reprisal of the subject of the earth's place in the heavens. He wanted to write a book about it. In fact he'd been planning such a book ever since his first published announcement of his discoveries. Their relations were quite cordial at this point, and Urban had long expressed interest in Galileo's work. However, the two men took opposing views of science. Galileo felt that God had created the universe in such a way that humans could figure it out from mathematical clues-that in fact figuring out the universe was humankind's mission on earth. But, from Urban's point of view, the universe would be a pretty puny affair if a mere mortal could figure it out. Much more likely that an Omnipotent Creator would fashion a different set of laws of physics for every individual phenomenon, and no matter how much evidence you gathered for any particular argument, you could not hold God to your argument. In other words, nothing short of Divine Revelation would suit.

As soon as Galileo's big book appeared, in 1632, it caused a furor. And the next time Galileo went to Rome, it was in answer to a summons to stand trial before the Inquisition. The trial resulted in Galileo's abjuration and public humiliation. His book was listed on the Index. He lived out his remaining years under house arrest, but he recovered his spirit sufficiently to continue a few observations and also to write another book, *Dialogues Concerning Two New Sciences*, which is considered to be his scientific masterpiece, though it isn't nearly as engaging to read as the *Dialogue on the Two chief Systems of the World.* That title stayed on the Index of Prohibited Books, along with Copernicus, for nearly two centuries.

Galileo is buried in the Church of Santa Croce, in Florence. At the time of his death, the Pope prohibited the Grand Duke from giving Galileo an important funeral oration or an imposing tomb. It took almost one hundred years for his devoted followers to move his remains to the much visited tomb.

Today, the moons that Galileo named for the grand duke of Tuscany are better known as the Galilean satellites of Jupiter-a name that Johannes Kepler gave them in 1610.

Galileo also has a crater named for him on Earth's Moon, as do Copernicus, Kepler, and even Bruno (on the far side). All moons belongs to science now.

The first direct evidence for the motion of the Earth came when English astronomer James Bradley, around 1730, discovered the aberration of starlight. Not long after that, in 1758, the Church lifted the general ban on teaching the heliocentric theory. But even at that point, certain specific works remained on the Index-including Copernicus's own book, Kepler's *Epitome of Copernican Astronomy* and Galileo's *DIALOGUE* - all expressly forbidden until the Index of 1835.

A little later in the 19th century, more definitive proof came to light regarding the Earth's motion: In Germany in 1838, Friedrich Wilhelm Bessel succeeded in measuring the parallax of the star 61 Cygni, and derived a distance to that star of 6 light-years. The universe was bigger by far than even Newton had imagined. Still more demonstrable proof for the Earth's motion came in the early 1850's, with Foucault's pendulum.

As for the official Church position on science, according to the 1893 Providentissimus Deus of Pope Leo XIII, it is held that the Bible does not aim to teach science. So, at least posthumously, Galileo won the argument.

But by then, at the end of the 19th century, the myth of Galileo as a martyr of science was firmly established. And it may never be dispelled. In popular perception, he is still pitted against the Church. Despite the fact that Pope John Paul II re-opened the Galileo case in the hope of taming its pervasive, divisive power, his public statement in 1992 acknowledged no wrong-doing on the Church's part. Instead, the Pope's statement blamed the theologians of the 17th century for failing to grasp the theological issues at play, and blamed Galileo for misunderstanding the limits of scientific certainty.

I had hoped that one outcome of the IYA might be a truce, in honor of Galileo, between science and religion. For, as his writings clearly show, he had no need to jettison his religious beliefs to perform first-rate science. Nor did Copernicus, nor Kepler, nor Newton.

Even today, there are working scientists who believe in a divine Creator, but they usually don't let on that they do. At least not in the laboratory or the observatory.

References

Galilei, Galileo, *Sidereus Nuncius or A Sidereal Message*, Translated from the Latin by William R. Shea. Introduction and Notes by William R. Shea and Tiziana Bascelli, Sagamore Beach, Mass.: Watson, 2009

Galilei, Galileo, *Sidereus Nuncius or The Sideral Messenger*, Translated with introduction, conclusion and notes by Albert Van Helden, Chicago: University of Chicago, 1989

Sobel, D., *Galileo's Daughter: A Historical Memoir of Science, Faith and Love*, New York: Walker, 1999

Galileo's Medicean Moons: their impact on 400 years of discovery
Proceedings IAU Symposium No. 269, 2010
C. Barbieri, S. Chakrabarti, M. Coradini & M. Lazzarin, eds.

doi:10.1017/S1743921310007210

The slow growth of humility

The Fault, Dear Brutus, lies not In the stars but in ourselves

Tobias Owen[1] and Scott Bolton[2]

[1]University of Hawaii,
Institute for Astronomy USA
email: owen@ifa.hawaii.edu

[2]Space Science Division, Southwest Science Institute,
San Antonio, Texas 78228
email: sbolton@swri.org

Abstract. Galileo's stunning discovery of the four largest satellites of Jupiter forced the over throw of the Earth-centered cosmology that had dominated astronomy for centuries. Such a fundamental transformation of the Western World's view of its importance in the cosmos could be expected to produce some humility in society. However, the deep desire for our uniqueness continues to struggle with the astronomical evidence.

Keywords. Galileo, History of Astronomy, Huygens, Cosmotheoros.

---

1. Introduction

This essay is a brief exploration of the social effects of the gradual growth in our knowledge about the configuration and dimensions of the universe. It is this aspect of Galileo's discovery of Jupiter's satellites that draws our attention.

The famous physicist Christaan Huygens (1629-1688) commented on the discovery and its consequences in a charming little book called *Cosmotheoros*. We will repeat some of his perceptions for their remarkable freshness.

What were society's responses to the sudden, stunning loss of our apparent importance in the cosmos that was created by Galileo'sobservations? That sense of importance was based on centuries of belief in an Earth-centered universe: a stationary Earth in the center of the solar system with the entire universe revolving around it. The importance of the humans who lived at the center of this system was self-evident. It was strongly reinforced by the religions that embraced it.

There is a latent paradox in this inquiry. The growing awareness of what a cosmic femtospeck we are in a universe over 90% of which is dark matter that we do not understand may indeed invoke a humbleness within us, yet our remarkable ability to achieve this awareness is at the very core of what makes us proud, even arrogant. We can find both attitudes manifested in human behavior.

2. A glimpse of ancient history

Aristarchos of Samos (310-230 BC) is generally credited with the first serious suggestion of a heliocentric solar system. Nevertheless, he also published a book supporting the geocentric configuration, by far the dominant model of the time. Much later *Hipparchos of Nicea* (~127 BC), the greatest astronomer of the classical period, expressed solid support for an Earth-centered solar system. His work strongly influenced Claudius Ptolemaeus, known in English as Ptolemy (90-168 AD) the final, definitive defender of geocentrism. Ptolemy's elaborate model could explain the apparent motions of the planets better than any previous work. The Islamic contributions to this problem culminated in the work of *Ibn Al-Shatir of Damascus* (1304-1375). Al Shatir developed a theory of the moon's motion identical to that of Copernicus.

3. Copernicus (1473-1543)

The great contribution of Copernicus was to substitute a sun-centered solar system for Ptolemy's geocentric model. He could not predict planetary motions better, but he showed that a heliocentric model had a simplicity and elegance that were a tremendous advance over geocentric models that required ever more moving parts to explain the improved observations. However, the Copernican scheme posed several deep philosophical and religious problems by removing the Earth from the center of the universe and by having it move through space. The Earth became just one more planet and its inhabitants could no longer claim an especially favorable connection with the cosmos.

4. The power of geocentrism

Before we consider Galileo's far-reaching discovery, we should ask why Geocentrism was universally accepted by leading scientists and intellectuals around the world for such a long time. Counting just from Hipparchos (127 BC) to Copernicus (1543) the span of acceptance was 1670 years. Not bad for a theory that was completely, gloriously wrong!! The obvious scientific reason was the ability of this model to satisfy the existing low-precision observations. But there were deep emotional reasons as well. Geocentrism = Egocentrism. The model satisfied a strong human urge to be exceptional, to have a position in the universe that was central and therefore unique. It can be argued that this desire is the natural consequence of an experience that every human being on the planet shares viz., being born! A baby is literally the center of the universe it knows, with all other visible bodies going around it (Figure 1).

Figure 1.

Our naked eye observations of the sun, moon and stars fully support this view. It is immediately obvious that the Earth, and we upon it are very, very special.
The reluctance to give up this simple, exalted place in the universe is so strong it has stayed with us in various forms despite the advances of science.

5. Galileo

Galileo's wonderful discovery of Jupiter's satellites is well known and is reviewed in detail at this symposium. Here we shall just give a very brief summary. The wonders he found with his telescopic survey of the night sky are documented in his fascinating book *The sidereal messenger.* This is how he relates his first, unknowing glimpse of Jupiter's satellites when he turned his telescope on Jupiter itself:

"...there were two stars on the eastern side and one on the west. The most easterly star and the western one appeared larger than the other. I paid no attention to the distances between them and Jupiter, for at the outset I thought them to be fixed stars, as I have said. But returning to the same investigation on January eight - led by what, I do not know - I found a very different arrangement. The three starlets were now all to the west of Jupiter, closer together, and at equal intervals from one another."

"Led by what I do not know". With those seven words, 1670 years of astronomical thought were about to be overturned. Human perception of the universe and our place in it would be forced into a mind-bending change.

How did Galileo feel when he recognized what he had found? Christiaan Huygens gives us an idea:

"Jupiter has his four moons which we owe to Galileo, and anyone can imagine that he was in no small rapture at the discovery." (Huygens, 1695)

This great "rapture" was felt by many of Galileo's contemporaries as well. When added to his other observations the elegant Copernican model for the solar system was now indisputable. What were the effects of the Copernican revolution on Society?

6. The post-Copernicus world

Even with Galileo's great discovery the seductive power of the geocentric model died a slow, difficult death. When Isaac Newton entered Cambridge University in 1661, the Ptolemaic model was still being taught. A survey carried out 344 years later in 2005 found that 20% of Americans still believe that the sun goes around the Earth.

Figure 2. An artist's view of our (Milky Way) galaxy showing the position of our solar system.

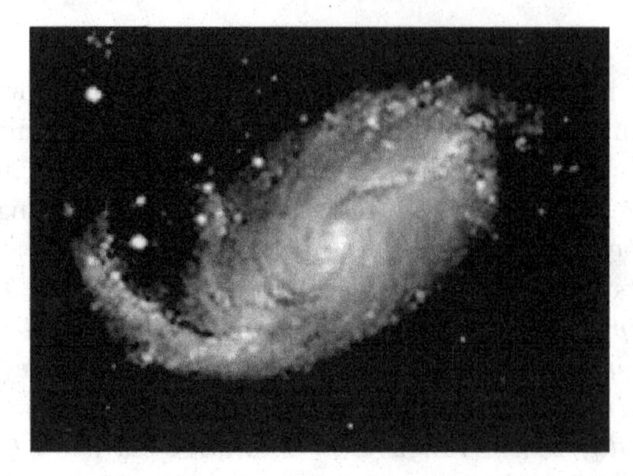

Figure 3. The barred spiral galaxy NGC1672 (NASA).

The comfortable feeling that everything in the universe moves around us, that we are the center of all activity in the cosmos, that conviction of our exalted importance was too wonderful to give up.
A quick look at history shows us there have been many examples of post-Copernican leaders who still felt the world should revolve around them.
For the rest of us, the sense of our uniqueness has continued to diminish. In what other ways could we be "special"? Perhaps it's our star.
No! There are at least 20,000 stars like the sun in our galaxy.
Maybe we occupy a special place in our galaxy - in the center?
No! We're not even near the central bar (Figure 2).

Other galaxies also have bars, some relatively larger and better decorated than ours (Figure 3).
So even our galaxy isn't special. It is certainly not the center of the universe - which has no center, and it is only one of thousands of billions of galaxies (Figure 4).

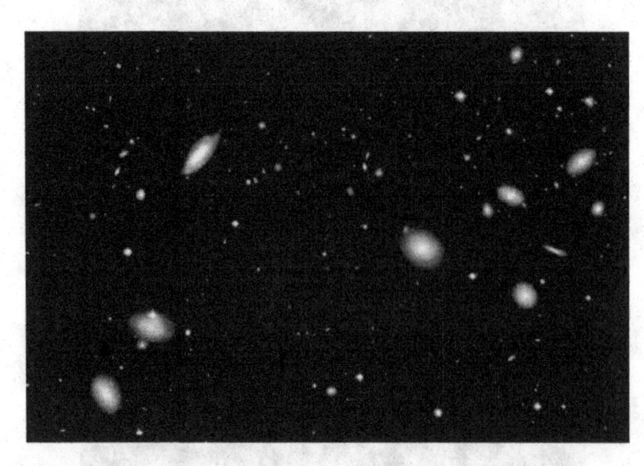

Figure 4. Part of the Coma cluster of galaxies. Almost every object you can see in this picture is a galaxy and the cluster contains many more. (NASA)

7. So how do we rate in the Cosmos?

There's nothing exceptional about our sun, our position in our solar system our position in our galaxy, or our galaxy itself. What else might satisfy this deep human longing to be first, best or even unique?

This quest takes us back to the Earth. This time we consider its intrinsic qualities, using criteria our ancestors could not appreciate. We can now compare the Earth as a planet with the other planets in the solar system and in that comparison our world is certainly unique.
It turns out that our position *is* important even though we are not at the center of the solar system. Our planet is in a nearly circular orbit at just the right distance from the sun to allow open bodies of water to exist on its surface for billions of years. The contrast between the arid, battered surface of the moon and the cerulean Earth with its thick atmosphere dappled with white, water-bearing clouds is beautifully illustrated in this famous Apollo 8 picture (Figure 5).

Figure 5. Earthrise at the limb of the moon as seen by the Apollo 8 astronauts.

But images like this also awakened some widespread humble feelings about our place in the universe. This cosmic jewel on which we live has been revealed as a fragile world whose well-being is essential for our continued existence. Without the good health of our planet we would disappear, regardless of all our proud accomplishments. This forces us to accept a humble position far lower than anything our ancestors could have imagined. There are probably other planets fundamentally like ours in the galaxy. Specialized telescopes that are designed to detect them are currently in orbit. These planets, if they exist, may or may not carry living organisms, possibly even intelligent life.
It is obvious that the existence of sentient life is surely the most unusual aspect of the world we inhabit. Yet the manner of life's origin, how extremely difficult it may be, remains hidden from us. The colorful ribbon of evolution leading from self-replicating molecules to art, music and poetry may unwind this far only rarely in the universe.
This possibility is also subject to experiment as scientists listen with powerful radio telescopes for intelligent signals from other worlds.
Should we really be undertaking such investigations or are we stretching our imagination too far?

The discovery of an advanced civilization on a different world would be the final challenge to our egocentrism. This was not a problem for Huygens who was quite taken with the idea. It was the main subject of Huygens' *Cosmotheoros.* He was a firm believer in the possibility that there were life forms on all the planets in our solar system and he described their probable characteristics. This was hardly a popular opinion in his day and he took pains to defend his arguments against both secular and religious skeptics. Galileo had shown the way in 1615, quoting a telling epigram from Cardinal Baronius (1538-1607):

"The intention of the Holy Ghost is to teach us how one goes to heaven, not how heaven goes"

Huygens offered more details:

The Secular Issue: Is it worthwhile to consider such a hypothetical problem? *"Conjectures are not useless because they are uncertain. In such noble and sublime studies as these, 'tis a glory to arrive at probability, and the search itself rewards the pains'.*

The Religious Issue: How could there be other beings if they were not mentioned in the Bible? *"God had no design to make an enumeration of all his works. Under the general name of stars or Earth are understood all the heavenly bodies, even the little gentlemen around Jupiter. So why must all those beings, which God has been pleased to place on these bodies, be excluded?" C. Huygens 1695.*

These are good arguments even today. We may be in the same intellectual position our ancestors were in the pre-Copernican world. Some scientists believe the emergence of intelligent life is so unlikely that we could be its only manifestation in the galaxy. A search for alien intelligence is accordingly a waste of time and money. Others are more sanguine. They are waiting for a modern Galileo to find evidence for other civilizations on other Earth-like planets orbiting other stars.

This will be the final proof that we are not the center of the universe. What would be the result of such a discovery? Will it kindle the humility that will finally make us kind to one another?

We know what we would like, but history doesn't make us optimistic.

We can give the last word to Huygens.

For Huygens this image of the Earth taken by the Cassini spacecraft from its orbit around Saturn (Figure 6) would have had a profound impact. It emphasizes the insignificance of our planet on a cosmic scale. 400 years after Galileo's epochal discovery of Jupiter's moons, 290 years after Huygens' *Cosmotheoros*, we can only echo the wish he expressed so well:

"This [new scale of the solar system] shows us how vast those Orbs must be, and how inconsiderable this Earth is, the Theatre upon which all our mighty Designs, all our Navigations, and all our Wars are transacted when compared to them. A very fit Consideration, and matter of Reflection, for those Kings and Princes who sacrifice the Lives of so many People, only to flatter their Ambition in being Masters of some pitiful corner of this small Spot." (Huygens, Cosmotheoros, 1695).

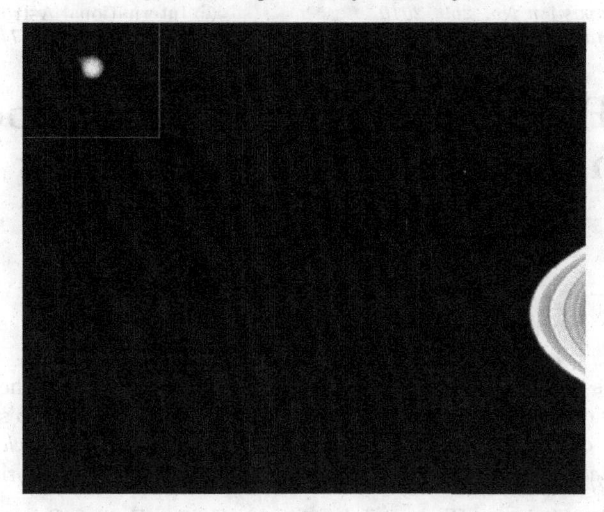

Figure 6. The Earth seen from Saturn by a camera on the Cassini Orbiter. To the right are a set of bright arcs from the edge of Saturn's famous rings, with the arcs of two more distant, hazy rings beyond them. Between these two hazy bands, one quarter of the frame from the right hand margin and one quarter down from the top is a small white spot. It is enlarged in the inset. This small spot is the Earth.

References

The Starry Messenger, Galileo Galilei (1615), trans. Stillman Drake.
Cosmotheoros, Christiaan Huygens (1695) (in 1757 trans. anon.).

Galileo's Medicean Moons: their impact on 400 years of discovery
Proceedings IAU Symposium No. 269, 2010
C. Barbieri, S. Chakrabarti, M. Coradini & M. Lazzarin, eds.
doi:10.1017/S1743921310007222

A new Physics to support the Copernican system. Gleanings from Galileo's works

Giulio Peruzzi

Department of Physics,
University of Padua, Italy
email: peruzzi@pd.infn.it

Abstract. Galileo's support to the Copernican theory was decisive for the revolutionary astronomical discoveries he achieved in 1610. We trace the origins of Galileo's conversion to the Copernican theory, discussing in particular the *Dialogo de Cecco di Ronchitti da Bruzene in perpuosito de La Stella Nuova.* Later developments of Galileo's works are briefly treated.

Keywords. History of Science, History of Astronomy, Galileo Galilei, Copernican System, Scientific Revolution.

---

1. Introduction

The use of the telescope alone doesn't explain the revolutionary astronomical discoveries achieved by Galileo from the end of 1609 onwards. To look doesn't mean to see, and the "sensate esperienze" must integrate observation and experimentation. Galileo looks and sees because in preceding years he had freed himself from prevailing convictions and he had progressively become aware that the facts he was studying both in the Heavens and on the Earth went in the direction of confirming the Copernican system.
It is well known that one of the first evidences of his adherence to Copernicanism lies in a letter to Kepler written on 4th August 1597 (*Opere*, vol. X, pp. 67-8, p. 68). Galileo is however well aware that the Copernican system, unlike the Aristotelian-Ptolemaic system, lacks a physics of its own. It is not by chance that in the years preceding the use of the telescope, his researches were devoted to both astronomy and the study of local motions. In this sense, it is emblematic that Galileo analyses the *Stella Nova* in the same year when he communicates to Sarpi his discovery of the law of falling bodies (letter on 16th October 1604, *Opere*, vol. X, pp. 115-6, p. 115).

2. The appearance of the *Stella Nova*

October 1604. The astronomers are fixing their eyes towards the region of the sky between the constellation of Sagittarius and that of the Ophiuchus or Serpentarius. They are observing quite a rare event, though cyclically recurrent and foreseeable: the celestial conjunction of three planets, Jupiter, Saturn and Mars. Many people are thus scrutinizing that part of the Heavens when, with great amazement, they suddenly see – some say on 9th and others on 10th October – a new source of light. The brightness of the new source of light increases during a couple of weeks and becomes equal to Venus. It then progressively decreases and finally disappears about one year and a half after its appearance.
Different kinds of emotions shake those who observe the phenomenon: a mixture of astonishment and fear, of superstition and curiosity emerges from letters and reports of that time. People recall a similar appearance and disappearance of a "stella nova" in

the constellation of Cassiopeia, observed in November 1572 by Tycho Brahe, which had raised some clamour also within the population.

What was going on? We know a lot today about these appearances. We can observe their remnants with our sophisticated instruments and we have at our disposal quite a satisfying theory of the stellar evolution, which enables us to catalog the appearance of these celestial bodies within the great class of *Variable Stars*. It is thus sure that the phenomena observed in 1572 and 1604 were *Supernovae* (the term was introduced by Fritz Zwicky and Walter Baade in 1934), catastrophic events within the stellar evolution during which the brightness of a star suddenly increases so that the star becomes visible from great distances.

In 1604, however, the knowledge was much different. The prevailing conception, supported by Aristotle's followers, sharply separated celestial phenomena and objects from terrestrial ones. Celestial bodies, created *ab inizio* by God, were made of a special substance, a highly perfect quintessence that did not undergo through any change; their perfection was mirrored by the perfection of their eternal circular motions. On the contrary, the sublunar region, including the atmosphere and the Earth, was the scene of changes, of life and death, of generation and corruption, and it hosted bodies made of the mixture of the four elements (earth, water, air and fire). These bodies, according to the proportion of their constituting elements, had their "natural" place at a given height or distance from the centre of the Earth: if they were in a different position, they moved (a "natural" motion) along a straight line, to go back to their natural place. The downwards motion of heavy bodies (towards the Earth's surface) and the upwards motion of flames were explained on the basis of this theory.

Such a conception of the Universe, imbued with theological and metaphysical elements, could not fit with the appearance of new stars: these appearances or generations had to be linked to entities or bodies located not in the celestial region but in the sublunar one, they had to be meteorological phenomena, though rare and strange. This is why the discussion on the new star focused on the position of the latter. The question did not involve only the explanation of an event, though such a peculiar one, but a millenarian conception of the Heavens based on a philosophy of nature that had become throughout the centuries more and more focused on the manipulation of bibliographies, the commenting of books and the research of an hypothetical consistency with the Holy Scriptures, forgetting little by little the importance of direct observation. A philosophy/theology of nature which tried to defend itself against attacks that, from the mid 16th century, had been more and more frequent. The scientific controversy thus involved consolidated powers and authorities both in the Church and in the academic community.

In Padua, where the *nova* was observed for the first time on 10th October, the controversy was very lively and involved the whole town, exciting curiosity and fears among the population and raising careful interest among scholars. Galileo, who was at the time professor of mathematics and astronomy at the University of Padua, particularly appreciated for his teaching capacities, had chosen "le teoriche dei pianeti" as the subject of his lessons for the year 1604-1605. It was thus natural that his friends and colleagues urged him to present his opinion about the phenomenon. He did so on three public lessons, which were probably held from the end of November and the first half of December 1604. The curiosity was such that more than thousand persons attended the lessons.

Unfortunately only some notes and a few fragments of the written texts of these lessons still survive in the archives (assuming that Galileo really completely wrote down his lessons).

Anyway, their main aim seems clear. As Galileo writes, though everybody was interested in knowing about "*de substantia, motu, loco et ratione apparitionis illius*", he only wanted at that time "*de motu et loco demonstrative constet*" (*Opere*, vol. II, p. 278). From other sources, it is known for sure that Galileo intended to write down and publish his lessons. This is quite clear in a letter written by Alessandro Sertini to Galileo on 16th April 1605 (*Opere*, vol. X, pp. 142-3, p. 143), and in an unfinished letter written in January 1605 by Galileo to an anonymous correspondent (maybe Onofrio Castelli or, more probably, Girolamo Mercuriale) (*Opere*, vol. X, pp. 134-5). In the latter, Galileo mentions reiterated requests to send "copia delle tre letioni fatte da me in pubblico" (*Opere*, vol. X, p. 134), and he says that the publication has already been repeatedly postponed and it is to be postponed again for a few more days, because the lessons have mainly dealt with the fact that the new star is much above the lunar orbit, while Galileo would now like to "mutarle in discorso et aggiugnervi circa la sustanza et generazione" (*Opere*, vol. X, p. 135) of the new star. Demonstrating that the star is much beyond the lunar orbit, Galileo writes, is quite "facile, manifesta e comune [...]; bisognò che io ne trattassi in grazia de i giovani scolari et della moltitudine bisognosa di intendere le demostrazioni geometriche" (*Opere*, vol. X, p. 134). But discussing the substance and generation of the *nova* was a much different matter. Galileo, in his letter, doesn't explicitly present his hypothesis on the subject (the autograph suddenly stops right with the sentence announcing a short summary of his ideas), he only explains that this hypothesis doesn't have evident contradictions and could thus be true, but he needs time to confirm it with observations, waiting for "il ritorno di essa stella in oriente dopo la separazione del sole, et di nuovo osservare con gran diligenza quali mutationi abbia fatto sì nel sito come nella visibile grandezza et qualità di lume [...]. Et perché questa mia fantasia si tira dietro, o più tosto si mette avanti, grandissime conseguenze et conclusioni però ho risoluto di mutar le letioni in una parte di discorso"†.

What was this "fantasia" rich in consequences Galileo was working on? First of all, though he did not take a definitive position about the nature of the *nova* (as he lacked indisputable evidences), Galileo started supporting, in those years, several hypothesis about the generation of the new star – that we will discuss later – that cancelled the difference between terrestrial and celestial physics. At the same time, Galileo hoped he could observe – but he did not succeed in this – the relative parallax of the *Stella nova* when the Earth was at opposite positions along its revolution orbit around the Sun, as he thought that the changing brightness of the *nova* was due to different distances from the Earth. This would have been a definitive proof that the Copernican system was true, against both the Ptolemaic system and the hybrid system proposed by Tycho Brahe. It is important to point out that in those very months, Galileo was working hard to study local motions, also in order to answer several objections to the Copernican system (an example: if the Earth is moving, why do we observe that a body falling from a tower arrives right at the base of the tower and not at a certain distance from it?). We can thus understand Galileo's emphasis about the consequences of his "fantasia".

† "the coming back of this star at east after the separation of the Sun, and observe again with great care what changes it [the star] shows both as for the position and the visible dimension and quality of light [...]. And as this *fantasia* of mine brings extremely important consequences and conclusions forth, I have decided to turn the lessons into a part of a *discorso*" (*Opere*, vol. X, p. 135).

3. Il dialogo di Cecco Rochitti

In Galileo's private correspondence, there are several letters from friends and acquaintances who sympathize with the antiaristotelian ideas which surely inspired the three public lessons held by Galileo. The Franciscan monk Ilario Altobelli, for instance, writes to Galileo on 3rd November 1604 (*Opere*, vol. X, pp 116-7), that "questo nuovo mostro del cielo" (*Opere*, vol. X, p. 117) seems to be there on purpose in order to "far impazzire i Peripatetici, ch'hanno creduto sin hora tante bugie in quella stella nova e miracolosa del 1572, priva di moto e di parallasse" (*Opere*, vol. X, p. 117). And Altobelli insisted on this point in a letter to Galileo written on 25th November 1604 (*Opere*, vol. X, pp. 118-20), where he repeated that the new star was clearly located on the fixed stars sphere and that "il suo sito rende possibile ogni impossibilità conietturata di Aristotile, distrugendo ogni sua imaginatione" (*Opere*, vol. X, p. 118) , in spite the "pertinacia" ("obstinacy") of "Peripatetici, o, per dir meglio, semifilosofi" ("Peripatetics or, to say it better, semiphilophers"), unable to confront the observation data (*Opere*, vol. X, p. 118). And Galileo, who carried out by himself observations and measures on the position and features of the new star, though probably not in a systematic way, acquired through this intense correspondence, further precious details not only on the 1604 *stella nova* but also on the previous appearances, in particular on the 1572 one, which he was studying by reading (and commenting) Tycho Brahe's works.

The antiaristotelian spirit of the three Galilean lessons raised a lively discussion in the Academic world, where scientific questions were mingled – as often they are - with personal, prestige and power questions. Cesare Cremonini in particular, authoritative scholar of Aristotle and holder of the first chair of Natural Philosophy at the University of Padua, openly criticised Galileo and supported the Aristotelian tradition. It is likely that Cremonini himself inspired, at least partially, the publication in Padua, at the end of January 1605, of the *Discorso intorno alla nuova stella* by Antonio Lorenzini da Montepulciano. The core of Lorenzini's argumentation was the strenuous defense of the celestial essence perfection: the immutability and incorruptibility of the Heavens had to imply that the *nova* was nothing else than a meteor located in the sublunar world. To support this conviction, Lorenzini mentioned Aristotle, according to whom the Heavens would stop moving if a new star was added in it; he then introduced a series of reflections about the fact that, as the Heavens was only made of a quintessence, the contrary elements necessary for corruption and generation could not be produced in it, and he concluded with the rhetoric question: in what way could the Heavens corrupt the Heavens to generate the Heavens? After this question, he proposed a long digression about parallax and questions, confused if not even wrong, about geometric-astronomical theorems, and he then presented ideas from the scholastic tradition about lunar spots and the Via Lattea, until a further discussion on the position of the *nova*. There were also a couple of chapters on the so called "judicial astrology", where Lorenzini discussed the influence of the *nova* on seasons and harvests, on public health and on physical and moral conditions of humanity.

An answer to Lorenzini arrived very quickly. Six weeks after the publication of the *Discorso intorno alla nuova stella*, a short booklet was published in Padua with the title *Dialogo de Cecco di Ronchitti da Bruzene in perpuosito della Stella Nuova.*

The marginal notes of this booklet contained precise references to Lorenzini's text (*Opere*, vol. II, pp. 310-34). Written in Paduan dialect, the *Dialogo* has two main characters, Natale and Matteo: the first one gives an account of the ideas of a Paduan "letterato" (Lorenzini) and the other one ribs these ideas by using Galilean inspired arguments presented in a simple way and with examples taken from everyday life. It is nowadays ascertained that the text was written jointly by Galileo and Girolamo Spinelli, a young Benedictine monk of Galileo's circle. This circle included intellectuals and churchmen – like the canon Antonio Querengo, to whom the *Dialogo* is dedicated – all interested not only in the new developments of science but also in the Paduan dialect and his great mentor, Ruzzante (alias Angelo Beolco). And not only the choice of the Paduan dialect is consistent with Beolco's ideas, but also the choice of the rough characters, who show how the wisdom *snaturale* can prevail on the book based culture.

The *Dialogo*, characterised by an irony particularly manifest in the original dialectal version, starts with a conversation on the hypothetical correlation between the drought of the countryside and the appearance of the new star. But if it is really a star, says Matteo, "as it is so far away", it will be difficult to prove that it is the cause of the drought. Natale observes that a Paduan "letterato" supports in a "librazuolo" that the *nova* is located in the sublunar region. Matteo then asks whether the author of the booklet is an expert of measures and, as he is told that the author "l'é Filuorico" ("he is a philosopher"), he reacts with indignation wondering "what has his philosophy to do with measuring?": the work of mathematicians is intended to carry out measures and they have to be asked about the position of the star. All right, answers Natale, the "letterato" also says that mathematicians carry out measures but they do not understand anything, because they have concluded from their measures that the star is far away and this implies an unacceptable generation and corruption of the Heavens. But this should not matter to mathematicians, answers Matteo upset, because they concern themselves with measuring and not with the essence of things or the substance of what they measure: "even if the star was made of polenta, they could nevertheless observe it". The readers of the *Dialogo* are thus warned: the controversy on the Stella does not concern the simple field of astronomical observation but it involves the core of philosophical tradition consolidated beliefs. And these beliefs are to be criticised and ribbed in the following pages of the booklet.

Here is the argumentation proposed. Of course, Matteo argues, it is not possible for the moment to prove that the new star is really a star like all the others, but at the same time one can propose a series of conjectures. For instance, as it is not possible that "all the stars in the Heavens could be seen" (a recall to Giordano Bruno's idea), some of them could have merged to give birth to a new visible star, or the *nova* could have been formed in the air and it could then have raised in the Heavens. As a matter of fact, though this star seems peculiar because of its sudden appearance, which suggests a forthcoming disappearance, who could support that the stars are not, like the Earth, slowly changing, with apparently unperceivable changes? All such arguments are based on the unity of the physics of the Universe, without any distinction between Earth and Heavens. Natale tries to answer to these reflections mentioning once more the "librazuolo", which says that according to Aristotle, the Heavens could not move any longer if a star was added. But in fact, as Matteo points out, this would not be such a big problem, because there are many people "ed anco di buoni" ("and good ones") who believe that the Heavens does not move at all. This evident reference to Copernicans is explicitly written down in a marginal note in the Paduan edition of the *Dialogo*.

This was more than enough to drive Galileo to publish the *Dialogo* under a pseudonym. Such a practice was common at that time, but here the issues were particularly delicate and they had already started shaking the consolidated powers within and outside the University.

4. The Copernican system at work

The content of the *Dialogo de Cecco di Ronchitti* and the studies on local motions enable Galileo to seize all the opportunities offered by the new instrument, the telescope. The trust in the observation without prejudices and the abandoning of the Aristotelian-Ptolemaic system are the base of his fundamental discoveries. At the same time, his growing trust in the Copernican system enables him to obtain in a natural way some of the consequences of these observations. In particular, as we can read at the end of the *Sidereus Nuncius*, the discovery of Jupiter's moons is a demonstration of the inadequacy of those who, though accepting at first the Copernican system, become anti Copernican because they do not accept the idea that the Moon revolves around the Earth while both revolve around the Sun in one year. As a matter of fact, we now see that four moons are revolving around Jupiter, and all these celestial bodies together revolve around the Sun in twelve years. We still do not know how this can happen, but it happens, Galielo says (*Opere*, vol. III, pp. 51-96, p. 95). Fourteen years later, in the letter to Francesco Ingoli of 1624 (*Opere*, vol. VI, pp. 509-561), with his famous metaphor of the ship (*Opere*, vol. VI, pp. 547-9), Galileo will provide "physics arguments" (*Opere*, vol. VI, p. 534) to support the impossibility to prove another of the paradoxes against the Copernican system: if the Earth moves, how can a stone fall perpendicularly to the Earth's surface? Staying on the Earth we cannot decide if the Earth (ship) is motionless or in motion. The physics of local motions can help in understanding questions related to celestial motions. A step forward in the construction of a physics for the Copernican system.

But let's go back to astronomical questions. In the "Postscriptum" of the tables on the *Costitutiones of the Medicee* added to the *Istoria e dimostrazioni intorno alle macchie solari* (*Opere*, vol. V, pp. 247-9, p. 248), Galileo concludes that, in order to explain the observed variations of the length of the eclipses of Jupiter's moons, it is necessary to take into account the fact that the shadow cone of the planet also depends on the annual revolution motion of the Earth [besides the dependence on the "diverse latitudini di Giove" ("different latitudes of Jupiter") and "dall'essere il pianeta che si eclissa de i più vicini o de' più lontani da esso Giove" ("on the fact that the planet that is eclipsed may be one of the closest or most distant from Jupiter")]. Once again the idea of a proof of the Copernican system. The same idea that will bring Galileo to hypothesise, in the same year, that the changing form of Saturn (sometimes with two satellites very close to the two opposite sides of the planet, sometimes alone) could depend on the relative position of the planet with regard to its source of illumination (the Sun) and to the observer (the Earth in its revolution motion)†.

† Galileo's hypothesis emerges from a letter written by Agliuchi to Galileo on 13th July 1613 in answer to a letter of Galileo now lost (*Opere*, vol. XI, pp. 532-5, p. 532).

The adherence to observed and experimental facts and the research of their explanation within the most advanced scientific knowledge make Galileo a modern scientist. A modernity that we can see also in his contrariety to use Pythagoric or Platonic arguments so current at that time (much diffused, only to mention another great scientist of that time, in Kepler's work). To those who tried to explain, with a-priori arguments, why the moons around Jupiter were right four and, on the basis of these arguments, proposed the existence of other moons around Jupiter or around other planets†, Galileo answers indirectly in the letter to Dini written on 21th may 1611 (*Opere*, vol XI, pp. 105-16, p. 115), reaffirming his adherence to facts: I have observed four (moons) around Jupiter and two moons around Saturn, "non posso negare né affermare cosa alcuna" ("I cannot deny or affirm anything") about whether others exist.

With good cause, many of his contemporaries greeted Galileo as a new Columbus or a new Amerigo Vespucci.‡ A similar acknowledgement was to be addressed in 1904, about three hundred years after the discovery of Jupiter's moons, to J. J. Thomson, the scientist who discovered the electron, the first elementary particle (Langevin, *The History of Modern Physics*).

References

Opere di Galileo Galilei, Edizione Nazionale a cura di Antonio Favaro (referred to simply *Opere* in the text)

Langevin, P., *"The Relations of Physics of Electrons to other Branches of Science"*, in K. R. Sopka (ed.), *Physics for a new Century, Papers Presented at the 1904 St. Louis Congress, The History of Modern Physics* 1986, vol. 5, American Institute of Physics, pp. 195-230, p. 195

† See for instance the letter of Altobelli to Galileo on 17th April 1610 (*Opere*, vol. X, pp. 317-8, p. 317), and the *Dissertatio* of Kepler (*Opere*, vol III, pp. 100-25).

‡ Galileo is compared to Columbus by Orazio dal Monte (letter to Galileo on 16th June 1610, *Opere*, vol. X, pp. 371-2, p. 372) and by Kepler in his *Dissertatio* (*Opere*, vol. III, p. 119), and to Amerigo Vespucci by Ottavio Pisani (two letters to Galileo, the first on 15th September 1613, *Opere*, vol. XI, p. 564-5, p. 564, and the second on 18th December 1613, *Opere*, vol. XI, p. 608-9, p. 608).

Galileo's Medicean Moons: their impact on 400 years of discovery
Proceedings IAU Symposium No. 269, 2010
C. Barbieri, S. Chakrabarti, M. Coradini & M. Lazzarin, eds.

doi:10.1017/S1743921310007234

The telescope in the making, the Galileo first telescopic observations

Alberto Righini

Dipartimento di Fisica ed Astronomia dell'Università degli Studi di Firenze
Largo E. Fermi 2
50125 Firenze, Italy
email: alberto.righini@unifi.it

Abstract. In the first part of this paper we briefly discuss some historical constraints useful for understanding when Galileo for the first time aimed his telescope to the Moon which most probably was the first astronomical object observed by the Scientist. In the second part we discuss the dates of the observations on which the etchings, published in the *Sidereus Nuncius*, are based. It results that the five etchings refer to observations performed in December 1609 and January 1610. The measurement of the position, of some peculiar structures of the lunar surface clearly represented by Galileo in the etchings, shows that he was very careful in trying to produce a faithful graphical rendering of what he was observing.

Keywords. History and Philosophy of Astronomy, Galileo Galilei, telescope.

---

1. Introduction

The year 2009 was devoted to astronomy, to celebrate two great events occurred in 1609, four hundred years ago. In that year, in Prague, the astrologer (and astronomer) Johannes Kepler published his *Astronomia Nova* in which he enunciated the first two of the three laws, on planetary motions which we, today, designate as *Kepler's Laws*; in Padua, a Professor of Mathematics (but he was also teaching Astronomy and Cosmography) Pisan in birth but Florentine by adoption, built a powerful tool for naval warfare and for observing the universe by transforming an optical toy, used by ladies and gentlemen for spying their secret loves in the Italian gardens. Professor Galilei son of the late Vincentio Galilei, nobleman in Florence (this was the name of that ingenious lecturer who desperately wanted to return to Florence) was a well known Professor in Padua, very skilled to occupy all available areas of academic power, and a good friend of influent personalities in Venice; he was also well introduced in many cultural circles of Europe. The intelligent use of his new optical instrument, which he defined as *cannon* or *optical cannon*, brought several advantages to Galileo: he obtained tenure for life at the University with a salary of 1000 florins per year, and the Venetian Senate asked him to build twelve telescopes for the Navy. Galileo was not only a great academic, very well acquainted with the University tribal uses, but also a skilled and careful researcher of things of nature, in the language of the time he was a *natural philosopher*. In this context we should consider the Galileo's almost spontaneous deed of aiming his precious tube to the sky: this was a deed of a physicist not of an astronomer, a deed of someone curious about the physical nature of the celestial objects; the Moon was most probably the first object to be observed, and therefore we may ask ourselves, once again, when this happened (A. Righini 2009).

2. About the date of the first Galilean observation of the Moon

Galileo himself tells us the story of his re-discovery of the telescope in a letter to his brother in law Benedetto Landucci dated August 29, 1609. In this letter Galileo states that *in Venice news arrived that a Fleming had presented to the Count Maurice [of Nassau] a glass by means of which distant objects could be seen distinctly. Upon hearing these news I returned in Padua and set myself to thinking about the problem. The first night after my return I solved it, and on the following day I constructed the instrument and sent word to some friend in Venice with whom I had discussed the matter the day before.....* (Favaro 1968); the telescope, that Galileo had perfected had 8-10 enlargements. We can therefore consider that in Venice the news of the Dutch *spyglass* goes back to the end of June or early July 1609, in fact, in that period Augusto Bartoli, Tuscan ambassador in Venice, writes to Belisario Vinta, State Secretary of the Grand Duke in Florence, that there was around in the *Serenissima Republica* (Venice) a person willing to sell the secret of a device to see far.

Galileo employs one or two months to improve the telescope and on August 25th, he submits the telescope, as an invention by his own, to the *Doge* Leonardo Donato. On August 29th; from Florence, Enea Piccolomini Aragona writes to Galileo on behalf of the young Grand Duke of Tuscany Cosimo II asking a telescope, or the recipe for building one in Florence. We have also another letter which reaches Galileo from his friend Don Antonio de' Medici in Florence asking him a telescope, dated September 19. We are sure therefore that Galileo until September 19, 1609, after the improvements of the telescope did not return to Florence, as he was accustomed to do every year before the beginning of the academic year. In early November we find in the Galilean correspondence a letter in which our scientist writes to Belisario Vinta that he just returned to Padua. Everything seems to indicate that Galileo was in Florence after September 19, 1609 and he was back in Padua before October 30,1609.

Given as certain the presence in Florence in the Fall of 1609, which is confirmed also by other sources (Drake 1978), let us examine the letter written by Galileo to Belisario Vinta on January 30, 1610 in which he announces that the *Sidereus Nuncius* (Starry Messenger) was in press. Galileo describes to the Tuscan Secretary of State the surface of the Moon and claims that he had already shown our satellite to the young Cosimo II Gran Duke of Tuscany, although with an imperfect telescope. We may therefore infer that the first observations of the Moon made by Galileo may be dated before October 30, 1609. From the analysis of lunar phases of that period we understand that the first days of October are those in which perhaps the Grand Duke of Tuscany has observed the mountains and valleys on the Moon, albeit imperfectly, with a telescope with a ten enlargements. Of course it is not reasonable to think that Galileo, in the presence of the Grand Duke decided to turn the telescope to the sky without knowing what Cosimo II would have observed and it is therefore logical to consider that the first observations of the Moon by Galileo date back at least to previous lunation. With these arguments we can conjecture that the first observations of the Moon, by Galileo were performed in Padua in early September of 1609 but with a telescope delivering no more than 10 enlargements, we discard the lunation of August because at that time the telescope was still very imperfect and we do not find any hint on Galileo's correspondence of an astronomical use of the new instrument.

We should note that the men of the Gran Duke of Tuscany secret service, had already aimed to the sky the imperfect Dutch telescopes they had in their hands, reporting that using those spyglasses some nebulae were resolved in a multitudes of stars which were not visible with the naked eye.

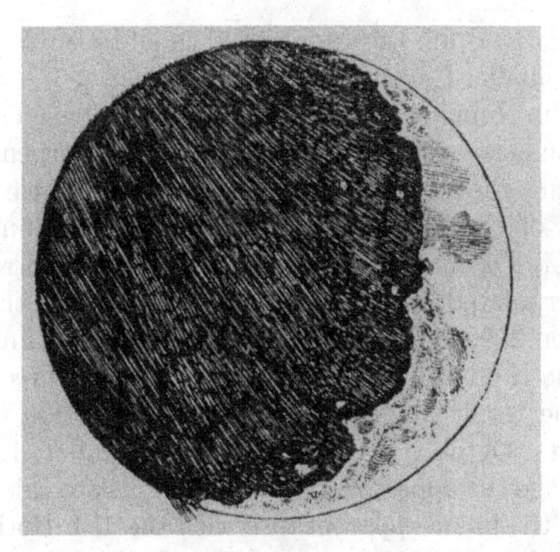

Figure 1. The first etching of the Moon in the *Sidereus Nuncius* (Starry Messenger). The moon shows an illuminated fraction of $(21 \pm 2)\%$.

3. The Lunar etchings in the *Sidereus Nuncius*

Discussing about the Galileo's first telescopic observations, we should discuss also about the epoch of the lunar observations which gave origin to the etchings of the lunar surface we find in Galileo's *Sidereus Nuncius* (Starry Messenger), the booklet published in March 1610, which announces to the entire world the new celestial discoveries. The idea to print the *Nuncius*, as we know it now, may be dated some days after January 7, 1610 when Galileo realized that those stars, that he (or some of his friends) had observed around Jupiter, were four Jovian moons. However Galileo probably had already in December 1609 the idea to print his observations of the Moon carried out in November and December 1609.

Proof that Galileo had already collected a considerable experience in observing the Moon is given by the letter he writes just on January 7, 1610 to Don Antonio de' Medici in Florence, in which he explains how difficult is to work with the telescope. In this letter Galileo examines various aspects of the lunar surface by enriching the words with many approximate drawings. It should be noted that we do not have the original letter but a copy of it and possibly the letter may be one he had never sent. In the letter closure, Galileo incidentally writes, that on that evening he noted Jupiter accompanied by three fixed stars, invisible with the naked eye for their smallness and aligned along the Ecliptic with the planet. In this letter to Don Antonio, Galileo claims to have a telescope of 20 enlargements and not 30 as he later states in the *Nuncius*.

A dating of the *Nuncius* etchings must take account of what was written by Guglielmo Righini (GR)(G. Righini 1978), who studied in depth the problem, with whom we disagree in part. The procedure adopted by GR is based on the measurement of the illuminated fraction of the Moon and on the position of the center of the drawing to compute the libration. This method implicitly admits that Galileo, graphically transposing telescopic observations, was able to maintain the relative positions of the observed structures. As known in the Galilean (or Dutch) telescope it is not possible to overlap a reference grid on the image due to optical structure of the instrument, however we know that on the occasion of observation of Jupiter satellites, Galileo was able to solve this problem watching with the eye, which was not applied to the telescope, a distant sheet of paper on which he draw a set of concentric circles or an orthogonal grid. The brain of the observer merges the two images by providing a reference scale, it may be that this device was used by Galileo also to draw the Moon. We should also consider that the field of Galileo's telescope was smaller than the Moon angular size.

The first etching of the Nuncius (Figure 1) refers to a phase between New Moon and First Quarter. In this image, simply tracing the chords perpendicular to line joining the cusps and evaluating the ratio of the chord lengths which lie in the bright and those in the dark part of the Moon we may evaluate the illuminated fraction which results to be $(21 \pm 2)\%$. Of course, the terminator, which correctly Galileo draws as irregular, must be interpolated. Using the data available on the web thanks to JPL project Horizon (Horizon 2009) it is possible to compute the illuminated fraction of the Moon observed in Padua from July 1609 to February 1610. Three dates of the year 1609 are compatible with the result about the illuminated fraction of the Moon and with the historical constraints: October 2, November 1, December, 1. To make a choice we superimposed on the image the appropriate Stonyhurst disk to calculate the selenographic latitude of the center of the lunar apparent disc with the JPL Horizon program, properly rotated so that the coordinates of the center of the Mare Crisium, clearly indicated by Galileo, fit at best. Than we selected four areas: Plana, Capella, Rabbi Levi and Baco changing the epoch of the observation trying to obtain the best fit. This procedure has led us to discard the dates of November 1,1609 and October 2,1609, and converging on December 1, 1609 with a mean absolute error of 2.5% in unit of lunar diameter in the positioning which should be compared with the value of 4.6% for the November date and again 2.5% for the October date. The December date is the more plausible for the following considerations: a) on October 2, Galileo was in Florence and did not have a proper telescope, b) the illuminated fraction on December 1st is more in agreement with the drawing, c) Galileo that day would have had more time to work, d) Galileo must have spent the whole month of November to improve the telescope, while producing the instruments commissioned by the *Serenissima* and only at the end of the month he had in his hand an appropriate instrument delivering 20 enlargements. We want once again to pay a tribute to Galileo physical and experimenter skill: the small error we found in assessing the coincidence of the lunar structures means that Galileo's drawings have clearly the meaning of measurements, and not only, as suggested by O. Gingerich Gingerich 2009, an attempt simply to give the idea of the mountains and plains observed on the Moon with his instrument.

The following two etchings (Figure 2, drawings 2 and 3) are dated by GR relying mostly on libration, the first of the two, which refers to the Moon in the first quarter is attributed to December 3, 1609 and the third, which represents the Moon at last quarter, to December 18,1609 with a libration in latitude -5.9. On these dating we do not object. GR dates the fourth drawing (Figure 2, drawing 4), representing a Gibbous Moon at an age between the Full Moon and the Last Quarter, a few days before December 18 (we would say the 17). It should remarked that in this case the etching in the Nuncius is not in chronological order with the previous one (Figure 2, drawing 3); but we could make the hypothesis, considering its position in the book, that this etching might refer to the day January 16, 1610, for that the astronomy is the same which supports the dating to December 17. The fifth etching is identical with the third one, it is again the Last Quarter and technically seems that the printer uses the same plate which printed the third image. GR suggests that this image is an editing error. Maybe! However we may consider that Galileo has really observed the Moon at the last quarter on January 18, 1610 (in this date the astronomical circumstances were similar to those of the December 1609 Last Quarter), and, perhaps, to speed up the publication of the booklet (and to save money) he deliberately duplicated the third etching.

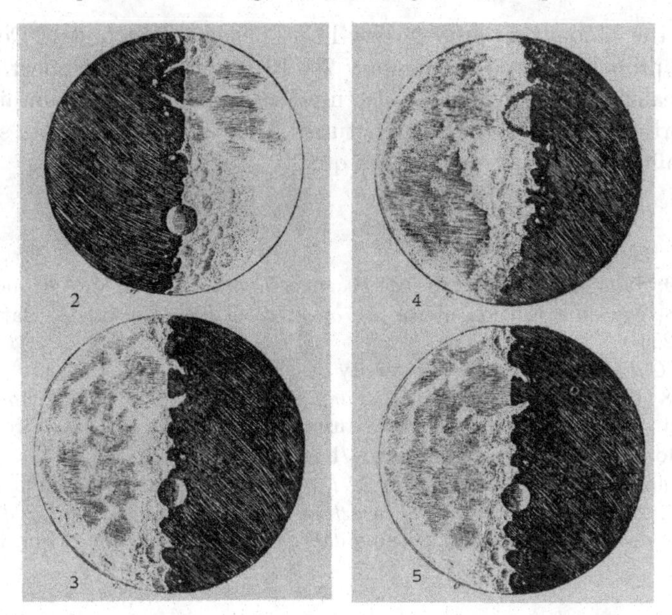

Figure 2. The other four etchings of the Moon in the *Sidereus Nuncius*. The dates of the observations of the etchings 2 and 3 are respectively: December 3 and December 18, 1609. It may be guessed etching 4 refers to an observation made on December 17, 1609 or on January 16, 1610. The etching 5 might be an editorial error because identical to 3 (or might also be inserted by Galileo to represent an observation performed on January 17, 1610 since the Moon in that night was very similar to that of the December 18, 1609).

Let us compare our results with those exposed in two recent papers one signed by E. Whitaker (Withaker 2009), which appeared in the catalog of the exhibition on Galileo held at *Palazzo Strozzi* in Florence to celebrate 400 years from the first telescopic observations, and the other signed by O. Gingerich (Gingerich 2009) on the puzzling copy of the *Sidereus Nuncius* owned by the Martayan and Lan Company in New York. Withaker proceeds to an accurate dating of the lunar drawings existing in the National Library in Florence and dates the etching in the hypothesis that those drawings were the originals supplied by Galileo for making the etchings. The first lunar etching in the Nuncius (Figure 1) is dated by Witaker November 30 and confirmed by Gingerich, we would prefer, for the considerations set above, December 1, the second and the fourth are dated December 3 and December 18 as already stated by Guglielmo Righini, dates on which we also agree. The third is assigned by Witaker to December 17 in agreement with our results, but Gingerich also suggests that it might refer to January 16, 1610. It is very interesting that among the Galilean materials, in the Florence National Library, Whitaker identifies an image which can surely dated to January 19, 1610 thanks to a nearby bright star recorded by Galileo, which might be the θ LIB occulted that night by the Moon, demonstrating that Galileo continued to observe the Moon at least in the two first decades of January 1610 while writing his Starry Messenger.

4. Conclusions

We are convinced (with many others) that Galileo, as soon as he had available a suitable instrument, aimed his 10 enlargements telescope to the sky and in particular to the Moon; this happened in early September of the year 1609. When he had available a more advanced telescope delivering 20 enlargements, thanks to experience gained in constructing telescopic ordered by the Doge, in November 1609, he undertook a systematic observation of the Moon, this project very likely began in the first of December of the year 1609 and probably ended on January 20, 1610 (as far as it concerned the Starry Messenger). Realizing his drawings Galileo used some artifices, to maintain in his drawings the proportions of the observed structures. And, as usual when we study in depth the work of Galileo, we are amazed by the quality of his work!

References

Righini, A. 2009, *Sulle date delle prime osservazioni lunari*, Giornale di Astronomia, 35, 3

Favaro, A. 1968, National Edition of the *Le opere di Galileo Galilei*, G. Barbera, Florence, Vol. X, p. 253

Drake, S. 1978, *Galileo at work*, The University of Chicago Press, Chicago

Righini, G. 1978, *Contributo alla interpretazione scientifica dell'opera astronomica di Galileo*, Supplemento agni annali dell'Istituto e Museo di Storia della Scienza, Firenze

JPL Program Horizon, http://ssd.jpl.nasa.gov/horizons

Gingerich, O. 2009, *The curious case of the M-L Sidereus Nuncius*, Galilaeana VI, 2009

Withaker, E. A. 2009, *Identificazione e datazione delle osservazioni lunari di Galileo*, in P. Galluzzi *Galileo Immagini dell' Universo dall'antichità al telescopio* Giunti, Firenze

Galileo's Medicean Moons: their Impact on 400 years of Discovery
Proceedings IAU Symposium No. 269, 2010
C. Barbieri, S. Chakrabarti, M. Coradini, & M. Lazzarin, eds.
doi:10.1017/S1743921310007246

The Appearance of the Medicean Moons in 17th Century Charts and Books—How Long Did It Take?

Michael Mendillo

Department of Astronomy, Boston University
725 Commonwealth Avenue, Boston, MA, USA
email: mendillo@bu.edu

Abstract. Galileo's talents in perspective and *chiaroscuro* drawing led to his images of the Moon being accepted relatively quickly as the naturalistic portrayal of a truly physical place. In contrast to his resolved views of the Moon, Galileo saw the satellites of Jupiter as only points of light (as with stars). He thus used star symbols in *Sidereus Nuncius* (1610) for the moons, in constrast to an open disk for Jupiter. In this paper, I describe methods used in subsequent decades to portray objects that could not be seen in any detail but whose very existence challenged the scholastic approach to science. Within fifty years, the existence of the moons was such an accepted component of astronomy that they were depicted in the highly decorative "textbook" *Atlas Coelestis seu Harmonia Macrocosmica* by Andreas Cellarius (1660). Other symbolic methods, ranging from the routine to the dramatic, were used in subsequent centuries to portray the moons. Actual photographs using ground-based telescopes were not possible until the 20th century, just years before cameras on spaceflight missions captured the true details of the Medicean Stars.

Keywords. Atlases, History and Philosophy of Science, Planets and Satellites: Jupiter

1. Introduction

Galileo's observations of the moons of Jupiter defined what we now call "discovery-mode" astronomy. Their presence in the night sky was a completely unanticipated effect. Transient astronomical phenomena had, of course, occurred before Galileo's use of a telescope on the night of 7 January 1610. Meteors, comets and novae had been chronicled since ancient times, along with major uncertainties about their locations in the sky (atmospheric or celestial), as well as origins (natural or divine). The Galilean moons were soon accepted as permanently present bodies, objects that had escaped detection, and thus they were added to the inventory of known celestial bodies. The fact that they were not transient events required a fundamental re-thinking of what had been considered a known and stable ensemble of heavenly bodies. This led to re-appraisals of past hypotheses and the contemplation of new physical mechanisms. Pivotal to these tasks were the evaluation of evidence and the determination of the proper roles of authority: What was actually being seen? How reliable were the observations? What were their implications?

2. A New Role for Imagery

The responses from various communities to Galileo's moons were also precedent-setting aspects for new modes of discovery. They set the stage for the modern scientific requirement of experimental reproducibility by independent researchers. Verification versus falsifiability, concepts introduced later but today routine components of the scientific method, saw their origins in the announcement of *Sidereus Nuncius.*

Among the many forms of scientific revolution launched by Galileo in 1610, the one treated here is the role of visualization in astronomy. Today it is routine to consider astronomy to be the most visual of the sciences—complete with a "picture of the day" website (http://apod.nasa.gov/apod/). Yet prior to Galileo, the pictorial component of astronomy was confined almost exclusively to how one portrayed circular motion: the paths of celestial objects either circled the Earth (Ptolemaic), the Sun (Copernican), or a blend of the two (Tychonic). Indeed, one of the most reproduced figures in the history of science (Barrow 2008), perhaps second only to Leonardo's 1490 drawing of the *Vitruvain Man* (with outstretched arms and legs within a circle), was the simple but elegant drawing of the heliocentric cosmology that Copernicus put into *De Revolutionibus* (1543).

Galileo introduced a new form of celestial imagery in *Sidereus Nuncius* that went far beyond details of organization. His drawings of the Moon, a place with mountains, valleys and patterns of shadows similar to those found on Earth, contradicted the Platonic-Aristotelian doctrine of the perfection of bodies beyond the terrestrial domain. The realistic portrayal of Jupiter's moons was quite a different matter. Galileo's telescope was far too primitive to show any details on the disks of the four jovian moons. Indeed, it would take over 300 years before the ground-based telescopes at Pic du Midi gave hints of surface features upon those moons (Murray 1975, Dollfus 1998), and images from satellite missions soon followed (as discussed throughout this volume).

Galileo coined the phrase *Medicean stars* for Jupiter's satellites, and portrayed them using actual symbols of stars in *Sidereus Nuncius.* He used the terms *planet* and *star* interchangeably, and both words were correct usage within the prevailing Aristotelian terminology (Van Helden 1989). Copernicus, of course, had no reason to address the issue since the moons were unknown in his day, and it was Kepler who came up with the word *satellite* to describe objects in orbits around planets.

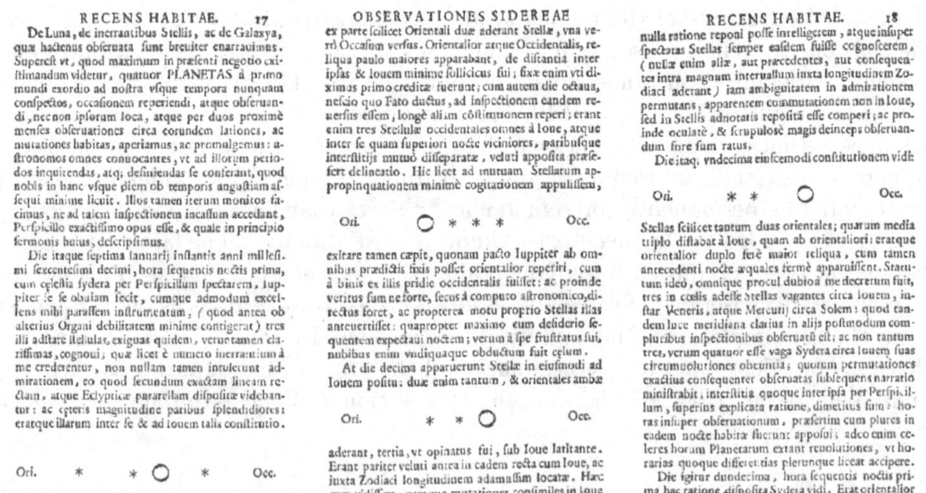

RECENS HABITAE. 17

De Luna, de inerrantibus Stellis, ac de Galaxya, quæ hactenus obſeruata ſunt breuiter enarrauimus. Supereſt vt, quod maximum in præſenti negotio exiſtimandum videtur, quatuor PLANETAS à primo mundi exordio ad noſtra vſque tempora nunquam conſpectos, occaſionem reperiendi, atque obſeruandi, necnon ipſorum loca, atque per duos proximè menſes obſeruationes circa eorundem lationes, ac mutationes habitas, aperiamus, ac promulgemus: aſtronomos omnes conuocantes, vt ad illorum periodos inquirendas, atq; definiendas ſe conferant, quod nobis in hanc vſque diem ob temporis anguſtiam aſſequi minime licuit. Illos tamen iterum monitos facimus, ne ad talem inſpectionem incaſſum accedant, Perſpicillo exactiſſimo opus eſſe, & quale in principio ſermonis huius, deſcripſimus.

Die itaque ſeptima Ianuarij inſtantis anni milleſimi ſexcenteſimi decimi, hora ſequentis noctis prima, cum cęleſtia ſydera per Perſpicillum ſpectarem, Iuppiter ſe ſe obuiam fecit, cumque admodum excellens mihi paraſſem inſtrumentum, (quod antea ob alterius Organi debilitatem minime contigerat) tres illi adſtare ſtellulas, exiguas quidem, veruntamen clariſſimas, cognoui; quæ licet è numero inerrantium à me crederentur, non nullam tamen intulerunt admirationem, eo quod ſecundum exactam lineam rectam, atque Eclypticæ pararellam diſpoſitæ videbantur: ac cęteris magnitudine paribus ſplendidiores: eratque illarum inter ſe & ad Iouem talis conſtitutio.

Ori. * * ○ * Occ.

E ex parte,

OBSERVATIONES SIDEREAE

ex parte ſcilicet Orientali duæ aderant Stellæ, vna verò Occaſum verſus. Orientalior atque Occidentalis, reliqua paulo maiores apparabant, de diſtantia inter ipſas & Iouem minime ſollicicus fui; fixæ enim vti diximus primo creditæ fuerunt; cum autem die octaua, neſcio quo Fato ductus, ad Inſpectionem eandem reuerſus eſſem, longè aliam cōſtitutionem reperi; erant enim tres Stellulæ occidentales omnes à Ioue, atque inter ſe quam ſuperiori nocte viciniores, paribuſque interſtitijs mutuò diſſeparatæ, veluti appoſita præſefert delineatio. Hic licet ad mutuam Stellarum appropinquationem minimè cogitationem appuliſſem,

Ori. ○ * * * Occ.

exitare tamen cæpit, quonam pacto Iuppiter ab omnibus prædictis fixis poſſet orientalior reperiri, cum à binis ex illis pridie occidentalis fuiſſet: ac proinde veritus ſum ne forte, ſecus à computo aſtronomico, directus foret, ac propterea motu proprio Stellas illas anteuertiſſet: quapropter maximo cum deſiderio ſequentem expectaui noctem; verum à ſpe fruſtratus fui, nubibus enim vndiquaque obductum fuit cęlum.

At die decima apparuerunt Stellæ in eiuſmodi ad Iouem poſitu: duæ enim tantum, & orientales ambæ

Ori. * * ○ Occ.

aderant, tertia, vt opinatus fui, ſub Ioue latitante. Erant pariter veluti antea in eadem recta cum Ioue, ac iuxta Zodiaci longitudinem adamuſſim locatæ. Hæc cum vidiſſem, cumque mutationes conſimiles in Ioue nulla

RECENS HABITAE. 18

nulla ratione reponi poſſe intelligerem, atque inſuper ſpectatas Stellas ſemper eaſdem fuiſſe cognoſcerem, (nullæ enim aliæ, aut præcedentes, aut conſequentes intra magnum interuallum iuxta longitudinem Zodiaci aderant) iam ambiguitatem in admirationem permutans, apparentem commutationem non in Ioue, ſed in Stellis adnotatis repoſitã eſſe comperi; ac proinde oculatè, & ſcrupuloſè magis deinceps obſeruandum fore ſum ratus.

Die itaq; vndecima einſcemodi conſtitutionem vidi:

Ori. * * ○ Occ.

Stellas ſcilicet tantum duas orientales; quarum media triplo diſtabat à Ioue, quam ab orientaliori: eratque orientalior duplo ferè maior reliqua, cum tamen antecedenti nocte æquales fermè apparuiſſent. Statutum ideò, omnique procul dubio à me decretum fuit, tres in cœlis adeſſe Stellas vagantes circa Iouem, inſtar Veneris, atque Mercurij circa Solem: quod tandem luce meridiana clarius in alijs poſtmodum compluribus inſpectionibus obſeruatũ eſt; ac non tantum tres, verum quatuor eſſe vaga Sydera circa Iouem ſuas circumuolutiones obeuntia; quorum permutationes exactius conſequenter obſeruatas ſubſequens narratio miniſtrabit; interſtitia quoque inter ipſa per Perſpicillum, ſuperius explicata ratione, dimetitus ſum: horas inſuper obſeruationum, præſertim cum plures in eadem nocte habitæ fuerunt appoſui; adeo enim celeres horum Planetarum extant reuolutiones, vt horarias quoque differentias plerunque liceat accipere.

Die igitur duodecima, hora ſequentis noctis prima hac ratione diſpoſita Sydera vidi. Erat orientalior

E 2 Stella

Figure 1. The pages of *Sidereus Nuncius* showing the first portrayals of the Medicean Stars on the nights of January 7th (left) and 8th (middle) and 10th (right), 1610. (Courtesy of the Houghton Library, Harvard University, *IC6.G1333.610sa.)

Here a brief survey is offered of how Jupiter's satellites were portrayed in the years after 1610. Attention is given to the amount of time it took for other astronomers to show in graphical forms that Jupiter was a center of motion, as well as to the types of symbols they used for Jupiter and the moons. The final topic addressed is the iconographic appearances of the moons in what might be considered the 17th Century "textbook" for astronomy, i.e., in a publication not reporting new observations but meant for summarizing the status of the field to both general and technical readers.

3. 17th Century Depictions of Jupiter's Moons

Discovery and Controversy

Galileo's announcement that Jupiter had four companions in the sky was done with the simple insertion of linear drawings embedded within his published text. That is, there were no separately engraved plates (see Figure 1). He used an open disk to portray Jupiter and star symbols for the moons, presumably because he did not want to portray them as larger and smaller versions of the same types of bodies. Their reality was not accepted by all interested parties. Some simply doubted (or denied) their existence and suggested that they were merely the products of defects within the lenses used (Drake 2001). This was not classical falsification, as defined by Karl Popper (see discussions in Kuhn 1962) and as practiced today, because the rejections were made without offering evidence. I could not find an account by an experienced observer, using a telescope equal to or better than Galileo's, publishing a map of the sky showing Jupiter and fixed stars (and hence reliability), but no objects associated with Jupiter. Thus, there are no artistic renderings attesting the non-existence of the Medicean stars.

Validation is a richer field of pursuit. Van Helden (1989) points out that, unknown to Galileo, when Jupiter became visible in the Fall of 1610, his Medicean Moons were spotted by Thomas Harriot (1560-1621) in England, and by Joseph Gaultier de La Vatelle (1564-1647) and Nicolas-Claude Fabri de Peiresc (1580-1638) in Aix-en-Provence. The first to publish a book with observations was Simon Marius (1573-1625), a German astronomer who had studied with Tycho Brahe in Prague and spent the years 1602-05 as a student in Padova. He was implicated in a plagiarizing scandal with one of Galileo's former students (Baldesar Capra) concerning the publication of one of Galileo's earlier works. Returning to safer quarters in Germany for a Lutheran under investigation in Veneto, Marius became an even greater annoyance to Galileo when he published *Mundus Jovialis* in 1614. Marius claimed that he had been observing Jupiter's "four planets" since 1609 (i.e., prior to Galileo) and thus claimed the right to name them, not after the Medici family but after Jupiter's love conquests: Io, Europa, Ganymede and Callisto (a suggestion he received from Kepler). Marius' *World of Jupiter* contained tables of satellite positions but only one figure that showed the moons using star symbols in circular orbits about a solid disk representing Jupiter (Figure 2(a)). Predictably, Galileo's response was strong. In *The Assayer* (1623), he wrote:

> *"Notice the craft with which he tries to show himself prior to me.... What he neglects to mention to the reader, that since he is outside our church and has not accepted the Gregorian calendar, the seventh of January of 1610 for us Catholics, is the same as the twenty-eighth day of December of 1609 for those heretics.*
> *So much for the priority of his pretended observations."*

Thus, even if Marius' first recorded observation (29 December 1610) is accepted, it was one day *after* Galileo's! (from Drake, 1960, and The Galileo Project, Rice University).

The astronomical community since that time has been far more gracious to Marius. Recognized as a good observer and a careful ephemeris maker, the International Astronomical Union (founded in 1919) decided in 1935 to honor him with the lunar crater named Marius, and then in 1979 by naming a feature on Ganymede photographed by Voyager *Marius Regio*.

Subsequent Findings

Rene Descartes (1596-1650) published his *Principia Philosophiae* in 1644, and in his treatment of the Copernican system of cosmology he updated the iconic image of a sun-centered system with Jupiter having four satellites. Descartes appears to be the first to use the same artistic symbol (open disk) for both the planet and its moons (Figure 2(b)) — implying, perhaps, a similarity in physical make-up (in *Il Dialogo* (1632), Galileo used an open disk for Jupiter and solid disks for the moons). Two years after Descartes, the Italian lawyer and astronomer Francesco Fontana (1602-1656) published in Naples what we might consider the first picture-book for astronomy. *Novae Coelestium Terrestrium Rerum Observationes* (1646) contained wood-cut prints of his telescopic appearances of the Moon, Mercury, Mars, Jupiter and Saturn. Fontana's observations of Jupiter and its moons started in 1630 and continued up to the year prior to publication (1643). Figure 2(c) shows an example of his jovian images: the disk of the planet is now resolved

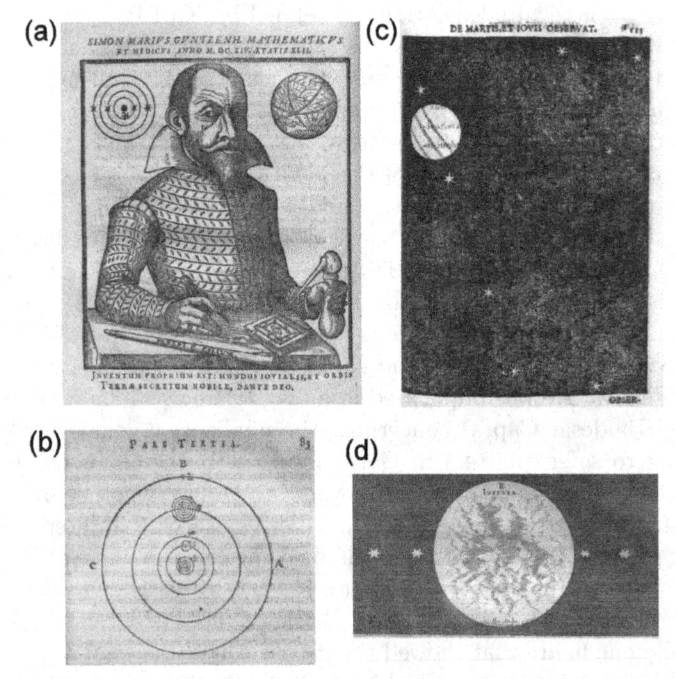

Figure 2. (a) Image of Simon Marius with Jupiter and its moons published in *Mundus Iovialis* in 1614. (b) Image from René Descartes' *Principia Philosophiæ* (1644) showing the solar system with Jupiter having four satellites. Note that both the planet and its moons are depicted using open disks. (c) Image of Jupiter and its moons from observations in 1630, published in *Novae Coelestium Terrestriumque* (New Observations of Things Celestial and Terrestrial) by Francesco Fontana in 1646. Jupiter has its atmospheric band structure and the moons are depicted using the same symbols as employed for stars. (d) Image of Jupiter and its moons from the *Selenographia* (1647) of Johnnes Hevelius. It shows a resolved disk with considerable structure, with the four moons depicted by star symbols in a symmetrical layout. (All courtesy of the Houghton Library, Harvard University: *GC6 M4552 614m, *FC6 D4537 644p, *IC6 F7342 646n, and f*GC6.H4902.647s.)

and Fontana shows that it has bands. For the moons, still points of light, star symbols are used. In his multiple plates of Jupiter spanning 13 years of observations, Fontana showed that there were changes in the banded structures (and, of course, in the positions of the moons).

Johannes Hevelius (1611-1687), the son of a wealthy merchant in Danzig (now Gdansk, Poland), created a personal observatory with some of the most advanced naked-eye and telescopic observing equipment of his time. While his *Selenographia* (1647) was primarily an atlas portraying the features of the Moon, Hevelius also used it as a way to publish his observations of the Sun and planets. Of interest here is his depiction of Jupiter and its moons (Figure 2(d)). This image shows a resolved disk of the planet that contains structure (curiously, with band-like features somewhat more meridional than the zonal patterns known to exist in its atmosphere). The four Galilean satellites are shown using star symbols — and arranged symmetrically, two to each side, and at the same distances. This image comes from a plate summarizing the appearances of three planets (Jupiter, Saturn and Mars), and thus it may well be intended to be more an illustration of characteristic patterns than a physical description on a specific night.

The famous Jesuit astronomer Giambattista Riccioli (1598-1671) championed the cosmological system of Tycho Brahe as the most acceptable to Catholic doctrine. The frontispiece to his *Almagestum Novum* (1651) is a wonderfully allegorical image that has been the subject of interpretation and speculations for centuries, mainly on the topic of hidden opinions and meanings in his portrayals of the three great world systems (Vertesi 2007). Riccioli does not include the jovian moons in any of his graphical portrayals of cosmology. However, in the upper right corner of the image [Figure 3(a)], we find Jupiter and its four moons — with the giant planet having banded structures and the moons depicted symmetrically by star symbols. The satellites are in orbits that appear polar rather than in their correct (but more difficult to show graphically) positions in the equatorial plane.

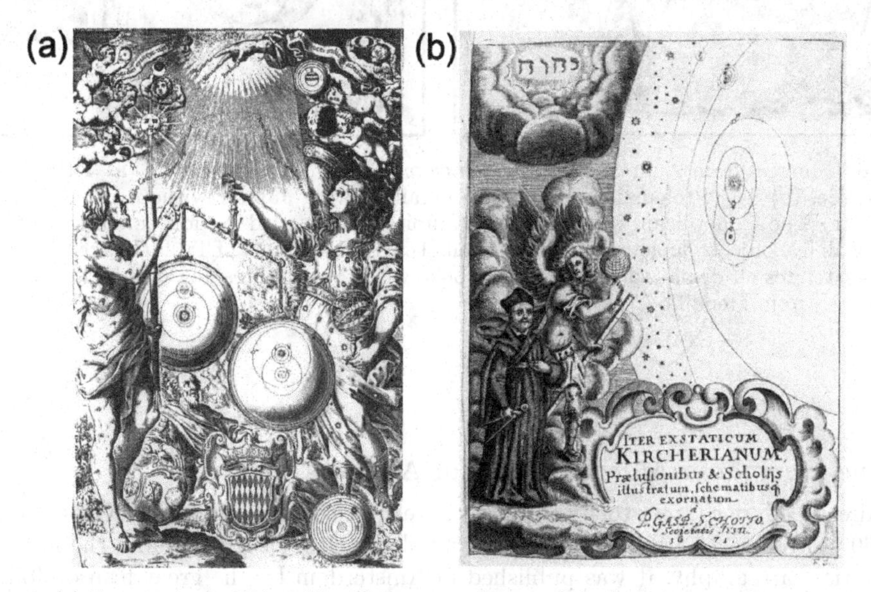

Figure 3. (a) Frontispiece of Riccioli's *Almagestum Novum* (1651). Jupiter and its moons appear in the upper right (from The Library of the Department of Astronomy, University of Bologna.). (b) Frontispiece of Athanasius Kircher's *Iter Exstaticum Coeleste* (1671) showing the system of Tycho Brahe, updated with the telescopic discovery of Jupiter's four moons, depicted as stars orbiting a resolved disk with banded structures (from Department of Special Collections, Stanford University, Green Library.).

Given the decorative nature and rich symbolism of this particular frontispiece, the jovian system's portrayal is consistent with the theme of a representational-only illustration.

Athanasius Kircher (1602-1680), that most remarkable and controversial polymath of the Society of Jesus (Findlen, 2004), offered his views on cosmology via a dream-dialogue in 1656, with later editions in 1660 and 1671 under the title *Iter Exstaticum Coeleste* (The Celestial Ecstatic Journey). The frontispiece (Figure 3(b)) depicted the Church-approved Tychonic system with a specificity that goes beyond Riccioli — dramatically having a telescope point towards Jupiter — with its revelation of four moons, portrayed by tiny star symbols (Godwin 2009). Father Kircher's intended message to the reader of his dream has long been suspected as being pro-Copernican (De Santillana 1955).

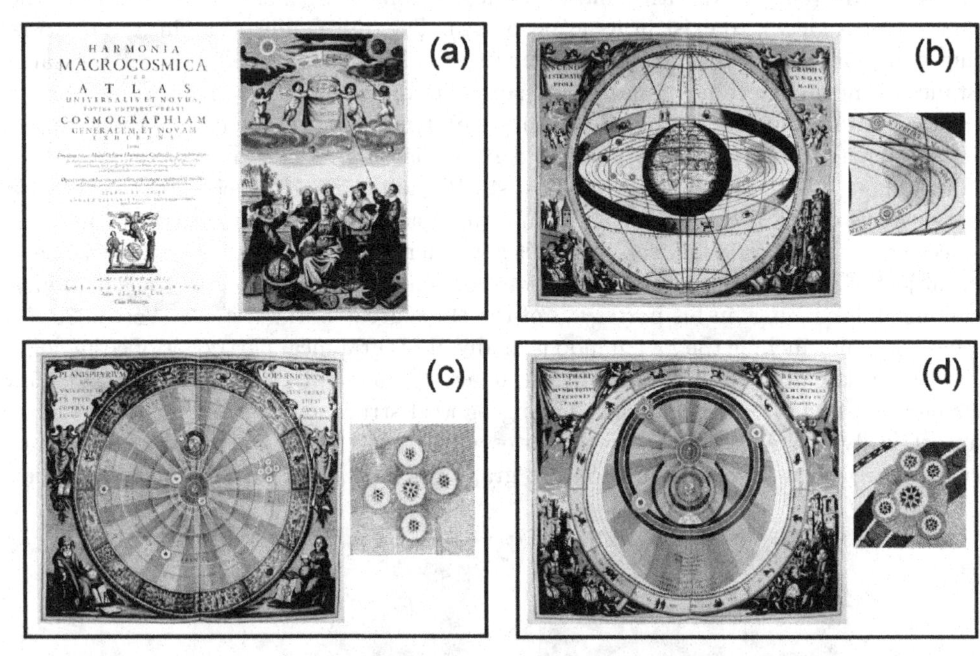

Figure 4. Images from Andreas Cellarius' *Harmonia Macrocosmica* (1661). (a) Title page and frontispiece, (b) The Ptolemaic system, with detail showing Jupiter without its four Galilean moons. (c) The Copernican system showing Jupiter surrounded by four moons, with detail showing all five objects depicted using star symbols. (d) The system of Tycho Brahe with Jupiter and its satellites all displayed using star symbols, with the satellites at variable distances from the planet. (from Mendillo Collection).

4. The *Harmonia Macrocosmica* of Andreas Cellarius

Andreas Cellarius (1596-1665) produced his exquisite "Harmony of the Great Cosmos" in 1660 (with a second printing in 1661), a volume that clearly signifies the high point of celestial cartography. It was published in Amsterdam by the great firm of Johannes Janssonius. Cellarius was an educator, and thus this elaborate atlas may be taken, not as the work of a practicing astronomer, but as a compendium of all knowledge about astronomy from antiquity to his day (Kanas 2007). Its publication date is, conveniently, 50 years after *Sidereus Nuncius*, and thus serves as an appropriate answer to the question posed for this study: it took 50 years for Galileo's starry message to get into the textbooks.

The *Harmonia* contained 29 copper engraved plates, individually hand colored, and often with gilding for stars or points of emphasis. Eight of the plates included information about Jupiter, with three showing its four moons. Figure 4 shows the title page, frontispiece and three of its plates. Panel (b) describes Ptolemy's geocentric system and, keeping with all celestial objects encircling only the Earth, Jupiter is shown without its satellites. Clearly an update would have been logically inconsistent. Panel (c) displays the famous plate *Planisphaerium Copernicanum.* The heliocentric system of Copernicus has golden rays extending from a humanized Sun to the stellar sphere indicated by the signs of the zodiac. Jupiter has its four moons in a representational fashion (i.e., with star symbols all equidistant from the planet). Elaborate cartouches frame the Latin title and astronomical instruments hang nearby. The two astronomers below appear to be Ptolemy (in an Eastern turban) and Copernicus, sitting with their globes and tools gazing out at the viewer.

Panel (d) of Figure 4 gives one of two depictions of the compromise cosmology of Tycho Brahe — ones that include Jupiter with four moons. The planets encircle the Sun, and the Sun with its brood circles the Earth. This modification of Ptolemy kept the system geocentric, but allowed for objects (other than the Moon) to orbit the Sun. Consistent with that approach, Cellarius has Jupiter depicted by a large star symbol with its four moons shown using smaller stars. Not only are these Medicean stars not all the same size, they are also at slightly different distances from the planet. This, then, is the most dramatic departure from Ptolemy that could be tolerated by the Church.

5. Later Depictions of Jupiter's Moons

Following Galileo's discoveries of 1610, astronomers continued to scan the skies for new objects, and cartographers continued to create atlases with updated astronomical content. Limitations of space allow only a few to be mentioned. First to come was the discovery of moons around another planet (e.g., Christian Huygens discovered Saturn's moon Titan in 1655, and G. D. Cassini found four more between 1671-1684). Even more astounding were the discoveries of new planets (e.g., William Herschel's discovery of Uranus (1781) and LeVerrier's discovery of Neptune in 1846). These advances were due to rapid developments in the technologies of telescope design and construction in the later half of the 17th Century and beyond. Yet, actual photographs of the disks of the jovian moons did not occur until the mid-years of the 20th Century (e.g., see images taken in 1962 in review by Dollfus (1998)). During the intervening centuries, prints and paintings had no choice but to show Jupiter with only tiny disks for its moons. Examples occur in *Four Systems of Cosmology* by Alexis Hubert Jaillot (Paris, 1690), *Descriptions de l'Universe* by Alain Manesson Mallet (Paris, 1683), Systems of Cosmology in *Atlas Historic* by Nicolas de Fer (Paris, 1705), and in the *Papal Astronomical Paintings* by Donato Creti (Bologna, 1711). The series of atlases by Dopplemayer and Homann (1720) contained many images of Jupiter with bands and of Saturn's rings, but the moons still remained as points of light. Examples can be seen in McCarroll (2005). Perhaps the most unusual depiction of Jupiter's moons in the 18th Century was offered by Carel Allard (1648-1709). His chart of the southern sky appeared in the *Atlas Minor* of N. Visscher in 1717, and later in an atlas by Covens and C. Mortier in 1759. Allard was known for innovative publications in which border decorations contained additional items of scientific information related to the central image, rather than the usual depictions of angels, philosophers or astronomers (as in the Cellarius images in Figure 2). One of Allard's side-bar inserts is shown in Figure 5. This rare, and perhaps unique, depiction of the four moons of Jupiter that Galileo named after the Medici family actually shows

Figure 5. A side-bar item from the star chart *Planisphærii Coelestis Hemisphærum Meridionale* by Carel Allard (1759) that has Jupiter's four Medicean moons depicted by four armed men—assumed to be the four Medici brothers, Cosimo, Francesco, Carlo and Lorenzo. (Courtesy of the Pusey Library, Harvard University).

the four Medici brothers (Cosimo, Francesco, Carlo and Lorenzo) in combat gear. Galileo never used their specific names in his writings about the moons, instead referring to them simply and Medicean moons I, II, III and IV. It is thought that the French astronomer Nicolas-Claude Fabri de Peiresc (1580-1637) may have made the suggestion to use their actual names in this way (see Galileo Project, Rice University).

6. Conclusions

Galileo's announcement of the Medicean Moons created a new way to do astronomy and, indeed, natural science. One of his innovations was the introduction of astronomical art— and visualization has played an increasingly prominent role in science and education ever since. In this brief summary that focused on how the Medicean Stars were portrayed in the half century following publication of *Sidereus Nuncius*, we pass from Galileo's simple drawings embedded within lines of text to the sumptuous illustrations of Cellarius. For centuries, no one knew what these bodies looked like, and yet they caused a sensation, led to a trial, and to the validation of Newtonian physics beyond Earth. They offered a scheme for finding longitudes in distant lands and upon the seas. Upon closer inspection, the images we now have challenge our models for volcanism, magnetism and, most surprisingly, for potential host sites for life.

Acknowledgements

At Boston University, Mr. Michael Hamilton of the Archaeology Department provided photographic images from my Cellarius atlas, Ms. Joei Wroten of Center for Space Physics provided assistance with the images and manuscript preparations, and Mr. Aaron Shapiro of the English Department provided assistance with Latin translations. The staff and directors of the Houghton Library and the Pusey Library at Harvard University graciously provided images from their collections suitable for publication. Much of this work was done in conjunction with the exhibition *Mapping Discoveries in the Heavens and Controversies on Earth*, at the Pusey Library (December 2009 to March 2010), for which I was Guest Curator. Sabbatical support from Boston University is gratefully acknowledged, as well as research funds made available through the Center for Space Physics. I thank Professor Cesare Barbieri of the University of Padova for organizing the fascinating meeting that generated this volume as its proceedings.

References

Barrow, J. 2008, *Cosmic Imagery: Key Images in the History of Science*,(New York: W. W. Norton & Co.)

De Santillana, G. 1955, *The Crime of Galileo*, (Chicago: University of Chicago Press)

Dollfus, A. 1998, *Planet. Space Sci.* 46(8), 1037-1073

Drake, S. 2001, *Galileo: A Very Short Introduction*, re-publication of 1980 and 1996 (Oxford: Oxford University Press)

Findlin, P. 2004, *Athanasius Kircher, The Last Man Who Knew Everything*, (New York: Routledge)

Godwin, J. 2009, *Athanasius Kircher's Theatre of the World*, (Rochester, VT: Inner Traditions)

Kanas, N. 2007, *Star Maps: History, Artistry, and Cartography*, (Chichester, UK: Praxis Pub. Ltd)

Kuhn, T. 1962, *The Structure of Scientific Revolutions*, (Chicago, University of Chicago Press)

McCarroll, S. 2005, *Celestial Images: Antiquarian Astronomical Charts and Maps from the Mendillo Collection*, exhibition catalogue (S. McCarroll, Curator), Boston University Art Gallery, (Seattle: University of Washington Press)

Murray, J. B. 1975, *Icarus*, 25, 397-404

Van Helden, A. 1989, *Sidereus Nuncius or The Sideral Messenger of Galileo Galilei*, (Chicago: University of Chicago Press)

Vertesi, J. 2007, *Stud. Hist. Phil. Soc.*, 38, 401-421

Galileo's Medicean Moons: their impact on 400 years of discovery
Proceedings IAU Symposium No. 269, 2010
C. Barbieri, S. Chakrabarti, M. Coradini & M. Lazzarin, eds.

doi:10.1017/S1743921310007258

Navigation, world mapping and astrometry with Galileo's moons

Kaare Aksnes

Institute of Theoretical Astrophysics
University of Oslo, Norway
email: kaare.aksnes@astro.uio.no

Abstract. Galileo realized that the four moons he discovered, besides supporting the heliocentric system, could also serve as a clock in the sky for longitude determination. Navigation at sea by this method did not prove practical but G. Cassini used it to improve land mapping. O. Rømer discovered that the interval between eclipses of the moons by Jupiter increased or decreased according to whether the Earth moved away from or toward Jupiter. He attributed this to the finite speed of light which he in 1676 determined with an error of about 25%. Timings of the eclipses by Jupiter have served to compute accurate orbits of the moons, notably by means of R. A. Sampson's theory of 1921. Beginning in 1973, light curves of mutual eclipses and occultations between pairs of moons have been made regularly at six years intervals. From these observations very accurate radii and positions of the moons have been obtained.

Keywords. Io, Europa, Ganymede, Callisto, eclipses, ephemerides, occultations, time.

---

1. The Galilean clock in the sky

Galileo discovered the four big satellites of Jupiter on 7 January 1610 and announced his discovery to the Prince of Venice (Figure 1). For the first time Galileo had proof that not all celestial bodies circle around the Earth as the geocentric world view claimed. This convinced him of the correctness of Copernicus' heliocentric theory. Galileo also realized that the satellites provided a clock in the sky which could be seen everywhere and give the longitude of the observer by comparison with local time. At that time chronometers had not yet been invented for determination of longitude at sea, and the available pendulum clocks could not keep accurate time in a rolling ship. It took more than 100 years before the first chronometer was constructed in 1735 by John Harrison.

Figure 1. Galileo and his letter to the Prince of Venice.

In 1612 the Frenchman Nicholas Peiresc tried to read the satellite clock based on the changing locations of the satellites, but they moved too slowly with too poorly known positions for timing. The same year Galileo for the first time observed an eclipse of a satellite by Jupiter or its shadow, and he pointed out that these frequent events could be timed fairly accurately. If one knew an eclipse time at a place of known longitude, the shift in longitude to another place where the eclipse was observed would be equal to the difference between that eclipse time and the local time which could be determined by sightings of the Sun or stars. But Galileo encountered two practical problems: He did not have sufficiently accurate eclipse tables, and the available telescopes needed to observe the satellites had only about 20 arcmin field of view, making it difficult to spot the satellites from an unsteady ship deck. He tried in vain to convince Spanish and Dutch ship navigators to use his technique and he worked for several years on perfecting predictions for the eclipse times.

When Galileo died in 1642, the best available eclipse tables were due to Simon Mayer (Figure 2) in 1614. He claimed to have discovered the four satellites independently of Galileo and at about the same time. But Galileo labeled him an impostor. That was really unjustified and it seems fair that the satellite names Io, Europa, Ganymede and Callisto, originally proposed by Mayer, were finally accepted by The International Astronomical Union in 1973.

Galileo's idea to navigate ships by means of *natural* satellites did not prove practical. But good ideas never die, and today ship navigation is performed to a very high accuracy by means of *artificial* satellites in the GPS system.

2. Giovanni Cassini's land mapping

It was left to Giovanni Domenico Cassini (1625-1712) (Figure 3), who became the first director of Paris Observatory, to perfect in 1668 the satellite eclipse tables to the extent that they could be used for longitude determination, not at sea but on land. Especially the innermost satellite, Io, with a period of only 1.7 days, lent itself for this purpose. By then, telescopes with convex lenses also offered much larger field of views than the Galileo telescope with a concave eyepiece lens. Cassini led in the 1670s the work to produce a more accurate map of France by means of eclipse observations. The map (Figure 4) was completed in 1679 and showed that the west coast of France on existing maps was a whole degree too far west. This is said to have made Louis XIV exclaim that he was losing more land to his astronomers than to his enemies! The satellite eclipse technique

Figure 2. Simon Mayer and his moon tables.

Figure 3. Giovanni Cassini and his moon tables.

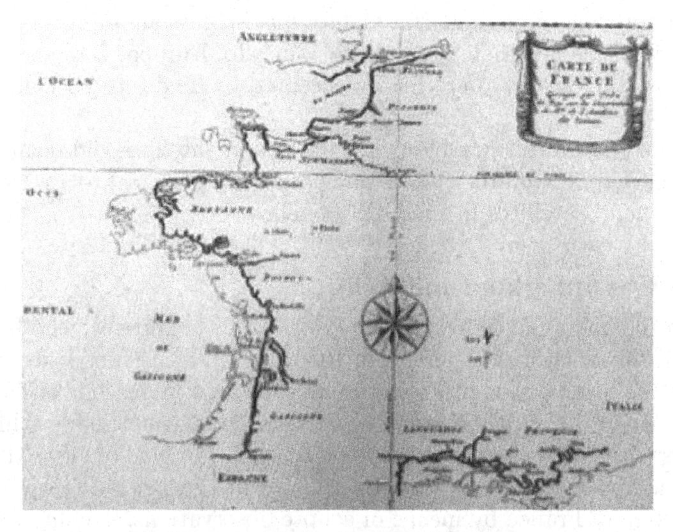

Figure 4. Cassini's map of France.

was even employed in the New World where Charles Mason and Jeremiah Dixon surveyed the border, known as the Dixon line, between Pennsylvania and Maryland.

3. Ole Rømer measures the speed of light

Attempts to measure the speed of light by, for instance, observing how long it takes for a light signal to pass from one mountain top to another, were of course doomed to failure. Equipped with Cassini's tables, the Danish astronomer Ole Rømer (1644-1719) (Figure 5) observed the Galilean satellites both in Denmark and at the Paris Observatory under Cassini's guidance. He discovered discrepancies of up to about 10 min between the predicted and observed times of the eclipses. Rømer correctly attributed this to the finite speed of light. The interval between consecutive eclipses will be shortened or lengthened according to whether the Earth in its orbit is moving towards or away from Jupiter. Rømer estimated that the light took 10 to11 min to travel from the Sun to the Earth. The correct figure is 8.3 min, so Rømer's value for the speed of light was about 25% too low, not bad since the best estimate until then was infinite speed.

4. Satellite orbit improvement from eclipses by Jupiter

It took a long time before a satisfactory theory of motion was developed for the Galilean satellites. Pierre-Simon Laplace (1749-1827) showed (Figure 6) that there are strong interactions between the three inner satellites Io, Europa and Ganymede because their periods or mean motions n_1, n_2, n_3 are in the nearly exact ratios 1:2:4. One consequence is that the four satellites can never be seen on the same side of Jupiter.

This gives rise to strong resonances, i.e. oscillations of large amplitudes and very long periods. A satisfactory orbital theory was not available until R. A. Sampson (1921) published a theory based largely on accurate visual timings of the satellites' entry into or exit from Jupiter's shadow (Figure 7). The interactions between the satellites also made it possible to estimate approximate masses of the satellites. Sampson's theory has been improved by Lieske (1987) at JPL, making use of better satellite masses determined from the Voyager flybys of the satellites. Occultations of the satellites by Jupiter's disk, or transits across the disk, cannot be timed as accurately because of the glare of Jupiter. Even for the eclipses, refraction and absorption of sunlight in Jupiter's variable atmosphere introduce considerable uncertainties despite the use of sophisticated models to define the shadow edges. The most accurate way to time the eclipses is by using a photometer to establish light curves of the satellites during shadow entry or exit. The time of eclipse is defined as the instant of half-light. Since 1975 members of the Association of Lunar and Planetary Observers (ALPO) have visually timed more than 10,000 satellite eclipses throughout the world.

5. Mutual eclipses and occultations among the satellites

The Galilean satellites move in near-equatorial orbits. Every six years when the orbital planes are seen nearly edge-on from the Earth, the satellites will occult or eclipse one another for periods of more than a year. In such a mutual occultation or eclipse, one satellite or its shadow will be seen to cover another satellite. Although timings of eclipses of the Galilean satellites by Jupiter have provided the bulk of observations for improving the satellites' orbits, the mutual events provide considerably more accurate position data. Because the satellites are airless, the beginning and end of a mutual occultation or eclipse are more precisely defined than is the case for an eclipse by Jupiter. Crude light curves of mutual events were made visually as early as 1926, but accurate light curves were

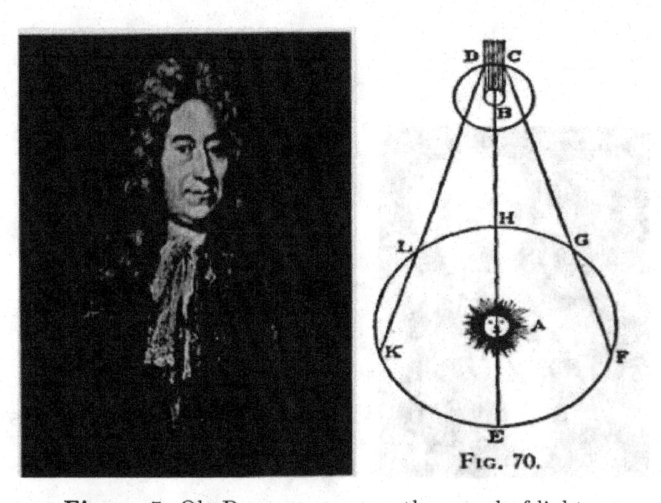

Figure 5. Ole Rømer measures the speed of light.

first recorded in 1973 when photometers with photomultiplicators had become available. That year a concerted world-wide observing campaign produced many hundred excellent light curves of mutual events.

Figure 8 is an example of an occultation of Europa by Io, that I observed at the Siding Spring Observatory in Australia. Notice the flat bottom of the curve which shows that Io covered Europa totally for about one minute.

In the six observing campaigns carried out world-wide between 1973 and 2003, many thousand observations of mutual satellite events have been recorded. This year there is an on-going seventh observing campaign. Most of the observations have been deposited in a databank (PHEMU) established at the Bureau de Longitude in Paris.

6. What have we learned from the mutual satellite events?

Essentially, the observed midtime of a mutual event depends on the longitudes of the satellites in their orbits, while the depth of the light curve depends on the latitudes and on the satellite radii. If I_1 and I_2 represent the brightnesses of the two satellites and A is the fraction of satellite *2* occulted by satellite *1*, then the joint brightness during the occultation can be represented as

$$I = 1 - A(1 + I_1/I_2) \qquad A = A(x, y, R_1, R_2) \tag{6.1}$$

The brightness ratio I_1/I_2 can be observed while A can be computed if we know the satellites' relative position x, y from an orbital theory and know the radii R_1 and R_2. By making adjustments in the radii and the relative position, the theoretical light curve can be fitted to the observed one through a least squares technique. This way the information of the entire light curve is made use of, and not only the midtime and the depth.

For the eclipses the light curve model is much more complex since one has to compute the light loss in both the umbra and the penumbra.

The radius of Io had been determined very accurately from a star occultation in 1971 giving a mean radius of 1821 km (O'Leary & van Flandern 1972). At that time, there was a spread of several hundred kilometers in the radii measured for the other three Galilean satellites with micrometers and radiometers.

$$n_1 - 2n_2 = n_2 - 2n_3$$
$$= 0.7395 \text{ deg/day}$$

$$n_1 - 3n_2 + 2n_3 = 0$$
$$l_1 - 3l_2 + 2l_3 \approx 180^\circ$$

n_1, n_2, n_3 = mean motions
l_1, l_2, l_3 = mean longitudes of Io, Europa, Ganymede

Figure 6. Laplace and the Galilean satellite resonances.

Table 1. Radii (km) of the Galilean moons.

Io	Europa	Ganymede	Callisto	Source
1821	–	–	–	Beta Scorpii occultation 1971
–	1533	2608	2445	1973 mutual satellite events
1822	1561	2631	2410	Galileo spacecraft 1995-2003

Aksnes & Franklin (1976) analyzed nearly 100 mutual events observations from 1973 and obtained the mean radii shown in Table 1for Europa, Ganymede and Callisto. These radii differ by at most 30 km from the more precise values determined later by direct photography from the Pioneer and Voyager spacecraft.

The satellite positions determined from the mutual events have an estimated mean uncertainty of only about 0.03 arcseconds which is an order of magnitude better than what can be obtained from eclipses by Jupiter or Earth based photography.

Just a few days before Voyager 2 arrived at Jupiter in July 1979, Peale, Cassen & Reynolds (1979) predicted, in an article in *Science*, that tidal stress in Io should produce volcanism, which was confirmed by dramatic pictures from the spacecraft of many active volcanoes on Io. Io's synchronous rotation and a slight ellipticity forced on Io's orbit by Europa cause the tidal bulge raised by Jupiter to oscillate across Io, flexing its surface. The energy dissipated is taken from Io's orbit which tends to shrink. This is counteracted by the tidal bulge which Io raises on Jupiter and which tries to push Io outward. It has been strongly debated which of these two opposing effects dominates, i.e., whether the time derivative dn_1/dt of Io's mean motion is positive or negative. Lieske (1987), using mainly eclipse observations covering more than a hundred years, finds a small negative value, while Aksnes & Franklin (2001), based on analysis of a large number of mutual satellite events, find a bigger positive value, according to which Io is now spiraling slowly

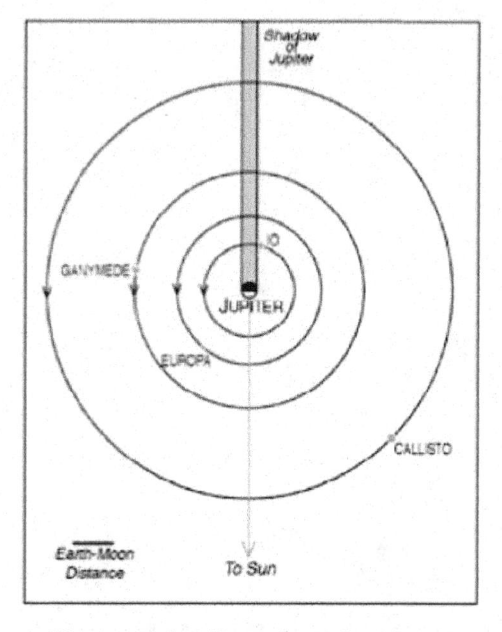

Figure 7. Satellite eclipses by Jupiter.

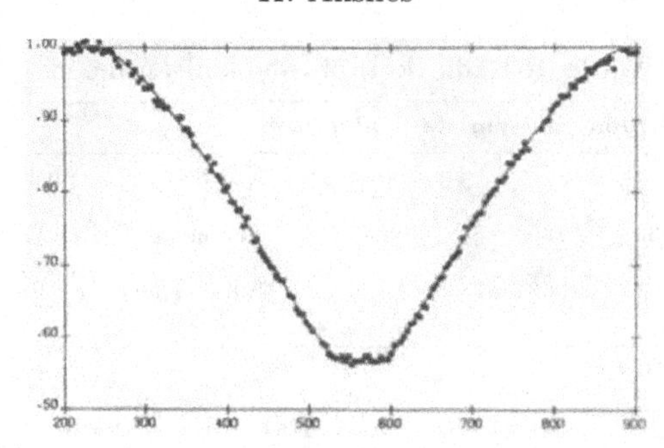

Figure 8. Total occultation of Europa by Io. Time in seconds after 10 September 1973, 12:35 UT.

inward. Recently, Lainey *et al.* (2009) also find a positive but much smaller value of dn_1/dt, based on tidal modeling and numerical integration over a time span of 116 years. The two last-mentioned articles both conclude that Europa and Ganymede are now moving outward, so that the three inner satellites are moving away from the present Laplacian resonance.

References

Aksnes, K. & Franklin, F. A. 1976, *AJ*, 81, 464
Aksnes, K. & Franklin, F. A. 2001, *AJ*, 122, 2734
Lieske, J. H. 1987, *A&A*, 176, 146
O'Leary, B. & van Flandern, T. 1972, *Icarus*, 17, 209
Sampson, R. A. 1921, *Mem. RAS*, 63, 1

Galileo's Medicean Moons: their impact on 400 years of discovery
Proceedings IAU Symposium No. 269, 2010

C. Barbieri, S. Chakrabarti, M. Coradini & M. Lazzarin, eds. doi:10.1017/S174392131000726X

Modern exploration of Galileo's new worlds

Torrence V. Johnson
Jet Propulsion Laboratory, California Institute of Technology,
4800 Oak Grove Dr. Pasadena CA, USA
email: Torrence.V.Johnson@jpl.nasa.gov

Abstract. Four hundred years ago Galileo turned his telescope to the heavens and changed the way we view the cosmos forever. Among his discoveries in January of 1610 were four new 'stars', following Jupiter in the sky but changing their positions with respect to the giant planet every night. Galileo showed that these 'Medicean stars', as he named them, were moons orbiting Jupiter in the same manner that the Earth and planets revolve about the Sun in the Copernican theory of the solar system. Over the next three centuries these moons, now collectively named the Galilean satellites after their discoverer, remained tiny dots of light in astronomers' telescopes. In the latter portion of the twentieth century Galileo's new worlds became important targets of exploration by robotic spacecraft. This paper reviews the history of this exploration through the discoveries made by the Galileo mission from 1995 to 2003, setting the stage for on-going exploration in the new century.

Keywords. History and philosophy of astronomy, space vehicles, planets and satellites: individual (Callisto, Europa, Ganymede, Io, Jupiter)

---

1. Introduction - The Era of the Telescope

Modern exploration of Galileo's new worlds began in a very real sense with the publication of *Sidereus Nuncius* itself, which announced the discovery of four moons circling Jupiter. Galileo's entire approach is recognizably modern - develop improved instrumentation (the telescope), make new observations of greater precision than ever before achieved, and, most importantly, make the results and the description of the methods used widely available as soon as possible for comment and confirming observations. The

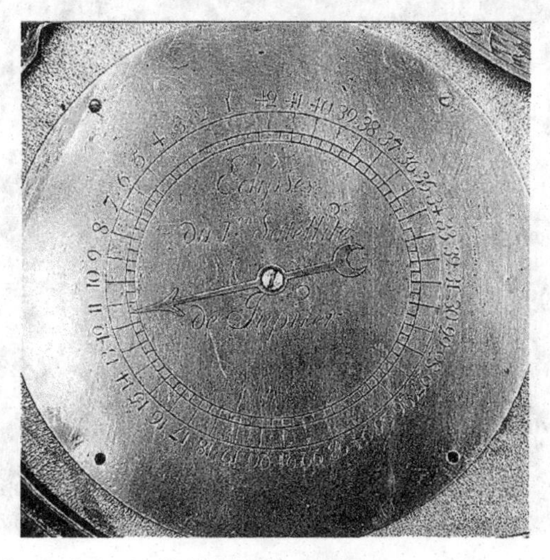

Figure 1. Io dial on planisphere in J. Paul Getty Museum collection. Note the ~42.5 marks around the circumference - approximately Io's orbital period in hours.

Galileo scholar Albert van Helden has pointed out in his translation that *Sidereus Nuncius* was an entirely new kind of astronomical publication, having more in common with a modern International Astronomical Union Circular than Ptolemy's *Almagest* (Galilei 1610).

The telescope remained the primary tool for studying the Galilean satellites for the next three and a half centuries. As the capabilities of telescopic optics improved, so did observing techniques and the instrumentation available to augment the 'Mark-I eyeball'. Before the end of the seventeenth century, accurate timing of Galilean satellite eclipses by Danish astronomer Ole Roemer (working at the Paris Observatory) provided the means to determine the first reliable value for the speed of light, a problem that Galileo himself had been unable to solve. Thus, within a few decades of Galileo's initial observations, his 'new worlds' were providing a key that would eventually lead to the modern world of relativistic physics.

These seventeenth century observations were accurate enough that they are still used in analyzing the long term variations of the orbits of the Galilean moons. Eclipse timings for Jupiter's moons also provided a cosmic clock which enabled accurate longitudes on Earth to be determined. A testament to the pervasiveness of Galileo's moons in the culture of the Enlightenment is found in a beautiful planisphere from the mid-eighteenth century. This device, now in the decorative arts collection of the J. Paul Getty Museum, includes a specialized dial which indicates the times of eclipses for Io, the innermost of the Galilean satellites (Wilson *et al.* 1996).

The next centuries saw the moons studied with ever increasing sophistication as the science of astronomy developed. The photographic plate, for precision measurements of position, the spectrograph, for analyzing the color of reflected light, and Michelson's stellar interferometer, for precision diameter measurements, were all applied to studies of Jupiter's four big satellites. In the twentieth century electronic instrumentation increased the power of telescopic observations many fold. The colors of the Galilean moons were measured, near-infrared photoelectric spectra identified water ice on the surfaces of Europa and Ganymede, and accurately timed stellar occultations determined the diameters of Io and Ganymede to a few kilometers.

Figure 2. Voyager's 'family portrait' of the Jupiter system. Montage of images from Voyager's cameras. PhotCourtesy NASA/JPL-Caltech.

Thus, at the beginning of the era of deep-space spacecraft exploration, in the 1970's, Galileo's new worlds, seen for so long only as points of light, were at last beginning to emerge as a reasonably well characterized, diverse set of planetary bodies. Their sizes were known, ranging from about the diameter of Earth's Moon, to somewhat larger than the planet Mercury. The available mass estimates and derived density for each satellite showed that while Io and Europa were mostly rocky objects, Ganymede and Callisto were likely an approximately even mixture of ice and rock. Io's highly reflective yellow-brown surface contrasted sharply with the less colored, icy surfaces of Europa and Ganymede and with the darker surface of Callisto. Clouds of both neutral sodium atoms and sulfur ions had been detected associated with Io. This was the sketchy but highly intriguing information which formed the basis for planning the Voyager mission's historic exploration of Jupiter's moons. The state of knowledge about the satellites at this time is reviewed in *Planetary Satellites* (1977).

2. Known Worlds - The Era of Spacecraft Exploration Begins

The Pioneer 10 and 11 spacecraft flew by Jupiter in 1973 and 1974, providing critical reconnaissance and scientific information about Jupiter's intense radiation belts, stunning views of Jupiter's clouds and glimpses of the satellites. Tracking data also improved estimates of the moons' masses significantly, confirming Io's rocky nature and Ganymede's low density and large ice content. By the time of the launch of Voyager 1 and 2 in 1977,

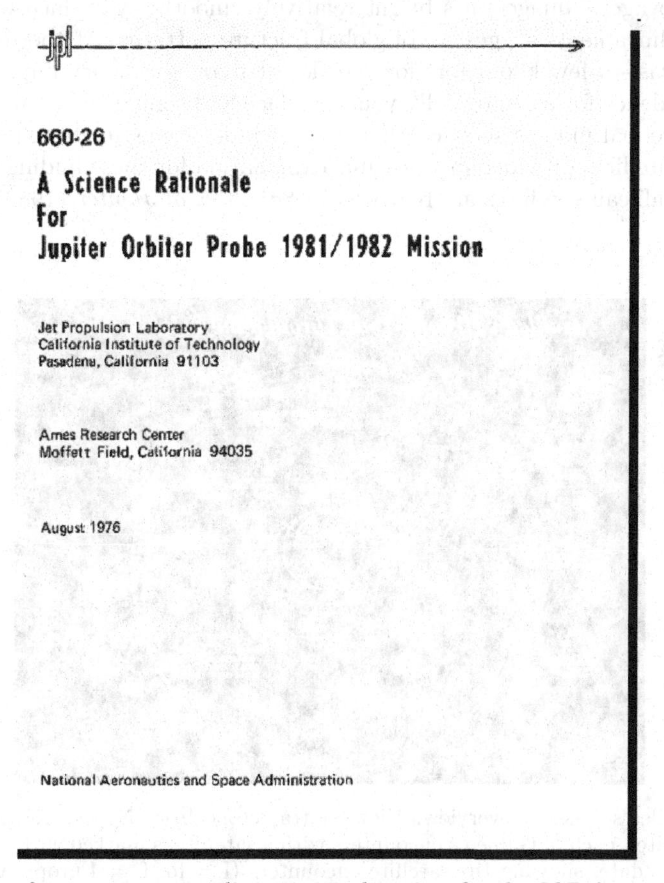
jpl

660-26

A Science Rationale
For
Jupiter Orbiter Probe 1981/1982 Mission

Jet Propulsion Laboratory
California Institute of Technology
Pasadena, California 91103

Ames Research Center
Moffett Field, California 94035

August 1976

National Aeronautics and Space Administration

Figure 3. Cover of report recommending science objectives for the JOP, later Galileo, mission.

the four Galilean satellites discovered in 1610 had become critical targets for investigation by the robotic explorers. The trajectories of the two spacecraft were carefully designed to allow each to have close encounters with several satellites on their way through the system. This 'celestial billiard shot' was enabled by the satellites' resonant orbits, which result in regularly recurring orbital configurations - a phenomena first explained and studied by Laplace in the eighteenth century.

Expectations were high for the first close-up views of these four planetary-scale worlds. Voyager's observations of the satellites from the multiple flybys in 1979 nevertheless exceeded even the most optimistic hopes of the mission's science teams. Images and other data from the two spacecraft revealed each satellite to have distinctly different characteristics and seemly different geologic histories. They had become, in the course of a few months, *terrae cognitae*, new worlds which could be mapped and studied in detail.
Io was found to be a young volcanic world with multiple active eruptions in progress during the Voyagers' visits. This activity confirmed the tidal heat source predicted by Stan Peale and his colleagues just months before the Voyager encounters, and the sulfurous emissions detected by spectrometers provided clues to the moon's bright, highly colored surface. In contrast, Callisto's dark surface was saturated with impact craters of all sizes, indicative of an ancient cold crust dating back to the near the formation of the solar system. Ganymede, the largest of the moons, appeared like a more youthful version of Callisto, with large regions of darker, heavily cratered terrain cut by bright bands of more recent origin. Europa, slightly smaller than Earth's Moon, was perhaps the biggest surprise. In Voyager's images, it's bright relatively smooth icy surface was crisscrossed by ridges and lineaments suggestive of global fracture patterns. At the resolution of the Voyager cameras - a few kilometers for the closest flyby - very few impact craters were detectable, evidence for a geologically young surface and some form of global resurfacing process. Theoretical models showing that liquid water oceans might exist under the icy crusts of the satellites provided one possible explanation for these findings. Voyager's results for the Galilean satellites are reviewed in *Satellites of Jupiter* (1982) and *Satellites* (1986).

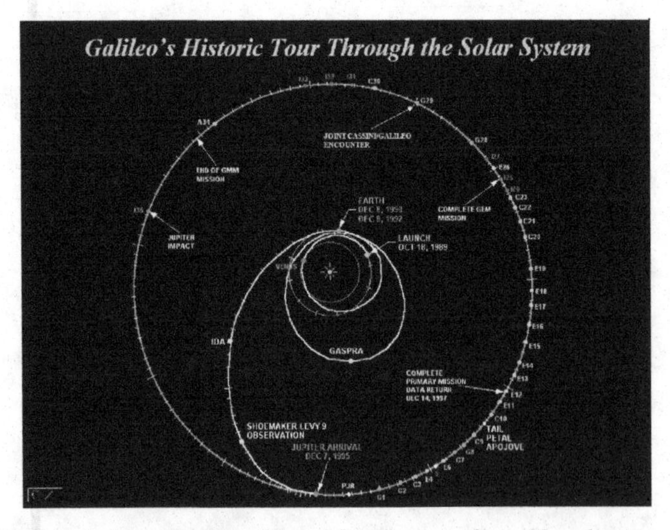

Figure 4. Galileo's mission overview. View of trajectory from N. ecliptic pole. After orbit insertion at Jupiter, each of Galileo's encounters with a satellite is marked with a dot at Jupiter's location for that date, showing the satellite encounters (I = Io, E = Europa, G = Ganymede, C = Callisto) and the orbit number.

3. Galileo's Namesake Explores Galileo's Worlds

Even while preparations were being made for the launch of the Voyager spacecraft, a return to Jupiter was being planned, this time with a spacecraft that would remain in the Jupiter system in orbit around the giant planet. Although Voyager was expected to accomplish much, several scientific committees recommended that this reconnaissance needed to be followed by an orbital mission capable of studying the whole system in detail over several years. This mission, initially dubbed the JOP (Jupiter Orbiter Probe) mission was begun by NASA in late 1977, with major international contributions from Germany. Given its goal of studying the 'mini-solar system' of Jupiter, its moons and radiation environment, it was fittingly renamed Galileo to honor the discoverer of the planet's satellite system.

The Galileo mission was delayed from its initial target launch date of January 1982, first by delays in the development of the space shuttle system and the necessary upper stage rockets needed to reach Jupiter, and then by the tragic loss of the Challenger and its crew in 1986. It was eventually launched on the space shuttle Atlantis in October of 1989, and used a gravity assist trajectory involving a Venus flyby and two passes by the Earth to acquire the extra energy needed for the trip to Jupiter. On December 7, 1995 the Galileo spacecraft arrived at Jupiter. In the span of few hours, Galileo conducted its first satellite flyby of Io, recorded data from deep in Jupiter's atmosphere as the entry Probe descended on its parachute, and fired its main engines to put the craft into orbit about Jupiter, becoming the first artificial satellite of a giant planet.

Once in orbit Galileo began what became a seven year long tour of Jupiter's system, making a close flyby of one of the Galilean satellites during each orbit. Each satellite encounter was design to set up the conditions for the next encounter, changing the spacecraft's trajectory using the gravity assist technique. This permitted multiple opportunities for scientific observations of the satellites from ranges up to a thousand times closer than Voyager had achieved. The first of these encounters after achieving orbit was with Ganymede, with the spacecraft skimming low over a dark, relatively older area known as Galileo Regio, appropriately enough.

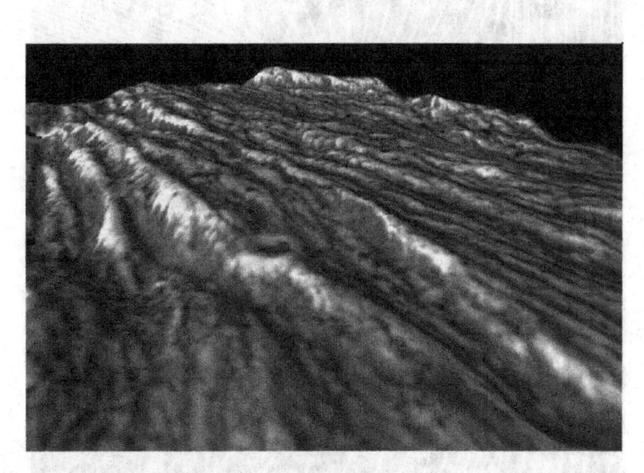

Figure 5. Galileo's view of Uruk Sulcus on Ganymede showing highly faulted surface in this shape from shading 3D rendering (approximately 3:1 vertical exaggeration. Courtesy NASA/JPL-Caltech.

High resolution pictures of the bright, younger appearing Uruk Sulcus on the southern border of Galileo Regio resolved one of the outstanding questions from the Voyager results. Based on Voyager's lower resolution images, it had been suggested that the sulchi were possibly smooth cryovolcanic flows of water and ice filling in faulted troughs. Galileo's images showed that, to the contrary, the entire surface of Uruk Sulcus was disrupted by many small, parallel faults, destroying older crust and craters through tectonic resurfacing.

Galileo's first close encounter with Ganymede also produced the first major discovery of the mission. Measurements by the magnetometer and plasma wave experiments showed that Ganymede was a 'moon with magnetism'. During the encounter, the magnetometer recorded strong distortions in the surrounding Jovian magnetic field while the plasma wave data showed a variety of signals indicative of high energy electrons interacting with a magnetosphere. Initial calculations suggested that Ganymede possesses its own, internally generated magnetic dipole field, strong enough to stand off the local Jovian field and create a 'magnetic bubble' around the satellite - a 'magnetosphere within a magnetosphere'. Subsequent encounters with Ganymede confirmed this model and Ganymede is now believed to be the only known moon with an active dynamo generated magnetic field. Gravity data support this interpretation, showing that Ganymede is differentiated, and probably has a core of iron or iron sulfide. This finding raises interesting and still unanswered questions about how Ganymede can maintain a hot, electrically conducting, convecting liquid core this long after its formation.

Volcanic Io was not ignored by the mission planners. Although close encounters with Io early in the mission were ruled out by concerns over the radiation environment that close to Jupiter, the spacecraft performed well in the Jupiter environment, and a series of close encounters were planned during the extended mission phase. Combined with more distant observations made during the primary mission, images and thermal data showed that Io's striking eruptions were continuing, apparently unabated since the Voyager encounters in 1979.

Figure 6. Ganymede's magnetosphere based on models by M. Kivelson *et al.* (Kivelson *et al.* 1997).

Many of the active volcanic sites discovered by Voyager were still erupting during the course of the Galileo mission and multiple new eruptions were also seen. One of these new eruptions, at a location in the northern region known as Tvashtar Catena, was observed to be still active some years after the end of the Galileo mission when the New Horizons spacecraft, on its way to Pluto, turned its cameras on Io during its flyby.

The composition of the volcanic flows on Io was an unresolved issue following the Voyager observations. The relatively low temperatures of some volcanic areas measured by Voyager's infrared experiment and the ubiquitous sulfur-colored deposits on the surface led to the suggestion that Io's volcanoes might be driven by liquid sulfur rather than the higher temperature molten silicates common in terrestrial basaltic eruptions. Telescopic infrared observations from the Earth had detected higher temperatures than seen by Voyager, showing that least some of Io's activity involved silicates, but the relative importance of the two styles of volcanism on the satellites remained uncertain. Galileo instrumentation included improved near-infrared capabilities for the CCD camera and the near-infrared mapping spectrometer. These near-infrared observations confirmed that most of Io's volcanic activity involves mafic or ultramafic silicate lavas, although there may be sulfur related flows in some areas.

Given theoretical models suggesting the possibility of a liquid water ocean beneath its icy crust, enigmatic Europa was a high priority target for Galileo observations. Images from Galileo confirmed the geologic youthfulness reported by Voyager. Even with much higher resolution images available, Galileo mapping shows only a handful of large craters (~10-20 km diameter) and many of the sprinkling of smaller craters seen at the highest resolution may be secondaries from the larger impacts. Estimates of the surface age from these results range from ~50-100 Myr, implying major resurfacing of essentially the entire surface in the recent geological past. In addition, Europa's surface seen at high resolution by Galileo resembles a frozen arctic landscape, with areas of disrupted 'chaos' highly reminiscent of broken sea ice in Earth's arctic regions. Gravity data suggests that the low-density outer layer of the moon is approximately 100 km thick.

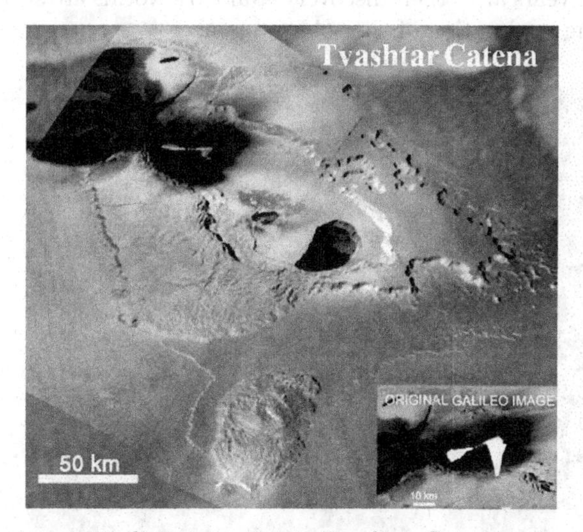

Figure 7. High temperature silicate eruption in Tvashtar Catena on Io. The infrared emission from the eruption site (marked in red) was so intense that the original CCD image was saturated in places (white "bleeding" in lower right inset). Courtesy NASA/JPL-Caltech.

This is consistent with models having a global liquid water layer under the surface, but the gravity measurements alone cannot determine how much of this layer consists of ice versus liquid water.

As with Ganymede, some of the most important information about the satellite's interior came from magnetic field measurements. In addition to detecting the strong intrinsic field of Ganymede, the magnetometer experiment made sensitive recordings of the effects of all of the satellites on the surrounding Jovian field during each flyby. For the three icy satellites these data showed that the field perturbations were consistent with a magnetic induction response to the changing Jovian field. Since Jupiter's dipole field is tilted about ten degrees from the planet's spin axis, each satellite experiences a varying field with a time period determined by Jupiter's rotation period and the satellite's orbital period. The perturbations in each flyby (after accounting for the intrinsic field in Ganymede's case) matched the predicted induction signature expected for an electrically conducting sphere the size of the satellite. The moons' icy surface materials and their tenuous ionospheres are calculated to have far too low a conductivity to account for the observed effect. However, liquid water oceans with dissolved salt, as in Earth's oceans, could easily provide the required electrical properties. Thus Europa, Ganymede, and Callisto may all have global liquid water layers, with the major difference that Europa's ocean would be in contact with its rocky mantle, while Ganymede's and Callisto's are predicted to be 'perched' oceans, with the water layer trapped between cold low density ice above and warm high density ice phases below.

Galileo's results are reviewed in Jupiter: *The Planet, Satellites and Magnetosphere* (2004).

4. Future Exploration

The results from the Galileo mission have laid the groundwork for future exploration of this complex system of moons. NASA and ESA have done studies of a joint mission to return to Jupiter, continue the exploration of the Jovian system and place spacecraft in orbit around both Europa and Ganymede to study their surfaces and interiors in greater detail. Four hundred years after their discovery, Galileo's worlds are still playing a central role in the continuing exploration of the solar system.

Figure 8. Ice rafts in Europa's Conamara Chaos region. Blocks are about 5 km across. Courtesy NASA/JPL-Caltech.

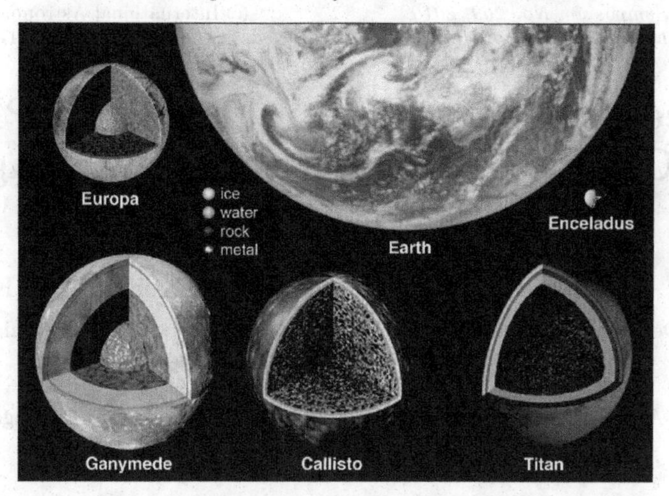

Figure 9. Planetary objects which may have oceans, to scale with Earth. Interior structures based on Galileo data and models (Galilean satellites) and Cassini/Huygens results (Titan and Enceladus). Art courtesy of Robert Pappalardo, JPL.

Acknowledgements This work was carried out at the Jet Propulsion Laboratory, California Institute of Technology under a contract from NASA.

Some of the materials used in this work were prepared for the author's presentation, The Galileo Mission: Exploring the Jovian System, at the Keck Institute for Space Studies Symposium "Challenging the Paradigm: The Legacy of Galileo" held at Caltech, November 19, 2009. http://www.kiss.caltech.edu/symposia/galileo2009/index.html

References

Burns, J. A. 1977, *Planetary Satellites*, Tucson: University of Arizona Press

Morrision, D. 1982, *Satellites of Jupiter*, Tucson: University of Arizona Press

Burns, J. A. & Matthews, M. S. 1986, *Satellites*, Tucson: University of Arizona Press

Bagenal, F., Dowling, T., & McKinnon, W. 2004, *Jupiter: The Planet, Satellites and Magnetosphere*, New York: Cambridge University Press

Galileo, G. 1610, *Sidereus Nuncius*, Chicago: The University of Chicago Press

Kivelson, M. G., Khurana, K. K., Coroniti, F. V., Joy, S., Russell, C. T., Walker, R. J., Warnecke, J., Bennett, L., & Polanskey, C. 1997, *Geophys. Res. Lett.*, 24, 2155

Wilson, G., Cohen, D. H., Ronfort, J. N., Augarde, J.-D., & Friess, P. 1996, *European Clocks in the J. Paul Getty Museum*, Los Angeles: The J. Paul Getty Museum

Galileo's Medicean Moons: their impact on 400 years of discovery
Proceedings IAU Symposium No. 269, 2010
C. Barbieri, S. Chakrabarti, M. Coradini & M. Lazzarin, eds.

doi:10.1017/S1743921310007271

Medicean Moons Sailing Through Plasma Seas: Challenges in Establishing Magnetic Properties

Margaret G. Kivelson[1,2], Xianzhe Jia[2] and Krishan K. Khurana[1]

[1]Department of Earth and Space Sciences, University of California, Los Angeles, CA 90095-1567,USA
email: mkivelson@igpp.ucla.edu

[2]Department of Atmospheric, Oceanic and Space Sciences, University of Michigan, Ann Arbor, MI 48109-2143

Abstract. Jupiter's moons, embedded in the magnetized, flowing plasma of Jupiter's magnetosphere, the plasma seas of the title, are fluids whose highly non-linear interactions imply complex behavior. In a plasma, magnetic fields couple widely separated regions; consequently plasma interactions are exceptionally sensitive to boundary conditions (often ill-specified). Perturbation fields arising from plasma currents greatly limit our ability to establish more than the dominant internal magnetic field of a moon. With a focus on Ganymede and a nod to Io, this paper discusses the complexity of plasma-moon interactions, explains how computer simulations have helped characterize the system and presents improved fits to Ganymede's internal field.

Keywords. Galilean moons, magnetic field, magnetosphere.

1. Introduction

Galileo would have been intrigued had he known that today's space explorers can tell us about the surfaces, interiors, atmospheres, and charged particle environments of his Medicean moons. The high resolution images provided by spacecraft have transformed his tiny points of light into worlds to be described by geologists and geophysicists. Relevant to this paper is the evidence that some of the moons are sources of magnetic fields, and that all of them are embedded in flowing, magnetized plasmas. Plasmas are highly ionized gases, on average electrically neutral but capable of conducting electrical currents. In the vicinity of a moon, the ambient plasma is diverted and otherwise perturbed. Those perturbations drive currents that can flow through the moons and their ionized environments and can also flow along Jupiter's background field to its ionosphere, its ionized upper atmosphere. The interactions are highly non-linear and are affected by properties of both local and remote parts of the system because of the currents that link them. This chapter emphasizes that the complexity introduced by the presence of plasma makes it challenging to characterize the internal magnetic properties of a moon from spacecraft measurements. Using Ganymede as an example, this paper describes a method of removing magnetic signatures generated within the ambient plasma from spacecraft measurements of the magnetic field in the vicinity of a moon. The remaining signature can be attributed to currents flowing within the moon and therefore provides information about the moon's interior.

In seeking to understand the interior structure of a moon, remote measurements provide little insight. Taking Ganymede as an example, Pioneer and Voyager were able to measure its mass, establishing an average density (1,940 kg/m^{-3} taken to be roughly

60% rock and 40% ice), but did not reveal whether the interior is relatively homogeneous or differentiated. Early models of the interiors of the icy moons typically consider only a rocky core and an ice-silicate shell (e.g., Schubert *et al.*, 1981). Only after Galileo acquired data from relatively low altitude flybys and established higher order gravitational moments did it become clear that the deep interior of Ganymede is differentiated into a metallic core and a rocky mantle (Anderson *et al.*, 1996). Although invaluable for establishing the distribution of mass within the interior of a moon, the gravitational moments are insensitive to the state of the matter forming the zones of differing density, and it is here that knowledge of magnetic properties becomes critical. Magnetic moments give insight into the distribution of conducting matter in the interior. For differentiated bodies like the moons of Jupiter, the dense cores are conductors, but only if they are partially fluid can dynamo action generate an internal field. Inductive fields can arise in any highly conducting layer exposed to time-varying external magnetic fields. The discussion below deals with both dynamo fields and induced fields and provides arguments supporting the view that induced fields (identified at Europa and Callisto as well as Ganymede) require warm (probably melted) layers within the icy outer shells. Thus the primary clues to the physical state of a moon's interior are provided by measurements of the magnetic moments and their temporal variation.

Because the magnetic moments provide insights that cannot be obtained from other measurements, it is important that their values be established accurately. The next section describes why refined estimates of the magnetic moments of a moon require knowledge of the contribution of plasma currents to the total field.

2. Problems in characterizing the internal magnetic field of a moon

A moon's internal magnetic moments are characterized by modeling the field measured in the vicinity of the moon as the sum of fields arising from internal and external currents. The problem is well posed if measurements are available over a current-free closed surface surrounding the moon. In this case, the internal and external multipole moments can be obtained by following a prescription that goes back to Gauss (Walker and Russell, 1995). However, measurements at the moons of Jupiter are sparse, and do not even approximately provide measurements over a closed surface. For example, at distances closer than 2 R_G (R_G is the radius of Ganymede, 2631 km) from the center of

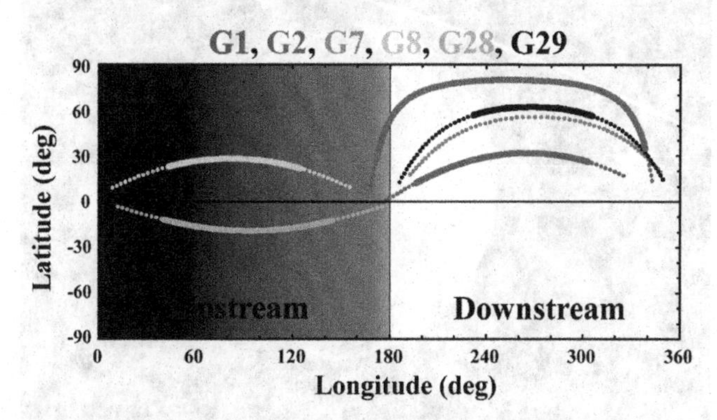

Figure 1. Latitude and longitude relative to a Ganymede-centered spherical coordinate system of the Galileo trajectories during the six close flybys of Ganymede. The portions of Galileo flyby trajectories closer than $2R_G$ to the center of Ganymede are represented by the solid traces. Latitude is measured relative to the spin equator and longitude is measured in a right hand sense relative to 0° in the Jupiter-facing meridian.

Ganymede, there are no measurements south of 20° latitude and in one quadrant there are no measurements at all (see Figure 1). Furthermore, strong currents flow throughout the near neighborhoods of the moons, so the available measurements are made in regions that are not current-free. Finally, the modeled sources referred to as internal include sources that are below the spacecraft trajectory but may be above the surface of the moon (see Figure 2). Currents flow above the surfaces because the gases that fill the immense Jovian magnetosphere are, for the most part, electrically charged but, because in any significant volume the charges balance, the system remains electrically neutral. Spatial changes of the magnetic field generate currents (through $\nabla \times \mathbf{B} = \mu_o \mathbf{j}$), while temporal changes produce inductive electric fields (through $\partial \mathbf{B}/\partial \mathbf{t} = -\nabla \times \mathbf{E}$) and associated currents. Here $\mathbf{B}(\mathbf{E})$ is the magnetic (electric) field, $\mathbf{j}$ is the current, and μ_o is the permeability of vacuum.

In considering the interaction of the gas of charged particles, a *plasma*, with the Medicean moons, it is useful to describe how currents arise and how particle flows and magnetic fields relate. Magnetospheric plasmas are tenuous; collisions are infrequent and particles interact principally through electromagnetic forces. Ions and electrons respond differently to the imposed forces and their differing motions produce net current. In response to the Lorentz force ($\mathbf{F} = q\mathbf{v} \times \mathbf{B}$ with q the particle charge and $\mathbf{v}$ its velocity) that acts perpendicular to the magnetic field, the plasma particles move across the background field on quasi-circular orbits (for a particle of mass m and perpendicular velocity $v_\perp$, the radius is $mv_\perp/qB$ with the sense of rotation depending on the sign of the charge). The Lorentz force has no component along the field, so charged particles respond only to field gradients in their field-aligned motion. The consequence is that charged particles bounce back and forth along $\mathbf{B}$ following spiraling paths with small transverse displacements. It is convenient to think of magnetic field lines loaded with spiraling charged particles as identifiable structures that can be followed as they move through space, and to assert

Figure 2. Schematic flyby of a moon. Currents flowing below the trajectory, whether inside the moon or above its surface, contribute as "internal" sources.

that the field and plasma are "frozen" together. It then follows that particle flows can displace and distort field lines. We will make use of this intuitive language in discussing the interaction of the magnetized plasma of the Jovian magnetosphere and the Galilean moons.

3. The plasma of Jupiter's magnetosphere

The structure of the plasma environment of a moon depends on the properties of the magnetized plasma in which it is embedded, but specifics of the interaction are controlled by the physical properties of the moon itself (Kivelson, 2004; Jia *et al.*, 2009b). The plasma relevant to describing interactions with the moons is the magnetospheric plasma trapped in Jupiter's strong magnetic field, dominated by a dipole moment tilted by about 10° relative to the spin axis. Over each Jupiter rotation period (~10 hours), the magnetic field imposed in the neighborhood of a moon is oriented dominantly southward but tilts radially inward and outward. The variation of the radial field occurs at the synodic period (the period of Jupiter's rotation as observed in the moon's rest frame), which is slightly longer than 10 hours because of the moon's orbital motion. The magnetospheric plasma in which the moon is embedded flows azimuthally, maintained in motion by field-aligned currents coupled to the planet. The angular speed is close to full corotation with Jupiter. Beyond ~2 R_J the speed of corotation is faster than Keplerian, so plasma overtakes all of the moons, flowing towards them from their trailing sides. [The Medicean moons move in nearly circular orbits in Jupiter's equatorial plane at distances ranging from ~6 R_J for Io to ~25 R_J for Callisto (R_J is Jupiter's radius, ~71,500 km).] Because plasma is concentrated close to the tilted magnetic equator, planetary rotation introduces additional slow north-south motions of the plasma relative to the rest frame of a moon.

If one of the moons of interest were inert and non-conducting, it would intercept and absorb the plasma flowing onto its cross section and a nearly empty wake would form in the region downstream in the plasma flow. This type of interaction approximates the situation at Earth's moon interacting with solar wind plasma. But the interaction of the plasma with Jupiter's moons and their surrounding ionospheres is quite different. Io, Europa, and Callisto (and their relatively dense ionized surroundings) are electrically conducting. Induced currents slow and divert much of the plasma flowing towards these moons so that only a small part of the upstream plasma flowing towards the cross-section of a moon actually reaches its surface. Because the field is frozen-in to the plasma, slowed flow on only a part of a flux tube (the part close to the moon) introduces a kink in the field. The associated field bending requires current to flow along the field and, in steady state, a bundle of perturbed flux tubes, referred to as an Alfvén wing, develops and moves through the plasma as if attached to the moon (Figure 3). Such structures and the currents that maintain them are of intrinsic interest (see, for example, Kivelson, 2004) but this paper focuses on plasma perturbations only to the extent that they mask the signatures of magnetic sources within the moons. This means that the distortion of the field produced by plasma currents, including those that form the Alfvén wings and others that close through the ionosphere or the solid portions of a moon, must be established in order to separate out the field perturbations generated by currents that flow completely within the moon.

4. Magnetic fields of the Medicean moons

Most of the Medicean moons have internally generated fields small compared with the field imposed by Jupiter at their orbital distances. Unique among the moons, Ganymede has a substantial internal magnetic field and a plasma interaction region dominated by internal sources of magnetism (Kivelson *et al.*, 1996, 1997). The existence and magnitude of the field were both generally unexpected. The scaling law of Busse (1976) had been used by Neubauer (1978) to predict plausible magnitudes of possible internally generated fields of all the Galilean moons, but before Ganymede's field was discovered it was generally thought that, over geological time, the moons would have cooled sufficiently for their interiors to have solidified and their dynamos to have shut down. Io, in a strong tidal resonance, was viewed as a possible exception, a case for which a planetary scale molten interior consistent with widespread volcanic activity and possible dynamo action could not be ruled out (Kivelson *et al.*, 1997). Yet Galileo ultimately found no evidence for a significant dynamo-generated magnetic field at Io (Jia *et al.*, 2009b) but discovered that Ganymede's internal field (of order 750 nT at the equatorial surface) is almost four times larger than had been suggested by scaling laws (Kivelson *et al.*, 1996, 1997). The large field appears to require dynamo action and a molten metallic layer in the deep interior. This discovery has led to suggestions that orbital migration took Ganymede through a tidal resonance at some time in the past billion years, with interactions strong enough to melt substantial portions of its metallic core (Showman and Malhotra, 1997).

Although Ganymede is the only moon with a large permanent magnetic field, the story differs when attention turns to temporally varying fields. If there is conducting material within a moon, the time-varying component of Jupiter's field (mostly radial in orientation) can induce an EMF that in turn drives a time-varying current that generates an (induced) magnetic field. Only if the conducting layer is global in scale and lies close to the surface will the induced field be comparable in magnitude to the time-varying field component that causes it. Galileo measurements detected magnetic perturbations consistent with inductive fields at Europa and Callisto (Khurana *et al.*, 1998; Zimmer *et al*, 2000). The amplitudes of the perturbation fields and their fall-off with radial distance provided evidence of the global scale of the conducting layers in which they were generated. A uniform time-varying field generates a dipolar response from a spherically symmetric shell. The amplitude of the field induced in a conducting body depends on its conductivity. The limiting case, for high conductivity, is that of a dipole field whose polar field is equal and opposite to the driving field at the surface of the conducting shell. Away

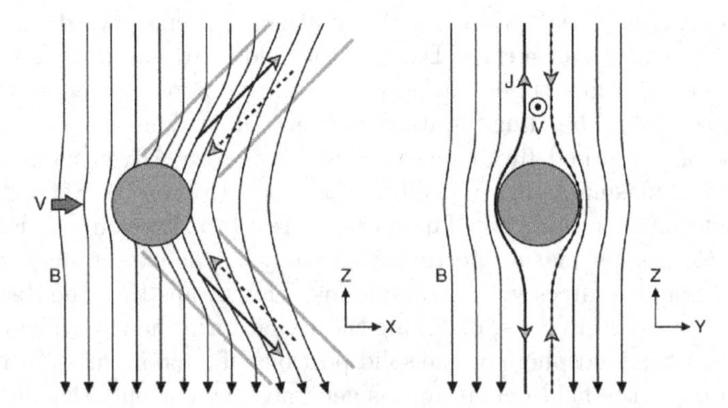

Figure 3. After Kivelson *et al.* (2001), Schematic of the Alfvén wing interaction. Lines with black arrowheads represent the magnetic field. Lines with gray arrowheads represent currents: those shown with dashed lines flow behind the plane of the image and those with solid lines flow in front of that plane. Light gray lines delimit the region in which perturbations linked to the Alfvén wing currents are significant. Left: A cross section through the center of the moon. Plasma flows to the right, transverse to the field. Right: A cross section orthogonal to the flow through the center of the moon.

from the conducting shell (of radius r_{shell}), the field strength falls as $(r_{shell}/r)^3$ where r is the distance from the center of the shell. The large amplitude of the perturbations observed at both Europa and Callisto required the conducting layers for both moons to lie close to their surfaces, and consequently to be buried within the outer ice layers. Solid ice is not a good enough conductor to account for the observations, but a liquid water shell with electrolytes to carry current is consistent with the magnetometer measurements in both cases. For Ganymede, the case for a sub-surface ocean is less clear as the large dynamo-driven field and the surrounding magnetosphere complicate the signature. However, results of the most recent analysis again lead to the conclusion that Ganymede also contains a conducting sub-surface ocean (Jia, 2009).

5. Interaction with the Jovian plasma

Each moon interacts with the Jovian plasma in a slightly different way. The interactions greatly perturb the field in the neighborhood of the moon and on flux tubes linked magnetically to it. The regions modified by the interaction may be highly extended and highly structured. Significant perturbations arise from ionospheric currents and the plasma currents that link to them. Some moons (notably Io and to a lesser degree Europa) release neutrals into their environments, much like comets. The ionization of neutral matter generates what are referred to as pickup currents, which also perturb the magnetic field.

The interaction region is particularly complex in the vicinity of Ganymede whose strong internal field carves out a large cavity within the Jovian magnetospheric plasma, a mini-magnetosphere whose boundary lies at distances greater than 1 R_G above its surface. The cavity develops because magnetized plasmas do not intermingle except on extremely small length scales (of order the ion gyroradius) so Jupiter's plasma is diverted well above Ganymede's surface. Ganymede's magnetosphere, embedded in the sub-magnetosonic plasma flow of Jupiter's magnetosphere, differs in shape from magnetospheres of planets (e.g., Earth, Mercury, Jupiter and Saturn) that form in the super-magnetosonic flow of the solar wind. Planetary magnetospheres are bullet-shaped cavities aligned with the solar wind plasma flow. Ganymede's magnetosphere is quasi-cylindrical, about an axis aligned almost perpendicular to the plasma flow velocity and nearly parallel to the externally imposed field as illustrated in Figure 4 (*a* and *b*).

A magnetosphere is often described in terms of distinct current systems that account for its structure. Especially in the vicinity of boundaries such as the magnetopause across which the field may reverse direction over distances of a few ion gyroradii, magnetospheric currents can be intense. Data acquired in current-carrying regions compromise the process of obtaining fits to a moon's internal field, as noted in Section 2. For this reason, we next discuss improved numerical simulations of the interacting system that provide good estimates of the contributions of the external current system, contributions that can be removed in order to improve the estimate of the internal sources of magnetic field.

6. Numerical simulations of the interaction region

The interactions of the Jovian plasma and the moons of Jupiter have been modeled using a variety of numerical approaches. Each approach focuses on modeling some aspect of the physical system particularly well. All start by assuming a model of the internal field and the field of Jupiter's magnetosphere imposed at the location of the moon. Some prescribe the magnetic field and calculate the interaction of the plasma and the moon's ionosphere using a realistic model of its conductivity (Saur *et al.*, 1998). The magnetic perturbations are calculated from the currents. This approach provides insight into the role of the ionosphere at the cost of self-consistency. Some follow ions and electrons separately, including small (ion gyroradii) scale contributions at the cost of higher spatial resolution (Paty and Winglee, 2006). Some use single fluid magnetohydrodynamics (MHD), ignoring multi-fluid and kinetics and simplifying the treatment of ionospheric conductivity (Schilling *et al.*, 2007, 2008; Jia *et al.*, 2008, 2009a). All have advantages and disadvantages and all can reproduce important features of the magnetic structure of the interactions.

Here we suggest that for a simulation to be useful for the purpose of establishing the internal field of a moon it must include:

- data-based upstream conditions;
- simulation boundaries that do not generate signals;
- physics-based boundary conditions at the moon's surface;
- appropriate handling of plasma resistivity to allow reconnection at a rate consistent with observations;
- output that faithfully reproduces all available measured properties of the system.

Challenges arise in implementing the requirements noted. Unperturbed plasma and field measurements are available only somewhat before and after the actual encounter, so upstream plasma/field conditions must be extrapolated to the location and time of the encounter. Downstream boundaries do generate signals unless the run is truncated, so a simulation must be run until it achieves quasi-steady state but not so long that signals

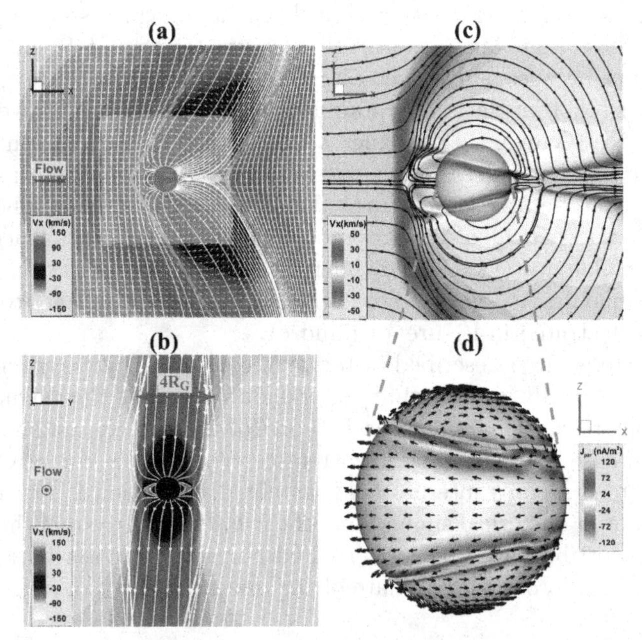

Figure 4. Figure from Jia *et al.* (2009a). (a) Flows and the projection of field lines (white solid lines) in the XZ plane at Y=0. Color represents the V_x contours and unit flow vectors in yellow show flow direction. A theoretical prediction of the Alfvén characteristics (orange dashed lines) is shown for reference. The projection of the ionospheric flow also is shown as color contours on a circular disk of r = 1.08 R$_G$ in the center. (b) As in (a) but in the YZ plane at X = 0. (c) A zoom-in of the highlighted area in (a). Flow streamlines are superposed with the color contours of V_x. Note a different color bar is used to better illustrate the relatively weak flow within the magnetosphere. (d) Field-aligned current density along with unit flow vectors shown on a sphere of radius r = 1.08 R$_G$.

return from the boundaries of the simulation space. The boundary conditions at the moon are not obvious, but systematic analysis of a range of boundary conditions can convincingly identify plausible results and improve the match to data.

The general form of the magnetic field data measured on several passes through Ganymede's magnetosphere can be obtained from the multifluid simulation of Paty and Winglee (2006). Their assertion that a multifluid treatment is needed to obtain good fits is contradicted by the results of the MHD simulation of Jia (2009). That simulation not only more closely represents the magnetic field measured on all passes by Ganymede, but also gives a good account of the plasma flow measurements. Focusing on the Jia simulation, we next comment on the assumptions and the modifications introduced that led to the desired good agreement with measurements.

The 3D MHD simulation model of Jia et al. The Jia (2009) simulation

- uses a resistive MHD code;
- uses a spherical grid that provides very high resolution near the moon, but generates unwanted signals where the spherical grid encounters the Cartesian downstream boundary;
- is run until it is quasi-stable but cut off before the signals get back to the interaction region;
- tests various inner boundary conditions (at 1.05 R_G) that differ in rather subtle ways but produce very different results. The acceptable inner boundary conditions are required to produce physically plausible flows and to predict field and plasma configurations consistent with observations.

The code employs a 131 × 132 × 128 × (r, θ, ϕ) non-uniform spherical mesh covering a calculation domain $0.5R_G < r < 40R_G$ with the grids ($\sim 0.02R_G$ or 50 km in characteristic lenght) near Ganymede. The resolution was increased in regions near the low

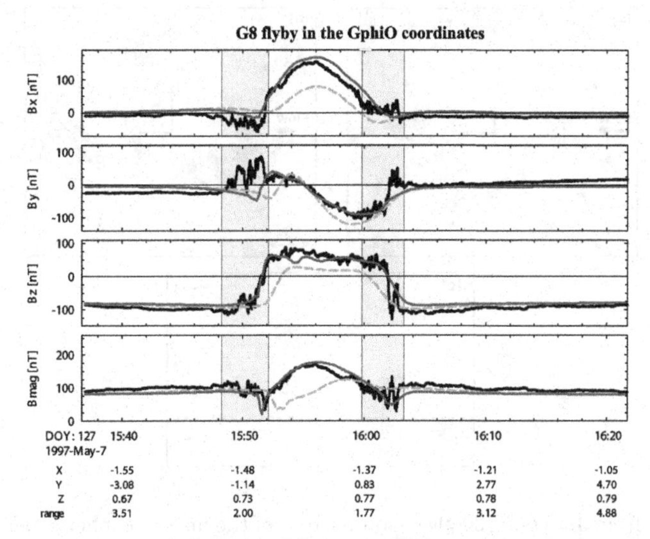

Figure 5. Figure from Jia *et al.* (2009a). Magnetic field comparisons between the simulation results and the Galileo observations for the G8 flyby. Black solid lines are the spacecraft measurements, green dashed lines are results extracted from the fixed boundary condition run and red solid lines are from the boundary condition allowing finite flow transverse to **B**. The locations where large amplitude magnetic fluctuations are observed during both inbound and outbound magnetopause crossings are marked with shading and have been found to correspond to regions with large temporal fluctuations in the simulation.

latitude upstream magnetopause boundary until the simulation reproduced the rapid field reversal observed as Galileo crossed that region. With inadequate spatial resolution, neither the duration nor the field magnitude on the two sides was properly reproduced.

The non-linear, coupled equations solved in the analysis are given in Jia *et al.* (2008). They include continuity, momentum, and energy density and allow for sources and sinks of ions. The magnetic field is derived from a vector potential that satisfies Faraday's law and Ampere's law is used to derive the current density from the curl of the field. Although the code in principle allows for a neutral source, the effects of such a source are not of great importance at Ganymede and have not been included.

Inner boundary conditions In matching the simulation results to the observations, Jia found that the solutions are extremely sensitive to the boundary conditions imposed on the plasma velocity at the inner boundary (Ganymede's ionosphere). In testing alternative ways of handling those boundary conditions, he focused on pass G8, an upstream pass at mid-Ganymede latitude. This pass was selected because the trajectory passed relatively close to Ganymede in a region also close to strong currents of the magnetopause, a particularly hard part of the system to model quantitatively. Inner boundary conditions (at 1.05 R_G) were tweaked until the magnetic field signature for pass G8 gave good agreement with the magnetometer data. Various familiar and seemingly plausible assumptions were tested: no flow (appropriate for a highly conducting obstacle), vanishing normal flow (plasma cannot penetrate the boundary) and either continuity of tangential flow or continuity of flow perpendicular to **B** (the latter form being consistent with equipotential field lines). Only the latter boundary conditions gave output that was consistent with the spacecraft data, and in particular gave a good correspondence to the observed location of the magnetopause.

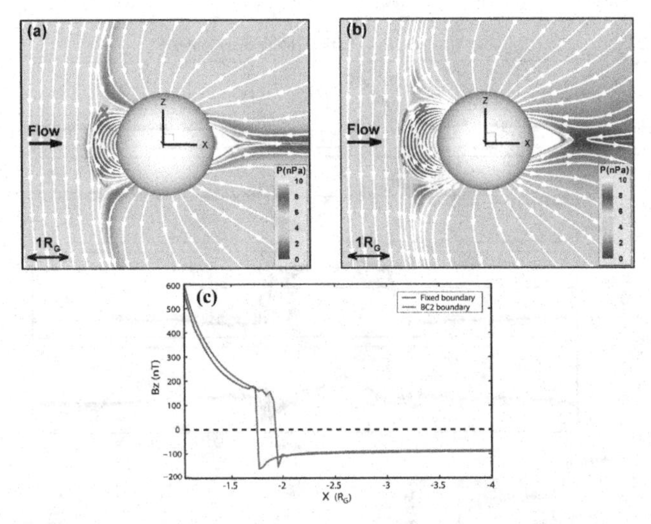

Figure 6. Figure from Jia *et al.* (2009a). Comparison of the magnetospheric configuration for (a) no flow at the ionospheric boundary and (b) no flow along **B** and continuous flow perpendicular to **B** at the inner boundary. In each plot, projected field lines are shown in white in the XZ plane at $Y = 0$ superposed with color contours of plasma thermal pressure. The intersection of the spacecraft trajectory with the XZ plane is represented by the magenta cross near the northern cusp. The inner boundary of the simulation is represented as a centered sphere. (c) The B_z component traced along the -X-axis on the upstream side (not along Galileo's trajectory): blue line for boundary conditions as in (a) and red line for boundary conditions as in (b).

Simulation results Adopting the inner boundary conditions consistent with equipotential field lines and making some additional adjustments to permit nonlinear increases of the resistivity in regions of large field gradients, Jia obtained extremely satisfactory agreement with the data for Galileo's G8 pass on May 7, 1997 as illustrated in Figure 5. The new treatment of resistance facilitates reconnection in the vicinity of the magnetopause and in the downstream return flow region. The large amplitude fluctuations in the data taken in the vicinity of the magnetopause crossings correspond to signatures of intermittent reconnection in the simulation, as evident from movies of the time series simulation data, so they are not likely to be either boundary waves (as suggested by Kivelson *et al.*, 1998) or spatial structures.

Changes of flow patterns associated with different boundary conditions are accompanied by changes in the shape and size of the magnetosphere; this is illustrated in Figure 6, a cut along the meridian containing the upstream flow. The appropriate boundary conditions enlarge the magnetosphere and move the magnetopause away from Ganymede.

Having developed a model that corresponds well with the magnetic field data from the G8 pass, Jia ran the simulations with no further changes other than to represent the upstream conditions appropriately for the remaining 5 Galileo passes by Ganymede. In all cases, the correct boundary conditions improved the fits and the simulated results gave a close approximation to the smoothed magnetometer data from all passes. (We believe that the fluctuations present in the actual data result from temporal fluctuations produced by intermittent reconnection occurring during the passes.)

Interpretations of the plasma data in regions of low flow velocity require assumptions about the mass per unit charge of the plasma ions. There has been some dispute as to whether the dominant ions within Ganymede's magnetosphere are protons (Frank *et al.*, 1997) or oxygen (Eviatar *et al.*, 2001). Jia's runs provide excellent models of plasma properties, as shown in Figure 7 for the G2 pass of Sept. 6, 1996, if ions are taken as oxygen.

In summary, the simulations developed by Jia (Jia *et al*, 2008, 2009a; Jia, 2009) provide excellent fits to data on both the magnetic field and the plasma within Ganymede's magnetosphere. They serve the objective desired for improving fits to the internal field of Ganymede.

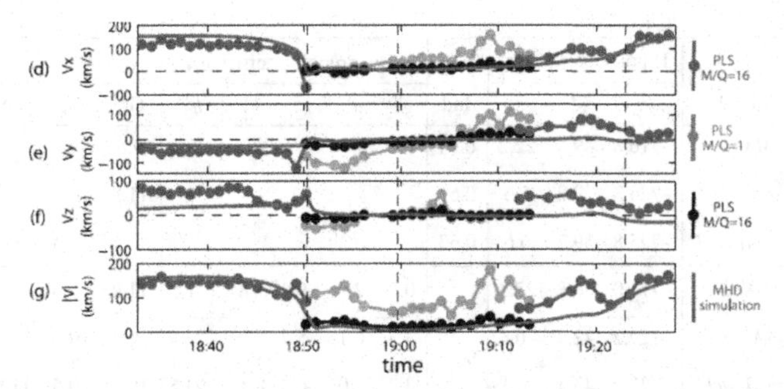

Figure 7. Figure from Jia *et al.* (2009a). Flow velocity inferred by the plasma investigation for the G2 pass of Sept. 6, 1996 with gray (black) dots relevant if ions are protons (O^+) (Frank *et al.*, 1997). Red curves are flow velocity from the Jia simulation.

7. Improved estimates of the internal field of Ganymede

Jia's simulation uses the best previous model of the internal field of Ganymede and adds to it a flowing plasma threaded by a uniform field representing the moon's surroundings in the Jovian magnetosphere. The output of the simulation differs from the input because of perturbation fields and their associated plasma currents. Thus, the difference between the output and the total input field (background field plus the assumed internal field) at each grid cell is taken to represent the contribution to the total field from currents driven through the plasma. By removing the sum of the modeled difference field and the input uniform field from the measured field, one obtains the approximate signature of the internal field along the spacecraft trajectory. The internal field should not vary over the few years of the Galileo encounters, so the difference signatures from all passes can be combined to obtain a model of the internal field. Evidently, if the new model differs markedly from that assumed in the simulation, the process should be iterated, but for the case of Ganymede, the changes between the assumed internal field and the new fit is rather small and the process has not been iterated.

Figure 8 (Jia, 2009) gives the model internal field obtained by fitting the difference signature from 6 passes to dipole plus quadrupole moments (8 independent parameters) or, alternatively, by fitting a permanent dipole (the same for all passes) plus an inductive response that varies with the phase of Jupiter's rotation and therefore differs from pass to pass (4 independent parameters). If there is a significant conducting layer, the inductive response can be characterized in terms of a single parameter, an efficiency factor, i.e., the fraction of the maximum response that would be found if the moon were a perfect conductor of radius r_{moon}.

Examining the table, one sees that:

- the new models differ rather little from the previously published ones (MK02, Kivelson *et al.*, 2002). However, the new models reduce induction efficiency compared with the earlier estimates;
- all new models give nearly same dipole moment: B_{suf} = 728 nT (719 nT), tilt= 177° or 176°, rotation ~4° or ~24° from the Jupiter-facing meridian;
- the ratio of quadrupole to dipole moments is small;
- the multipole model and the dipole plus induction model represent the data equally well, but the former requires fewer free parameters. Although the new fits were obtained by a refined process, they still do not enable us to establish the presence of an inductive response with any more certainty than did earlier treatments.

If the induced field is present, the relatively large value of induction efficiency listed in the table requires both that the conducting layer lie not far below the surface, well

		Dipole coefficients				Quadrupole coefficients					*rms* error				
	Model	g_1^0	g_1^1	h_1^1	Ind	g_2^0	g_2^1	g_2^2	h_2^1	h_2^2	G1	G2	G8	G28	total
Dipole	*MK02*	–716.8	49.3	22.2	0.84						9.8	7.5		15.6	11.5
+	*SimModel1*	–726.3	39.2	5.0	0.63						11.9	10.4		10.6	11.1
Induction	*SimModel3*	–725.8	39.7	6.6	0.63						11.6	10.8	11.3	10.7	11.1
Dipole	*MK02*	–711.0	46.8	22.3		0.9	27	–0.4	1.8	–11	11.9	15.6		13.2	13.5
+	*SimModel2*	–729.2	42.9	–0.6		4.3	6.1	–5.7	8.0	–23.7	10.4	10.0		9.8	10.1
Quadrupole	*SimModel4*	–727.9	43.4	1.0		3.3	7.0	–6.6	8.1	–21.8	10.4	10.5	11.9	9.5	10.4

Figure 8. Parameters of internal field models for Ganymede from Jia (2009). *Ind* denotes the induction efficiency

within the 800 km of ice that forms the outer shell, and that the conductivity be of order 1 S. The conductivity of ice is not sufficiently high to produce a response as large as 63% of the maximum possible response. However, a liquid layer buried within the ice layer could well be expected to have conductivity close to or somewhat below that of terrestrial sea water (2.5 S). The data do not constrain the precise depth of the ocean below the surface nor the precise conductivity, but reasonable combinations can be made consistent with the known thickness of the ice shell, probable thermal profiles, and sensible values of ocean conductivity.

8. Summary and a glimpse into future developments

Although the interaction of the Jovian plasma with the Medicean moons is itself of considerable scientific interest, this chapter describes the interactions mainly to emphasize that they compromise the reliable identification of the internal field of a moon, whether permanent or induced, from the field measured along spacecraft trajectories. An approach, recently developed, characterizes the perturbation of the field by use of a well calibrated MHD simulation. The perturbation field inferred from the simulation plus the unperturbed external field are subtracted from the measured field, yielding a good approximation to the field that would be measured along the spacecraft trajectory in the absence of external sources.

Using data from all the passes by Ganymede, the differenced field has been modeled either as a permanent internal field with dipole and quadrupole components or a permanent dipole plus an inductive response to the driving field imposed from Jupiter's magnetosphere. The inductive response is characterized by the efficiency of the inductive response, a parameter that depends on the conductivity of the responding layer and its depth. The inductive response model uses only 4 free parameters, but achieves a fit that is as good as the fit to the 8 parameters of the dipole plus quadrupole model. Although one cannot rule out the possibility that there is no inductive response, the rule of economy is appealing and it seems plausible that the inductive response is real. The new fits reduce the response efficiency from 0.83 previously suggested to 0.63 (see Figure 8). This level of response can be understood if the conducting layer is a buried ocean (at a depth of about 150 km) with conductivity lower than seawater. Elsewhere (Jia *et al.*, 2010) the implications of the imperfect response is discussed in greater detail.

The fact that the internal fields of Ganymede, Europa, and Callisto can be fit quite well by assuming inductive responses to the time-variations of the field imposed upon them is consistent with there being global shells of conductivity somewhere within their icy outer layers. For these icy moons, the conducting material is likely to be a buried liquid water ocean. Io is known to hide molten lava beneath its surface. Conducting magma is a plausible source of an inductive signature. Models of the internal field of Io based on the field measured in Galileo flybys of Io have not previously considered the possibility of an inductive response. However, Khurana, Jia, and Kivelson (2010) are using the technique described here for the case of Ganymede to examine Io's response to the time-varying field of Jupiter. The conductivity of melted magma is high enough to sustain an inductive response if the layer is thick enough and forms a nearly complete spherical shell. It may turn out that molten shells are ubiquitous among the Medicean moons, but confirmation must await additional analysis.

References

Anderson, J. D., Lau, E. L., Shogren, W. L., Schubert, G., & Moore, W. B. 1996, *Nature*, 384
Busse, F. H. 1976, *Phys. of the Earth and Plan. Int.*, 12, 350
Eviatar, A., Vasyliunas, V. M., & Gurnett, D. A. 2001, *Planet. Space Sci.*, 49, 327
Frank, L. A., Paterson, W. R., Ackerson, K. L., & Bolton, S. J. 1997, *Geophys. Res. Lett.*, 24, 2, 151
Jia, X., Walker, R. J., *et al.* 2008, *J. Geophys. Res.*, 113, A06212, doi:10.1029/2007JA012748
Jia, X. 2009, *Ph.D. Dissertation*, Dept. of Earth and Space Sciences, UCLA
Jia, X., Walker, R. J., Kivelson, M. G., Khurana, K. K., & Linker, J. A. 2009a, *J. Geophys. Res.*, 114, doi:10.1029/2009JA014375
Jia, X., Kivelson, M. G., Khurana, K. K., & Walker, R. J. 2009b, *Space Sci Rev*, doi:10.10.1007/s11214-009-9507-8
Jia, X., Walker, R. J., Kivelson, M. G., Khurana, K. K., & Linker, J. A. 2010, *J. Geophys. Res.*, doi:10.1029/2010JA015771
Khurana, K. K., Kivelson, M. G., *et al.* 1998, *Nature*, 395, 777
Khurana, K. K., Jia, X., & Kivelson, M. G. 2010, *A magma ocean at Io?*, in preparation
Kivelson, M. G. 2004, *Adv. Space Res.*, 33, 2061
Kivelson, M. G., Khurana, K. K., *et al.* 1996, *Nature*, 384, 537
Kivelson, M. G., Khurana, K. K., Russell, C. T., & Walker, R. J. 1997, *Geophys. Res. Lett.*, 24, 2127
Kivelson, M. G., Khurana, K. K., *et al.* 1997, *Geophys. Res. Lett.*, 24, 2155
Kivelson, M. G. *et al.* 1998, *J. Geophys. Res.*, 103, 19963
Kivelson, M. G., Khurana, K. K., *et al.* 2001, *J. Geophys. Res.*, 106, 26, 121
Kivelson, M. G., Khurana, K. K., & Volwerk, M. 2002, *Icarus*, 157, 507
Kivelson, M. G., Bagenal, F., *et al.* 2004, *Chapter 21 in Jupiter: The Planet, Satellites and Magnetosphere* edited by F. Bagenal, T. Dowling and W. McKinnon, Cambridge University Press
Neubauer, F. M. 1978, *Geophys. Res. Lett.*, 5, 905
Paty, C. & Winglee, R. 2006, *Geophys. Res. Lett.*, 33, L10106, doi:10.1029/2005GL025273
Saur, J., Strobel, D. F., & Neubauer, F. M. 1998, *J. Geophys. Res.*, 103, 19, 947
Schilling, N., Neubauer, F. M., & Saur, J. 2007, *Icarus*, 192, 41
Schilling, N., Neubauer, F. M., & Saur, J. 2008, *J. Geophys. Res.*, 113, A03203.doi:10.1029/2007JA012842
Schubert, G., Stevenson, D. J., & Ellsworth, K. 1981, *Icarus*, 47, 46
Showmann, A. P. & Malhotra, R. 1997, *Icarus*, 127, 93
Walker, R. J. & Russell, C. T. 1995, *in Introduction to Space Physics* edited by Kivelson, M. G., & Russell, C. T., New York, Cambridge U. Press
Zimmer, C., Khurana, K. K., & Kivelson, M. G. 2000, *Icarus*, 147, 329

Galileos Medicean Moons: their impact on 400 years of discovery
Proceedings IAU Symposium No. 269, 2010
C. Barbieri, S. Chakrabarti, M. Coradini & M. Lazzarin, eds.

doi:10.1017/S1743921310007283

Aurora on Jupiter: A Magnetic Connection with the Sun and the Medicean Moons

Supriya Chakrabarti[1] and Marina Galand[2]

[1]Center for Space Physics, Boston University
725 Commonwealth Avenue, Boston, MA 02215, USA
email: supc@bu.edu

[2]Dept. of Physics, Imperial College London,
Prince Consort Road, London, SW7 2AZ, United Kingdom
email: m.galand@imperial.ac.uk

Abstract. Observational astronomy began in Padova four hunderd years ago, when Galileo Galilei pointed a newly invented instrument towards Jupiter. After only one week of observations he discovered four moons circling Jupiter. In the intervening four centuries, technical progress in instrumentation and novel observational approaches have revealed much about the connection between these Medicean moons with Jupiter, none more revealing than the auroral emissions. In this paper we review observations of ultraviolet aurora made by earth-orbitting spacecraft as well as those that flew by the Jovian system.

Keywords. Medicean Moons, Aurora, Ultraviolet.

---

1. Introduction

Four hundred years ago today, observations of the night sky from this historic town revolutionized science. On this day, Galileo Galilee built his version of a *toy* invented by opticians, pointed it at the heavens and discovered four previously unknown moons orbiting Jupiter. This was truly a triumph of technology as these objects lied beyond the reach of the best observational tools available at the best observatories of the time. In an early demonstration of *faster, better, cheaper*, a scientific instrument that was shorter than a typical walking stick and took only weeks from the decision to develop to science results, established its superiority over naked-eye, castle-like observatories, that was the custom of the day. Since that auspicious day, creative minds have developed various telescopes and tools that have revealed many unexpected phenomena and surprising characteristics including those that are truly invisible to the human eye.

Four centuries of advances in technology have allowed us to answer many of the original questions on these Medicean moons. Our instruments on remote mountain tops, on space platforms near Earth, as well as those that have orbited Jupiter and flew past the Jovian system have given us a closer look of these bright points of lights. We now have a fairly good understanding of the major physical attributes of these moons, know how different they are from our own and from each other and why.

In this paper we will review observational progress made in our understanding of Jupiter and the Galilean moons and their magnetic connections as revealed through aurora. Auroral emissions and their morphology contain information on atmospheric composition, plasma interactions, magnetic field configuration, and magnetosphere-ionosphere coupling processes. Jupiter and its moons provide us with a rich wall-less laboratory environment that is accessible to probing by our spaceborne and ground-based observatories.

This is a chronicle of the triumph of technology. In the past two decades, new instruments behind large telescope have obtained conclusive evidence of planets around distant stars. We can now set our sights on these planets and, if the technical feats continue at their past or current pace, it would not be unexpected for another toolmaker to discover moons orbiting around one of these planets within this century.

2. Aurora: Background

On Earth, brilliant and dynamic auroral displays have conjured many colorful folklores and have been the subject of fascinating mythology (Eather (1980)). High occurrence frequency in certain geographic locations (now called *auroral zones*) and east-west extension of auroral displays which are correlated to large magnetic fluctuations established their geomagnetic connection. Even though people have studied aurora as long as we have studied the sky, there have been no less than three relatively recent instances when it was necessary to define aurora (Galand & Chakrabarti (2002), Clarke (2004) and Bharadwaj (2005)).

Pierre gassendi of the *Transit of Mercury* fame, coined the term *Aurora Borealis* (Northern Dawn) in 1621. Aurora observed at Earth was first defined by its temporal, spatial, spectral, and magnitude characteristics. However, these characteristics vary greatly from one Solar System body to another, as they depend on atmospheric composition, magnetic field configuration, energetic particle origin and acceleration mechanisms, and plasma interaction involved in their production. As a result, a more general definition based on their origin is needed for defining what aurora is beyond the Earth. Galand and Chakrabarti (2002) were the first to propose a definition applicable to bodies in the Solar System, followed later by Clarke (2004) and Bhardwaj (2005). A definition allows to make a classification, yielding a deeper understanding of plasma interactions in the Solar System.

Galand & Chakrabarti (2002) defined aurora to be *optical manifestation of the interaction of extra-atmospheric energetic electrons, ions, and neutrals with an atmosphere.* Clarke (2004) defined aurora to be *atmospheric emissions produced by the impact of external energetic particles* - a definition similar to that of Galand & Chakrabarti (2002) However, he was conflicted on the issue of emission produced by energetic neutrals, which do not follow magnetic field lines. He noted that *while this (the requirement of a magnetic field or localized region of particle precipitation) has been part of the historic definition of the aurora, it would require altering the literature as it has naturally evolved.* Bharadwaj (2005) required the presence of a magnetic field for optical emissions to be classified an aurora and accepted that such emissions are generally localized in the high latitude (presumably magnetic latitude) regions. He noted, *whether the magnetic field is intrinsic (natural) or induced, or of a nearby magnetic body (i.e., of planets for planetary satellites), or the interplanetary magnetic field, is a separate question.* He added, *even in cases of neutrals causing an aurora at low latitudes, they have been under the influence of a magnetic field until they got converted to neutral in the charge-exchange reaction.* Fortunately, for the cases of auroral emissions on Jupiter and the Medidean moons, the thread that connects them is the magnetic field and the emissions are produced by energetic ions or electrons, thus, avoiding any confusion.

While the physical and chemical processes that result in optical emissions in aurora are governed by the same underlying physics, the spatial, temporal and magnitude scales, and energetics vary enormously because the atmospheric composition, energy source and magnetic field geometry are often unique to each type of object. Thanks to such rich and diverse auroral processes, a comparative approach can be applied to develop a comprehensive understanding of one of the complex interactions in the solar system.

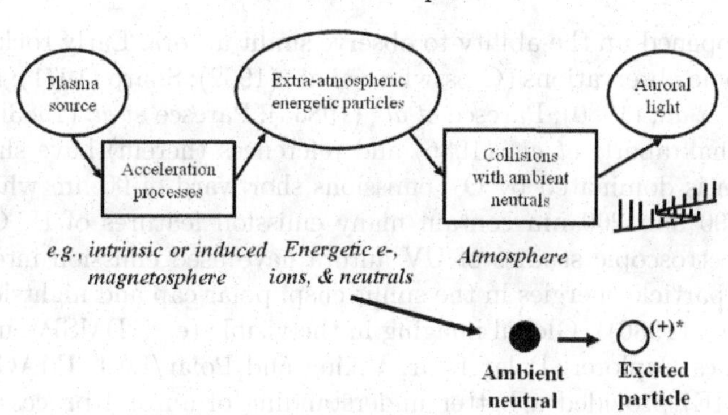

Figure 1. A cartoon depiction of processes leading to auroral emissions.

The general picture that is consistent with all these descriptions has been summarized in Fig. 1.

In this paper we will review the magnetic connections between Jupiter and the Medicean moons as manifested through ultraviolet aurora. While auroral observations have been made from IR to X-rays and beyond, we chose to restrict the discussions to UV emissions only. The UV emissions allow examination of spectral lines of many of the common atmospheric species in a compact wavelength region. Furthermore, the excitation mechanisms are simple and since the energy threshold are high (> 10 ev), complex photochemistry and albedo from lower atmospher, which complicates interpretation of visible emissions, can be ignored. However, the observed UV emissions are sometimes altered by radiative transfer processes, which need to be carefully accounted in the models.

3. Observational and Instrumentation Issues

The progress in optics, detector and space technology made our understanding of the complex processes that govern plasma interactions along magnetic fields leading to auroral emissions, which are tracers of solar wind-magnetosphere-ionosphere-atmosphere coupling. The Earth's aurora, being the most accessible of solar system objects, has been studied extensively. Terrestrial aurora has been scientifically studied since the 18th Century (see for example, Mairan (1733) and other listed in Chamberlain (1995)). Key observational milestones include the first spectroscopic study (Angstrom (1869)) and the first all-sky auroral morphology study (Gartlein (1947)). Since then, every advance in optics, detector and related technology, just like astronomical studies after Galileo, have been implemented in auroral observations.

Ground-based photometric, spectroscopic and imaging observations have established auroral spectra in the visible portion of the electromagnetic spectrum and their morphology over a given site (Vallance Jones [1974). Until recently, ground based observations have been limited to night time conditions centered around new moon. Recent advances in optical technology have demonstrated that selected auroral emissions could be observed round-the-clock even under fully sunlit conditions (Chakrabarti & Pallamraju (2006)).

With the availability of array detectors and space platforms, our understanding of the aurora increased dramatically. Ultraviolet observations, possible only from space-based platforms, have opened up the ability to observe sunlit aurora. Early rocket and satellite based spectroscopic observations (Crosswhite *et al.*, (1962); Sharp (1971); Eastes & Sharp (1987); Huffman *et al.*, (1980); Paresce *et al.*, (1983a); Paresce *et al.* (1983b); Chakrabarti *et al.*, (1985); Chakrabarti *et al.*, (1986) and references therein) have shown terrestrial auroral spectrum is dominated by O^+ emissions shortward of 90 nm while the spectral region between 90 and 200 nm contain many emission features of H, O, N, O^+, N^+, H_2, and N_2. Spectroscopic studies of UV aurora have used emission intensity ratios to characterize the particle energies in the sunlit cusp, polar cap and nightside aurora (e.g., Chakrabarti *et al.*, (1986)). Global imaging in the visible (e.g., DMSP), ultraviolet (e.g., Kyokko, Dynamics Explorer, Polar Bear, Viking and Polar/UVI, IMAGE) and X-rays (e.g., Polar/PIXIE) provided a better understanding of auroral processes. The size of the auroral oval derived from global images has been related to solar wind parameters (Siscoe (1991)). Evolution of the global aurora at up to 50 km spatial and 1 minute temporal resolutions have been examined to understand the substorm onset and poleward expansion [e.g., Craven & Frank (1987) and Shepherd *et al.*, (1987)].

Satellite borne ultraviolet spectrographs using microchannel plates, a key component of the night vision technology (Bowyer *et al.*, (1981)), established the geographic morphology and relationship with geomagnetic and solar activities for terrestrial aurora (Paresce *et al.*, (1983a); Paresce *et al.* (1983b)). Microchannel plates allowed observations of very low light levels as it enabled photon counting - building images or spectra one photon at a time. An imaging UV spectrograph aboard the International Ultraviolet Explorer (IUE) observed auroral emissions from Jupiter in the 120 – 170 nm band (Clarke *et al.*, (1980)), which was quickly followed by Voyager 1 observations in the 60 – 160 nm band (Broadfoot *et al.*, (1981)). Voyager observations of Jupiter's UV aurora were found to follow closely to the magnetic field lines connected to the Io Torus.

Charged Couple Devices (CCDs) and Intensified Charge Coupled Devices (ICCDs) have been used in auroral imaging and spectroscopy applications. With the spectra established, spectacular images and indeed movies of Jovian aurora in selected UV wavelength bands have been obtained by the Hubble Space Telescope (HST) which vividly showed the magnetic connections with the moon (Fig. 2).

Similar to Jupiter, the Medicean Moons have atmospheres. They are also embedded in Jupiter's magnetic environment where energetic particles are present, primarily from the Jovian system and with a small contribution from the Sun. As a result, we expect to see aurora, the optical manifestation resulting from complex and dynamic interactions between them, on each of these objects. The composition of the atmospheres of Jupiter and its Medicean moons, their magnetic field strength and geometry are very different from that of the Earth and from each other, which results in vastly different auroral signatures (see Bharadwaj & Gladstone (2000) for a comprehensive survey of auroral processes on Jovian planets). Through observations of UV aurora and comparing their brightness characteristics as well as spatial and temporal morphology against those observed in X-rays and IR, one can infer atmospheric composition and distributions, energy of the particles and trace the magnetic field configuration and plasma interactions.

4. Determination of Physical Properties and Processes from Auroral Emissions

Several techniques have been employed in inferring emission sources, key physical processes in the aurora and the environment they are produced; two common ones are pictorially depicted in Fig. 3 Fox *et al.*, 2008.

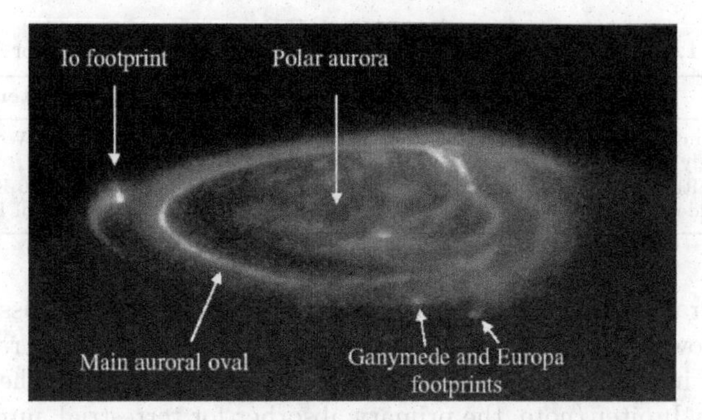

Figure 2. HST observations of UV aurora (false color) on Jupiter showing the presence of a main oval and polar cap aurora, similar to Earth (though at Jupiter, the main oval is associated with the breaking from co-rotation, unlike at Earth). The polar cap aurora was very dynamic and influenced by solar wind conditions. In addition, unlike at Earth, Jovian aurora also shows features magnetically connected to its moons. They represent the magnetic footprints of the moons in the jovian atmosphere. Note that Callisto footprint cannot be seen as its location is predicated within the main oval. Clarke *et al.* (1980) Photo Credit: NASA

The first one of these techniques is based on color intensity ratio. Noting that higher energy particles penetrate deeper in an atmosphere, the intensity ratios of selected emission lines observed from a vantage point above the atmosphere have been used to infer atmospheric composition as well as characteristic energy of the incident particles (see Table 1). Similar technique has been applied for the case of the Earth's sunlit cusp, polar cap as well as nightside aurora to characterize incident particle energy by examining intensity ratios of selected emission lines (Chakrabarti (1986)).

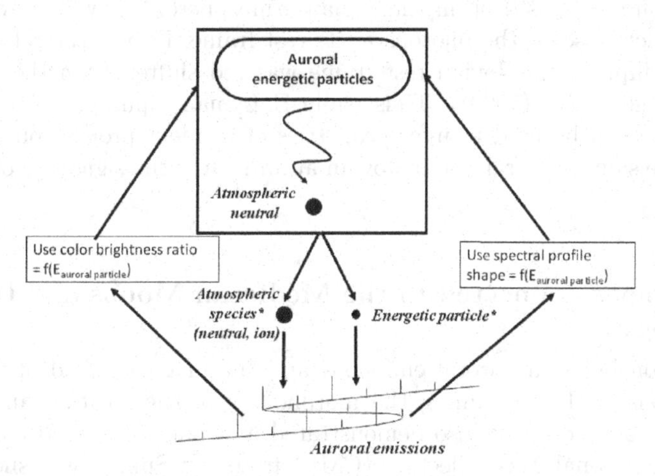

Figure 3. Two common methods of obtaining physical parameters from auroral emissions.

As an example, intensity ratios of two far ultraviolet windows covering H_2 bands at 123.0-130.0 nm and 155.7-161.9 nm have been used for Jupiter. Emissions in the shorter

Table 1. Example of Auroral Spectroscopic Analysis Using Color Ratios

Color Ratio	Earth	Jupiter
Two Spectral Bands Chosen in	N_2 LBH	H_2 Lyman and Werner bands
One band Strongly Abosorbed by:	O_2	CH_4
Electron Energy Range Covered	0.2 – 20 KeV	10 – 200 KeV
Type of Aurora identified:	Discrete Electron Aurora	Hard e- component of Electron Aurora

wavelength spectral window is strongly absorbed by methane, while emission in the longer wavelength window is absorbed only weakly. As a result, this intensity ratio is a function of the energy of incident electrons in an aurora as represented by the altitude of the methane layer on Jupiter (note, the primary absorber for terrestrial auroral emissions in the UV is O_2). Pallier & Prange (2004) and Rego *et al.*, (1999) used this technique with HST/STIS data to infer the energy of the electrons in Jupiter's aurora. Similar analyses have also been applied to spectroscopic data obtained by IUE, HUT, Galileo/EUVS, Cassini/UVIS as well as by New Horizon/Alice.

The second technique is based on the spectroscopic analysis of emission line spectra observed at high spectral resolution. The shape of the same spectral line can carry the signature of different physical processes. For example, the hydrogen Lyman alpha emission line in a terrestrial aurora as seen from space was observed to be red shifted by 0.13 nm. The spacecraft which made these observations was located at 600 km altitude, significantly above the peak emitting region. Galand *et al.*, (2002) interpreted these observations as emissions produced by precipitating protons and their interactions with the atmospheric neutral species.

The line shape of the H Lyman Alpha line in Jovian aurora observed by the HST/GHRS is significantly more complicated. The emissions show a self-reversed core which is red shifted from the geocoronal rest frame. While core reversal is common is solar and stellar spectra, such a feature had not been seen in planetary atmosphere and was first discovered in Jovian aurora. Further analysis of the observed red shift and the asymmetric line profile revealed that the core reversal is located at the Jovian rest frame thereby indicating that the emission source is deep in the Jovian atmosphere, below the atomic hydrogen atmosphere, which absorbs the photons at its rest frame. Due to the relative motion of the Earth and Jupiter, the Jovian rest frame was red shifted from the geocornal rest frame (see Prange *et al.*, (1996)). Thus while H Lyman alpha emission line shape in terrestrial aurora can be used to infer properties of incident protons on the Earth, the line shape of the same spectral line in Jovian aurora carry the signature of precipitating electrons.

5. UV Emissions connected to the Medicean Moons and their Morphology

The connection between auroral emissions and the magnetic field of the planet has been well established by examining the morphology of the location and geometry of terrestrial aurorae. They have also demonstrated a strong solar wind connection. In a study involving Coronal Mass Ejection (CME) from the Sun, it was shown how such an event leaves a profound signature through auroral emissions; four different spacecraft recorded auroral emissions due to a single CME event on the Earth (Polar spacecraft), Jupiter (Galileo and Cassini spacecraft) and Saturn (HST) (Prange *et al.*, (2004)).

Fig. 2 shows an image of the FUV (115 – 170 nm) aurora recorded by HST. From these and other multiwavelength observations, it has also been established that several auroral processes are active on Jupiter (Vasavada *et al.*, (1999)). The main oval observed in the UV, visible, hard X-ray, and near IR and produced by magnetospheric electrons while the soft X-ray are produced by energetic oxygen and sulfur ions. The polar cap emissions, which are very dynamic, have been seen in UV are responsive to solar wind conditions. Finally, localized emissions seen at the magnetic footprints conecting the Medicean moons and believed to be due to energetic electrons Thus, auroral emissions, at least emissions in the auroral zone, connect the Solar Wind to the magnetosphere, neutral atmosphere and ionosphere of a planet and its moon through diverse magnetic field geometry.

The auroral emissions and their temporal morphology have been studied in detail using ground- and space-based instruments. Prange *et al.*, (1996) established that the FUV auroral "spot" due to Io on Jupiter was located in a region consistent with the IR emissions (Connerney (1993)); however, both were seen poleward of model footpoint of the magnetic field lines (Hill *et al.*, (1983)) that connected Io. The emission brightness were found to represent 25% of the energy produced due to the motion of Io across Jovian magnetic field lines — a value that corresponds to an energy flux that equals the entire terrestrial auroral energy flux concentrated in a 10^4 sq. km area.

The location of the magnetic footprints observed in the UV has been used to constrain the topology of the magnetic field (Grodent *et al.*, 2008). Grodent *et al.* (2008) extracted the location of the magnetic footprints of the three Medicean moons, Io, Europa, and Ganymede, from individual UV images taken by HST in order to derive observed, reference contours. By comparison of these contours with the moon footprint locations extracted from the VIP4 magnetic field model (Connerney *et al.*, 1998), they inferred the presence of a magnetic field anomaly in the northern hemisphere of Jupiter.

Recently, Wannawichian *et al.*, (2010) have examined a decades of FUV emissions recorded by the HST at the footprint of the Medicean moons Io, Ganemede and Europa in terms of the electrodynamic interactions between Jovian magnetosphere and the atmospheres of the moons. After correcting for limb brightening effects, the authors found a factor of 8 – 10 brightness variations at the footprint of Io as a function of the location of Io (System III longitude). The intensity showed a two peak variations, with the second being approximately a factor of two dimmer than the first. Both peaks are observed when Io is near the plasma equator and suggests a strong connection between the brightness and the plasma density of the torus where the moon is located, thereby indicating a strong control by mass pick up processes.

The absolute brightness of the footprint of Ganymede is approximately a factor of two dimmer than that of Io. Ganymede also shows a variation of brightness that is correlated with the System III longitude and a peak near the equator. However, due most likely to its lower brightness, similar relations could not be established for Europa. Unlike Io, plasma densities at the locations of Ganymede and Europa are significantly lower. The brightness variations at these two locations suggests a stronger solar wind connection. The Jupiter and its Medicean moons presents a wall-less laboratory where the complex plasma processes manifest in bright UV emissions in the auroral zone.

6. Aurora on the Medicean Moons

Two Jovian Moons, Io and Ganymede, have substantial extended atmospheres. They are also immersed in Jupiter's magnetosphere. The location of the bright footprints seen near the auroral oval of Jupiter and the temporal variation of their brightness correlate well with the magnetic field geometry. It is therefore reasonable to expect that just like their footprints on Jupiter, some aurora-like features should be present on these moons.

Observational evidence of UV auroral features on Io was reported by Roesler *et al.*, (1999). Using the Space Telescope Imaging Spectrograph (STIS) the authors observed bright OI 135.6 nm and SI 138.9, SI 142.9 and SI 147.9 nm emissions near Io's equator that shifted positions with the orientation of Jupiter's magnetic field. While the authors also observed time-varying HI Lyman alpha emissions from Io's polar region, they are not believed to be due to aurora-like excitation process. More recently, Retherford *et al.* (2007) reported the observations of Io's aurora in eclipse by New Horizons ultraviolet spectrometer. These observations revealed the influence of plumes on Io's electrodynamic interactions with Jupiter's magnetosphere with the volcanoes supplying up to 3% of the dayside atmosphere.

After the detection of atmospheric atomic oxygen emissions by Hall *et al.*, (1998), Feldman *et al.*, (2000) used a similar approach on Ganymede and confirmed bright spots in OI 135.6 nm emissions on the trailing side near its poles (latitudes greater than 40°). These STIS observations showed temporal variations as well as inter hemispheric variations, which the authors attributed to aurora-like excitation mechanism. The location of auroral emissions near the poles is explained by the presence of an intrinsic magnetic field at Ganymede discovered by the Galileo spacecraft [Kivelson *et al.* (1996)]. Recent 3D multi-fluid simulations have confirmed the auroral nature of these emissions Paty & Winglee (2004).

7. Concluding Remarks

Ultraviolet aurora on Jupiter has revealed the complex magnetic connection with the galilean moons. The moons have left bright ultraviolet footprints on Jupiter. The magnetic field lines on the moons also glow in UV light. These emissions and their spatial and temporal morphology have been used to reveal the physical processes involving magnetospheric plasma and neutral atmospheric species.

The next 400 years will see a steady progress in optical technology. Already interferometric tools are being developed (See for example, Hicks *et al.* 2009 and references therein), that will be capable of directly imaging Jupiter-like planets around Sun-like stars within a decade. It should not strain anyone's imagination to envision a time within the next century when another Galileo will point an instrument towards such an exoplanet and find their own *Medicean moons.*

Acknowledgements

The work at Boston University was supported in part by NASA grants NNX06AC04G and NNG05WC17G.

References

Angstrom, A. J., 1869, *Spectrum des Nordlichts, Pogg. Ann*, 137, 161

Bharadwaj, A., 2005, *EOS Trans. AGU*, 86(11), doi:10.1029/2005EO110007

Bharadwaj, A. & Gladstone, G. R., 2000, *Rev. Geophys.*, 38, 285

Bowyer, S., Kimble, R., Paresce, F., Lamptom, M., & Penegor, G., 1981, *Appl. Opt.*, 20, 477

Broadfoor, A. L., Sandel, B. R., Shemansky, D. E., McConnell, J. C., Smith, G. R., Holberg, J. B, Atreya, S. K., Donahue, T. M., Strobel, D. F., & Bertaux, J. L. 1981, *J. Geophys. Res.*, 86, 8259, doi:10.1029/JA086iA10p08259

Chakrabarti, S., 1985, *J. Geophys. Res.*, 90, 4421

Chakrabarti, S., 1986, *J. Geophys. Res.*, 91, 8065

Chakrabarti, S. & Pallamraju, D., 2006, in: Marc Duldig. (Ed.) *Advances in Geosciences, Volume 2: Solar Terrestrial (ST)*,World Scientific Co., Pte. Ltd., Singapore, p. 201

Chamberlain, J. W., 1995 *Physics of Aurora and Airglow*, (American Geophysical Union, Washington, D. C.),

Clarke, J. 2004, *EOS Trans. AGU*, 85(52), 567

Clarke, J., Moos, H., Atryea, S., & Lane, A. L., 2004, *Astrophys. J.*, 241, L179, doi:10.1086/183386

Connerney, J. E. P., baron, R., Satoh, T., & Owen, T. 1993, *Planetary Science*, 262, p. 1035.

Craven & Frank, 1987 *J. Geophys. Res.*, 92, 4565

Crosswhite, H., Zipf, E., & Fastie, W., 1962, *Opt. Soc. Am.*, 52, 643

Eastes, R. & Sharp, W., 1987, *J. Geophys. Res.*, 92, 10095

Eather, R. H., 1980, *Majestic Lights*, American Geophysical Union, Washington, D.C.

Feldman, P., McGrath, M., Strobel, D. Moos, H., Retherford, K., & Wolven, B. 2000, *Astrophysical L.*, 535, 1085

Fox., J. L., Galand, M. I., & Johnson, R. E. 2008, *Space Sci. Rev.*, 139, 3-62, doi:DOI 10.1007/s11214-008-9403-7

Galand, M. & Chakrabarti, S. 1998, in: M. Mendillo, A. Nagy, & J. H. Waite (eds.), *Atmospheres in the Solar System*, (Washington, D. C., AGU Monograph), 130, p. 55.

Galand, M., Lummerzheim, D., Stephan, A., Bush, B., & Chakrabarti, S. 2002, *J. Geophys. Res.*, 107(A7), 10.1029/2001JA000235

Gartlein, C. W., 1947, *National Geographic Magazine*, 673, Nov.

Hall, D., Feldman, P., McGrath, M., & Strobel, D. 1998, *Astrophys. J.*, 499, 475

Hicks, B., Cook, T., Lane, B., & Chakrabarti, S. "Monolithic achromatic nulling interference coronagraph: design and performance", *Appl. Opt.*, **48**, 4963–4977, 2009.

Hill, T. W., Dessler, A. J., & Goertz, C. K. 1983, in: A. J. Dessler, ed., *Physics of the Jovian Magnetospheres*, (Cambridge Univ. Press), p. 353.

Huffman, R. E., Leblanc, F. J., Larrabee, J. C., & Paulsen, D. E., 1980 *J. Geophys. Res.*, 85, 2201, doi:10.1029/JA085iA05p02201

Kivelson, M. G., Khurana, K. K., Russell, C. T., Walker, R. J., Warnecke, J., Coroniti, F. V., Polanskey, C., Southwood, D. J., & Schubert, G., 1996 *Nature* 384, 537

Mairan, J. J. D. de, 1733 *Imprimerie Royale*, Paris

Pallamraju, D. & Chakrabarti, S., 2005 *Geophys. Res. Lett.*, 32, L03S10,doi:10.1029/2004GL021417

Pallier, L. & Prangé, R., 2004 *Geophys. Res. Lett.*, 6701, doi:10.1029/2003GL018041

Paresce, F., Chakrabarti, S., Bowyer S., & Kimble, R., 1983a *J. Geophys. Res.*, 88, 4905

Paresce, F., Chakrabarti, S., Kimble, R., & Bowyer S., 1983b *J. Geophys. Res.*, 88, 10247

Paty C. & Winglee, R. 2004 *Geophys. Res. Lett.*, 31, L24806, doi:10.1029/2004GL021220

Prangé, R., Rego, D., Southwood, D., Zarka, P., Miller, S., & Ip, W., 1996 *Nature*, 379, 323

Prangé, R., Pallier, L., Hansen, K. C., Howard, R. Vourlidas, A. Courtin, R., & Parkinson, C., 2004 *Nature*, 432, 78

Rego, D., Prangé, R., & Ben Jaffel, L., 1999 *J. Geophys. Res.*,104, 5939

Retherford, K. D. & 20 co authors, 2007 *Science*, 318, 237

Roesler, F., Moos, H. W., Oliverson, R., Woodward, Jr., R., Retherford, K., Schreb, F., McGrath, M., Smyth, W., Feldman, P., & Strobel, D., 1999,*Science*, 283, 353

Sharp, W., 1971 *J. Geophys. Res.*, 76, 987

Shepherd *et al.*, 1987 *Geophys. Res. Lett.*, 14, 395,doi:10.1029/GL014i004p00395

Siscoe, G., 1991, in, C.-I. Meng, M. J. Rycroft, & L. A. Frank eds. *Auroral Physics*, 159

Vallance Jones, A., 1974 *Aurora, D. Reidl Publishing Co.*, Dordrecht Boston

Vasavada, A. R., Bouchez, A. H., Ingersoll, A. P., Little, B., Anger, C. D., & the Galileo SSI Team, 1999 *J. Geophys. Res.*, 104, 27133, doi:10.1029/1999JE001055

Wannawichian, S., Clarke, J. T., & Nichols, J. D., 2010 *J. Geophys. Res.*, 115, 2206, doi:10.1029/2009JA014456

Galileo's Medicean Moons: their impact on 400 years of discovery
Proceedings IAU Symposium No. 269, 2010
C. Barbieri, S. Chakrabarti, & M. Coradini, eds.

doi:10.1017/S1743921310007295

Io's Escaping Atmosphere: Continuing the Legacy of Surprise

Nicholas M. Schneider[1]
[1]LASP, University of Colorado, Boulder CO 80309 USA
email: nick.schneider@lasp.colorado.edu

Abstract. The discovery of Io and her fellow Medicean Stars clearly altered the course of science as a whole. It is equally clear that the discovery of Io's tidal heating has altered the course of planetary science. One of the most directly observable consequences of Io's tidal heating is the prodigious escape of a ton per second of volcanically-supplied gases. I will review how studies of Io's escaping atmosphere since 1972 have advanced our deep understanding of Io itself, and helped formulate our perspective on planetary evolution in our solar system and beyond.

Keywords. Planets and satellites: Io

1. Io's unique nature revealed in sodium cloud observations

Galileo's experience of a new instrument rapidly discovering radical new phenomena has been repeated countless times in the Jovian system, most notably with Io. In 1972, Robert A. Brown was testing a new spectrograph with observations of "known objects", namely the Galilean moons of Jupiter. (These observations took place after Bigg's unexplained discovery of Io-correlated radio emissions from Jupiter (Bigg, 1964) but before the Voyager flyby of Jupiter in 1979.) Brown expected to confirm the performance of the instrument in the form of a simple reflected solar system (Binder and Cruikshank, 1966), but was surprised to find enhanced emission at the location of the sodium D lines near 589nm (Brown, 1972).

The emission was initially thought to be confined to Io's disk, like a glowing atmosphere, but followup observations by Trafton *et al.* (1974) showed the emission to be extended in a region up to tens of arcseconds from Io. This broad extent explains the discrepancy between Brown's spectra and those of prior observers : Brown's instrument possessed a wide entrance aperture which accepted light from Io and the surrounding region, which enhanced the emission feature relative to Io's reflected continuum. An instrumental design feature implemented to simplify the guiding process had enabled the discovery of the extended, escaping atmosphere of atomic sodium.

Several groups soon confirmed that resonant scattering was the emission process by correlating the brightness of the cloud with the Io's orbital phase. When Io was at its maximum velocity relative to the Sun at maximum elongation, co-moving sodium atoms appeared brighter because they experienced a higher solar flux due to their Doppler shift away from the minimum of the deep solar sodium D-line absorption (Bergstralh *et al.*, 1975; Macy and Trafton, 1975).

Narrow-band imaging then joined high-resolution spectroscopy as an important diagnostic tool (Matson *et al.*, 1978; Goldberg *et al.* 1980; (Pilcher *et al.*, 1984). These works enabled a new generation of physical models of escape processes whose parameters could be tuned to match observations (e.g., Smyth and McElroy, 1978 and subsequent works). Mendillo *et al.* (1990) then discovered the cloud extended hundreds of jovian radii away from the planet. These large distances, and the evidence of rapidly-changing features near Io (Schneider *et al.*, 1991) led to a new realization that high-speed escape processes, linked to plasma sweeping past Io at 70 km/sec, were involved.

In parallel, observations and theory were building up an understanding of the plasma torus created by the ionization of these neutral clouds. The plasma torus, in turn, flows onto and around Io's atmosphere, thereby causing escape and completing a positive feedback loop. The structure and behavior of the plasma torus is reviewed in Thomas *et al.* (2004) and Schneider and Bagenal (2007).

Together these observations have given a solid explanation of the many phenomena exhibited by Io's sodium clouds over a wide range of spatial scales (Figure 1). First, low-speed atoms ejected by sputtering make the slow "banana cloud" which extends insides and slightly forwards from Io (e.g, Smyth *et al.* 1978). Second, molecular ions are stripped out of the ionosphere, picked up and carried downstream in the torus, where they dissociatively recombine and create the fast sodium "stream" (Wilson and Schneider, 1994). Finally, a jet of sodium atoms is ejected in the anti-Jupiterward direction, created by sodium pickup ions undergoing charge-exchange as they rise out of the atmosphere (Burger *et al.* 2000). These pheonomena continue to offer a direct means to study Io's escape processes through groundbased observations. Contemporaneous observations of the infrared flux from Io's volcanoes suggests that high volcanic activity enhances energetic escape processes (Mendillo *et al.* 2004).

2. The Pervasive Effects of Io's Escaping Atmosphere

Nearly five decades of observations have stitched together a long list of cause-and-effect relationships which permeate the Jovian system (Figure 2). In the first step, Io's volcanoes release lava, pyroclastic debris and atmospheric gases. The gases form the most spatially and temporally variable atmosphere in the solar system, enhanced in regions of high volcanic outgassing and/or frost sublimation, and minimized in cold polar regions due to condensation. This molecular atmosphere, only weakly held by Io's gravity, begins to escape to space. Subjected to magnetospheric particles and UV radiation, the molecules are broken down into their constituent atoms as the escape continues.

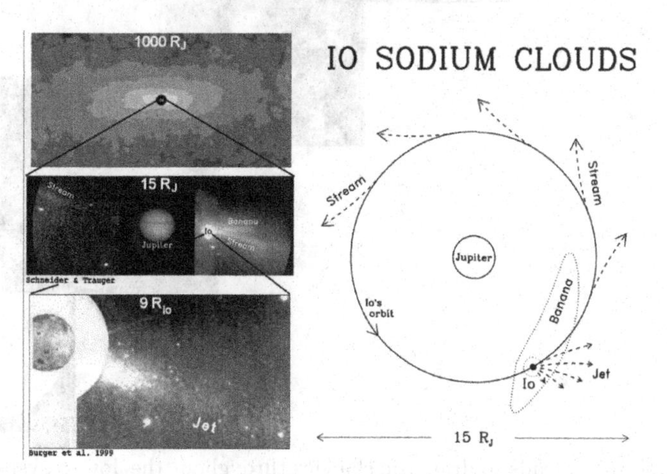

Figure 1. (Left) Io's sodium cloud on three spatial scales, as imaged by groundbased observations of sodium D-line emission. (Right) The features observed at left are explained by the three atmospheric escape processes shown schematically. The Ôbananaõ, jet and cloud are morphological features created by three distinct atmospheric escape processes and are explained in the text. From Schneider and Bagenal (2008) after Thomas *et al.* (2004)

Next the escaping gases are ionized and swept into a ring surrounding Jupiter, confined by the jovian magnetic field and co-rotating with the planet. The plasma diffuses inwards and outwards, becoming energized by unknown processes. Some of the plasma becomes capable of traveling along field lines all the way to Jupiter's polar regions, causing aurora. The auroras are sufficiently powerful to dominate sunlight as the prime energy input at high latitudes, driving chemical reactions which create the hydrocarbons which darken the poles at UV wavelengths. Some of the currents driven by precipitating particles give rise to the powerful Jovian radio emissions. Ironically, the driving cause of these phenomena, Io's volcanism, was discovered last, and its most distant effect, the radio emissions, were discovered first. Figure 3 summarizes the profound effects caused by the combination of orbital resonances, tidal heating, atmospheric escape and magnetospheric interactions.

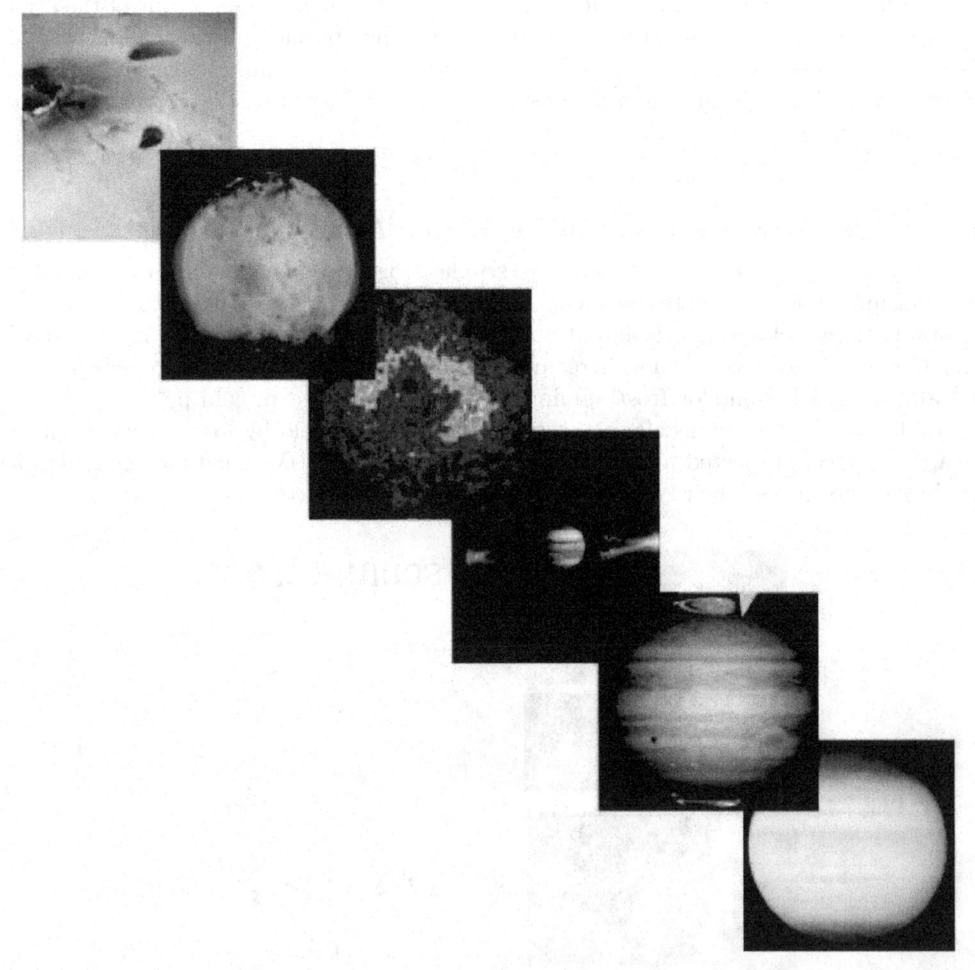

Figure 2. Io's volcanism leads to dramatic changes throughout the Jovian system. The images, respectively, are Io's Tvashtar volcano (imaged by Galileo), an artist's conception of Io's patchy atmosphere overlaid on Io's globe; a Hubble Space Telescope image of Io's corona; a ground-based image of the Io plasma torus, a Hubble Space Telescope image of Jupiter with its UV aurora, and another Hubble Space Telescope image of Jupiter showing polar darkening by auroral energy.

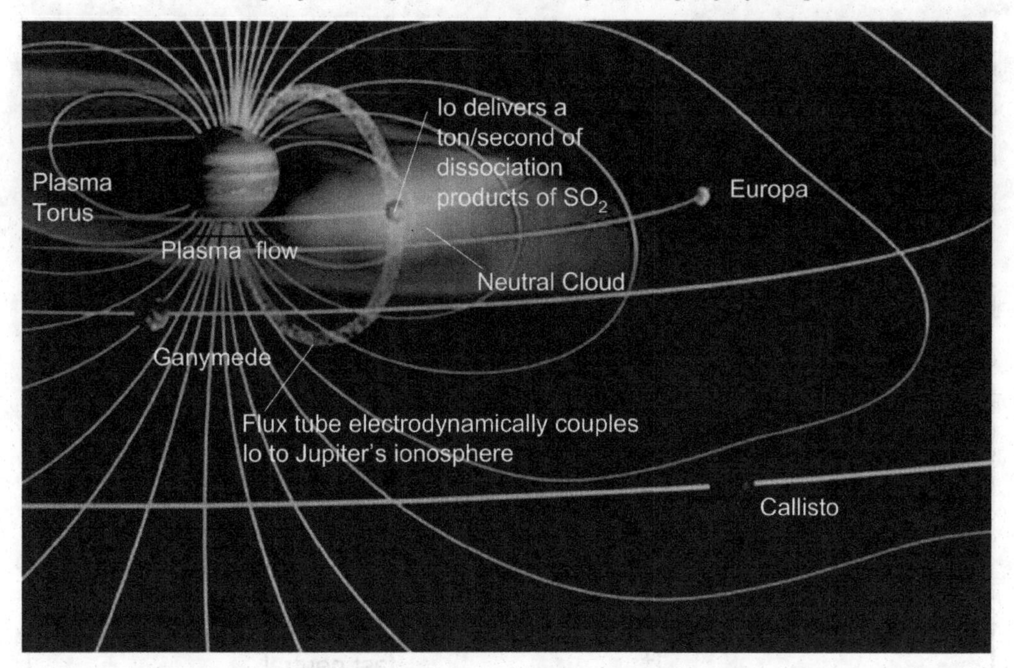

Figure 3. Io's Influences on the Jovian system. Without orbital resonances and tidal heating, these worlds would not affect each other to this dramatic degree.

3. How Io guides our understanding of atmospheric evolution

One of the surprises of planetary exploration has been the revelation that planetary atmospheres bear little resemblance to their primordial states. Virtually all objects apart from the jovian planets have undergone significant atmospheric escape. The escape on most worlds acts on timescales of millions or billions of years, and can only be studied by orbital spacecraft and theory. Smaller bodies, by virtue of their lower gravity, undergo more rapid and prominent escape. Io is arguably the archetype of atmospheric escape, losing a ton per second in the current epoch and perhaps 2km of its radius over the age of the solar system. Figure 4 shows many of the escape processes that have studied at Io with remote and *in situ* methods. New escape processes continue to be discovered at Io, and these new processes may prove to be important on other worlds or at at other times. The case of Mars is especially intriguing: NASA's upcoming MAVEN mission ("Mars Atmosphere and Volatile EvolutioN") seeks to measure many of the same non-thermal escape processes observed at Io, with a goal of extrapolating backwards in time to quantify the integrated atmospheric loss of the age of the solar system.

4. Were the Medicean Stars destined to change planetary science?

On the occasion of this Galilean anniversary, it's appropriate to speculate on whether the objects within the grasp of Galileo's telescope would inevitably be the ones to profoundly alter our understanding of the planetary processes described above. For example, if Galileo had discovered the moons of Mars, the impact on the Copernican revolution would have been the same. But the lasting impact on planetary science would have been different and arguably diminished. Phobos and Deimos, fascinating objects in their own rights, do not profoundly affect Mars, nor do they serve as the archetype of a transformative process like tidal heating.

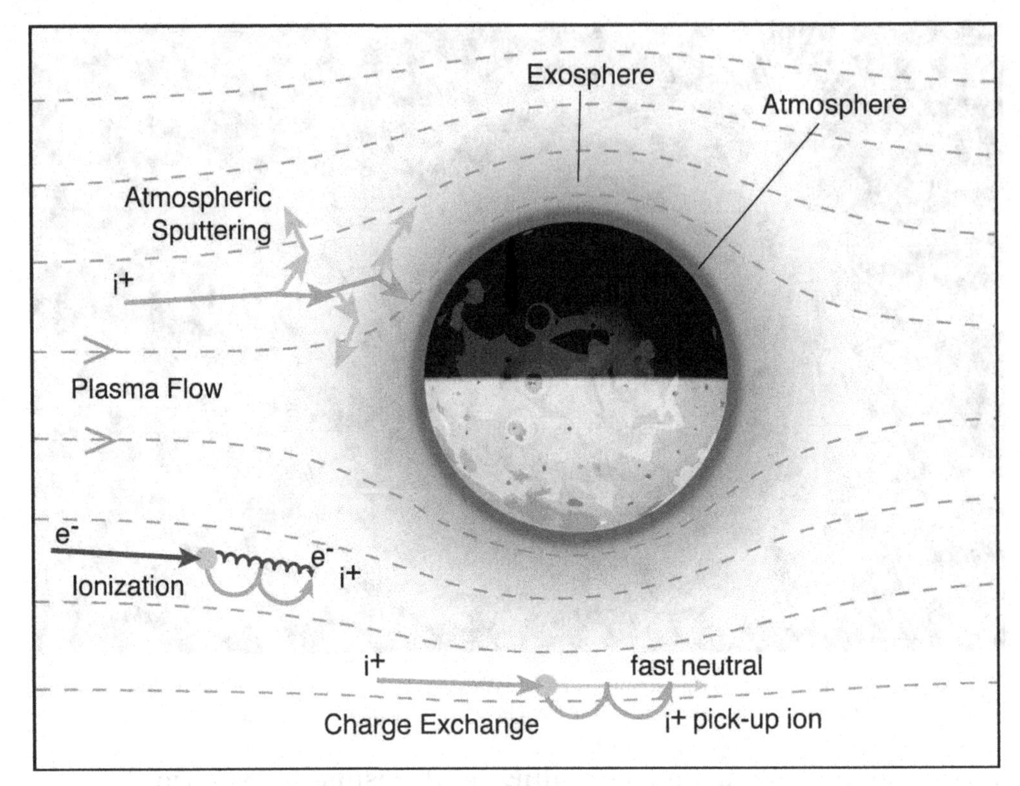

Figure 4. Important plasma/atmosphere interactions near Io. The scale of the gyromotions have been greatly exaggerated: the gyroradius of a pick up oxygen ion is 5 km, much less than Io's radius, and that of an electron is about 40,000 times smaller than the ions. Atmospheric escape occurs at such high rates at Io that many processes are detectable in groundbased observations.

But Galileo's telescope could not have discovered satellites as faint as Phobos and Deimos. His instrument and observing method were most sensitive to bright, fast-moving, numerous moons far enough from their planet to be resolved. The Medicean stars clearly meet these requirements: they are among the largest and most reflective moons in the solar system, and by virtue of their orbit around Jupiter are rendered brighter than satellites of Saturn, Uranus or Neptune. Their location relatively near Jupiter, and Jupiter's great mass, combine to give orbital periods of order days, making their motion easy to detect over the course of a few nights. The fact that there are four moons in the same orbital plane enhanced their detectability, both in their organized appearance and their orbital motions. The Medicean stars could perhaps have been detected even more easily if they orbited closer and faster, but if their orbits had been too small, Galileo's telescope could not have resolved them from Jupiter's glare.

Of all the moons in the outer solar system, only Titan shares any of these criteria for detectability. While its size rivals Callisto's, its greater distance and lower albedo render it roughly ten times fainter than the Medicean stars. And its large orbital distance about a lower-mass planet give a sixteen day period. These facts conspired to prevent Titan's discovery by Huygens until 45 years after Galileo's discovery of the Medicean stars. We can conclude that it was no accident that Jupiter's moons were discovered first.

One question remains: do the properties that lead to the great detectability of the Medicean stars inevitably lead them to have such fascinating properties? Specifically, is a system of several large, fast-moving satellites relatively close to a very large planet destined to undergo orbital resonances and tidal heating? The answer appears to be "yes". Orbital resonances by definition require two or more satellites, and the Medicean stars offer three. Were Io, Europa and Ganymede destined to become locked in an orbital resonance? Again the answer appears to be affirmative. These objects did not form in resonant orbits, nor is the resonance a coincidence. Each of these objects is large enough and close enough to Jupiter to raise tidal bulges on the planet. Strong tidal torques exerted by Jupiter evolved each of their orbits outwards, with Io moving outwards until achieving resonance with Europa, then the pair evolving outwards until locking into resonance with Ganymede. With eccentricities for all three moons pumped up by their orbital resonance, tidal heating was inevitable. And while their sizes are each large enough to enhance the magnitude of tidal heating to geoloigcally significant levels, the gravity of the moons is too weak to prevent the substantial atmospheric escape which powers the magnetosphere and profoundly alters the Jovian system. While it is conceivable that Jupiter might form with a satellite system containing fewer or smaller moons that might evade resonances and tidal heating, such a system might not have been noticed by Galileo. We are led to conclude that a satellite system within the capabilities of Galileo's telescope in 1610 was destined to undergo the transformative processes of orbital resonances, tidal heating, rapid atmospheric escape and magnetospheric interactions. Surprisingly, the first hints of these effects weren't known for another 350 years, and our understanding of these worlds in many ways remains rudimentary four centuries after their discovery.

This 400th anniversary celebrates the evolution of Galileo's "points of light" into worlds in their own right. By coincidence, recent years have also presented us with new "points of light" to consider: extrasolar planets (Figure 5). These points too will become worlds in their own right in coming years, and some may be controlled by the same processes of orbital resonances, tidal heating, rapid atmospheric escape and magnetospheric interaction. These worlds are sure to continue Galileo's legacy of surprising discoveries.

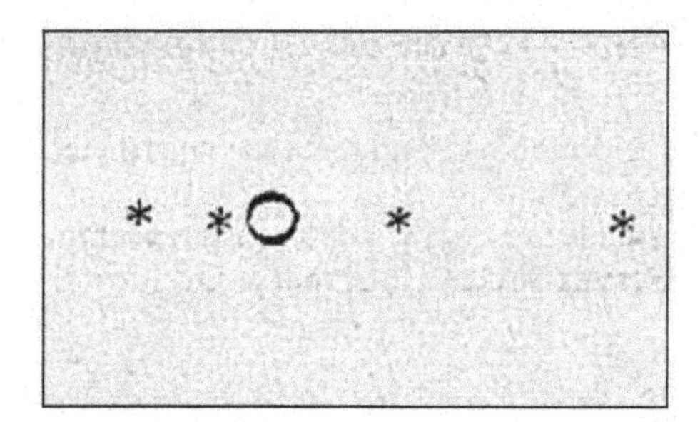

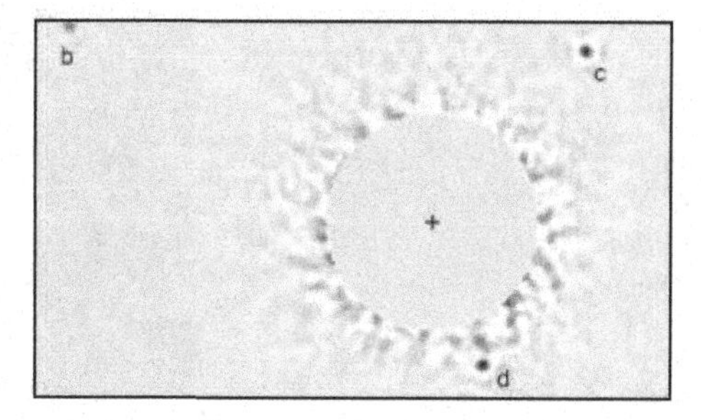

Figure 5. "Points of Light", then and now. The Medicean Stars (left) began as points of light, as shown in a figure from Siderius Nuncius. At right, an infrared image of the region surrounding the star HD8799 reveals three planets, now no more than points of light. Adapted from Marois *et al.* (2008).

References

Bergstralh, J. T., Matson, D. L., & Johnson, T. V. 1975, *Astrophys. J. Lett.*, 196, L131
Bigg, E. K. 1964, *Nature*, 203, 1008
Binder, A. B. & Cruikshank, D. P. 1966, *Astrophys. J.*, 144, 1240
Brown, R. A. 1974, *Exploration of the Planetary System*, 65, 527
Burger, M. H., Schneider, N. M., & Wilson, J. K. 2000, *Geophys. Res. Lett.*, 27, 1081
Goldberg, B. A., Mekler, Y., Carlson, R. W., Johnson, T. V., & Matson, D. L. 1980, *Icarus*, 44, 305
Goldberg, B. A., Garneau, G. W., & Lavoie, S. K. 1984, *Science*, 226, 512
Lellouch, E., Paubert, G., Moses, J. I., Schneider, N. M., & Strobel, D. F. 2003, *Nature*, 421, 45
Macy, W. W., Jr. & Trafton, L. M. 1975, *Icarus*, 25, 432
Marois, C., Macintosh, B., Barman, T., Zuckerman, B., Song, I., Patience, J., Lafrenière, D., & Doyon, R. 2008, *Science*, 322, 1348
Matson, D. L., Goldberg, B. A., Johnson, T. V., & Carlson, R. W. 1978, *Science*, 199, 531
Mendillo, M., Baumgardner, J., Flynn, B., & Hughes, W. J. 1990, *Nature*, 348, 312
Mendillo, M., Wilson, J., Spencer, J., & Stansberry, J. 2004, *Icarus*, 170, 430
Pilcher, C. B., Fertel, J. H., Smyth, W. H., & Combi, M. R. 1984, *Astrophys. J.*, 287, 427
Schneider, N. M. & Bagenal, F. 2007, *Io After Galileo: A New View of Jupiter's Volcanic Moon*, 265
Schneider, N. M., Wilson, J. K., Trauger, J. T., Brown, D. I., Evans, R. W., & Shemansky, D. E. 1991, *Science*, 253, 1394
Smyth, W. H. & McElroy, M. B. 1978, *Astrophys. J.*, 226, 336
Thomas, N., Bagenal, F., Hill, T. W., & Wilson, J. K. 2004, *Jupiter. The Planet, Satellites and Magnetosphere*, 561
Trafton, L., Parkinson, T., Macy, W., Jr 1995, *Astrophys. J. Lett.*, 190, L85
Wilson, J. K. & Schneider, N. M. 1994, *Icarus*, 111, 31
Wilson, J. K. & Schneider, N. M. 1999, *J. Geophys. Res.*, 104, 16567

Galileo's Medicean Moons: their impact on 400 years of discovery
Proceedings IAU Symposium No. 269, 2010
C. Barbieri, S. Chakrabarti, M. Coradini & M. Lazzarin, eds.
doi:10.1017/S1743921310007301

The Jovian Rings

Wing-Huen Ip
Institute of Astronomy
National Central University, Taiwan
email: wingip@astro.ncu.edu.tw

Abstract. A comparison of the Jovian and Saturnian rings is made by reviewing the recent advances in planetary spacecraft exploration and theoretical study. Two main issues are addressed, namely, the different structures of these two planetary ring systems and the water ice composition of the Saturnian rings. It is suggested that answers might be found by invoking tidal capture of Trans-Neptunian Objects with highly differentiated structures even though catastrophic breakup of pre-existing satellites in the ring regions remains a real possibility. Erosion mechanisms such as meteoroid impact, photo-sputtering, orbital instability of charged dust particles and thermal evaporation acting at different time scales could lead to the preservation of the Saturnian ring system but not the Jovian ring system of large mass originally.

Keywords. Jupiter, the Jovian rings, the Saturnian rings, erosion, gardening.

---

1. Introduction

At this conference, we celebrated the 400th anniversary of Galileo's discovery of the four Medicean moons of Jupiter in early January 1610. Because of the important philosophical and intellectual impacts on the world view by this unique observational result, it has been generally regarded as the apex of Galileo's fundamental contribution to astronomy and science. His other exciting discovery, namely, that of the Saturnian rings in the middle of July of the same year was generally not well publicized. In fact, that Saturn was observed to be not round and single but rather triple must have come as a real surprise to Galileo (van Helden, 1984a, 1984b). This puzzle was finally solved by Christian Huygens (1629-1695) who showed that Saturn was surrounded by a flat disk. Another breakthrough came when Giovanni Domenico Cassini (1625-1712) detected the existence of a gap dividing the disk into two parts which was later named after him. Ground-based study of the Saturnian rings was pursued until the very eve of the first close-up imaging observations by the Voyager 1 spacecraft (Dollfus, 1979a, b). Before the flyby of the Saturnian rings of Voyager 1 the monopoly of Saturn in owning a ring system was already broken by the serendipitous discovery of a system of narrow rings around Uranus by Elliot *et al.* (1977). The same technique of ground-based occultation measurements was used in the early 80's of the 20th Century to detect Neptune's narrow rings and a series of ring arcs (Manfroid *et al.*, 1986; Hubbard, 1986). For Jupiter, the crowning moment came when Voyager 1 obtained a most exquisite image of the Jovian rings (see Figure 1). Since then a lot of important observations and theoretical studies have been dedicated to understand the structure and dynamics of the Jovian rings which are mainly composed of small dust particles (see review by Burns *et al.* (2004) and references therein). The Jovian ring detected by Voyager 1 has an outer edge at 1.81 R_J (Jovian radii) which is also the orbital distances of two tiny satellites, J15 Adrastea and J16 Metis. Recent observations have shown clearly that the main ring detected by Voyager 1 is at the inner edge of yet another broad, tenuous disk of dust extending all the way to 3.1 R_J. The material can be traced to two small satellites, Amalthea and Thebe (Showalter *et al.*, 2008). The formation of the tenuous Jovian rings as a result of ejection of surface material from some parent bodies has interesting counter part in the Saturnian rings. That is, the G ring, the Janus/Epimetheus ring and thePallene ring as shown in Figure 2, are in principle very similar to the Jovian rings if the main Saturn rings inside the orbit of Janus could be taken away.

If Galileo had been able to see these different ring systems, the first question he would ask would probably be why only Saturn has a sizable ring system and not others? To answer this fundamental question, we will argue in the following that a comparison of the similarities and dissimilarities of the Jovian and Saturnian rings might give us some hints.

2. Thick Disk vs. Narrow Rings

In comparison with the Jovian rings with an average optical depth (tau) of ∼a few 10^{-6} the rings of Saturn is much thicker and more massive. The total mass can be estimated to be on the order of 7.0×10^{22} g assuming a water ice composition. If all the mass is collected into a single body, it would be of the size (diameter) of 225 km. It is very unlikely that this ring system is of primordial origin. This is because the inner region of the satellitary nebula surrounding Saturn should have a temperature on the order of 1000 K during formation of the planetary satellites (Ayliffe and Bate, 2009), and such a high temperature would not allow the co-accretion of Saturn and the icy ring system. The breakup of an inner satellite due to catastrophic impact with a large projectile has been postulated to be a possible origin of the ring system (Pollack, 1975; Ip, 1988). An alternative possibility is to invoke the tidal disruption of one or more large comets during their close passages inside Saturn's Roche limit (Dones, 1991). Charnoz *et al.* (2009) revisited this issue and reached the conclusion that either mechanism is a possible origin of the Saturnian rings in spite of some intrinsic difficulties of each process. According to their dynamical study, the ring formation event might have taken place during the so-called Late Heavy Bombardment Event (LHB) at which time (∼3.8 billion years ago) an intense flux of TNOs should have been injected into Saturn-crossing orbits from outside. There is, however, an important conundrum concerning the lifetime of the Saturn rings. That is, a number of studies have shown that the Saturn rings are subject to

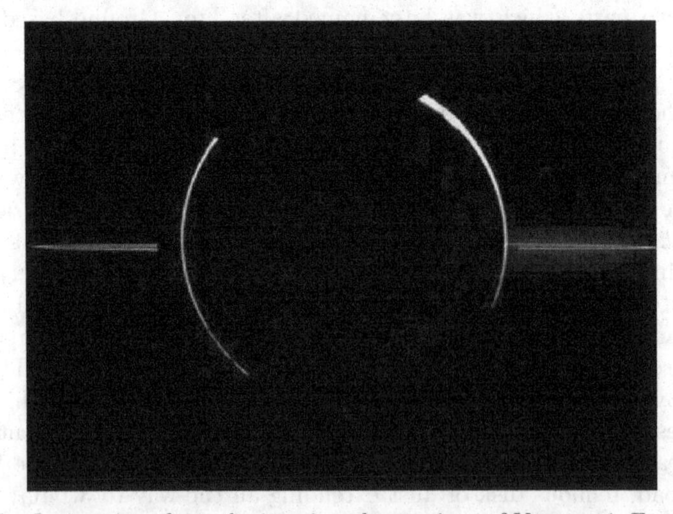

Figure 1. The Jovian rings from the imaging observations of Voyager 1. From NASA JPL.

rapid erosion effects as a consequence of orbital decay, interplanetary meteoroid impact, plasma siphon flow and photosputtering (Morfill *et al.*, 1980; Ip, 1983, 2005; Johnson and Quickenden, 1997; Johnson *et al.*, 2006). The estimates of the ring lifetime have been given to be between 200 million years and one billion years (Morfill *et al.*, 1983; Ip, 1989; Farmer and Goldreich, 2007). Barring the event that these parameter studies are wrong by orders of magnitude, it is reasonable to cast in doubt the primordial nature of the Saturnian rings. In this event, the existing ring material must have been replenished since LHB. This could be achieved by continuous impact capture of incoming TNOs (Trans-Neptunian Objects) over the eons.

In the model calculations of Charnoz *et al.* (2009), the difficulties of both tidal disruption of TNOs and satellite breakup mechanism were discussed. The problem posed by the exogenous origin has to do with the fact that among the outer planets, Jupiter had the biggest chance to acquire a ring system via tidal disruption/capture of TNOs while Saturn, Uranus and Neptune had a much smaller probability of doing so. For the satellite origin, several inner icy satellites (like Janus and Mimas) of Saturn should have been destroyed during the episode of intense bombardment according to this mechanism. However, these small satellites could have reformed via re-accretion since they are outside the Roche limit.

Assuming that Jupiter had a thick ring system like that of Saturn immediately after LHB - whatever the production mechanism, its disappearance could be understood in terms of the much faster erosive effects by a factor of 4-5 (i.e., thermal sublimation, photosputtering, meteoroid impact and orbital instability of the charged small grains) mentioned earlier because of its closer distance to the Sun. Amalthea which has a density as small as 0.86 g cm^{-3} (Anderson *et al.*, 2005) could be the remnant of such a process. It could have been pushed outward because of the tidal effect. The same can probably be said of Janus in the case of the Saturnian ring system.

Figure 2. An image from the Cassini mission showing the Saturnian rings with the faint rings associated with small satellites indicated. From NASA JPL.

3. Gardening on the Rings

The remaining issue has to do with the composition of the Saturn rings which are mainly water ice. The average densities of large TNOs and Enceladus are comparable and of the order of 1.5- 2 g cm^{-3} basically suggesting that they must contain a significant amount of rocky materials. We must then accept the scenario that the original ring composition should be a mixture of ice and silicate material at the very beginning. The question is how to hide the rocky part from view. Where has it gone? Or rather where has all the water ice come from?

Recent Cassini imaging observations of the small Saturnian moon, S12 Helene, which is orbiting at 6.25 R_S (Saturnian radii) showed that its very smooth surface might have been shaped by the deposit of icy grains from the E ring. Can similar mechanism be operational at the ring system? Noting that the present-day production rate of H_2O molecules from Enceladus is of the order of 10^{28} molecules s^{-1}. (Waite *et al.*, 2006; Richardson and Jurac, 2004), the total mass emitted over a time interval of 10^7 years would be 10^{20} g assuming that the outgassing rate could be kept constant. With about 10% of the E ring material being intercepted by the outer edge of the ring system (Jurac and Richardson, 2007), the accreted mass would be on the order of 10^{19} g which is comparable to the mass of a small satellite with a radius of 13 km. Even though this mass is much smaller than that of the ring mass, it could change the outlook of the ring system completely if the volatile ice can be added to the top layer of the ring particles thus disguising their true nature.

Another mechanism is to invoke recycling of the icy material of the original population of the broken pieces produced by tidal disruption of a TNO or catastrophic fragmentation of a satellite. The surfaces of planetary bodies like asteroids and planetary satellites are subject to collisional bombardment by interplanetary meteoroids and dust particles. Such hypervelocity impact would produce craters and eject surface materials to different regions depending on the emission speed. By the same token, the surface layer up to a certain depth (i.e., the regolith layer) is made up of fragments coming from various parts of the object. This process is called surface "gardening". It is expected that similar mechanism should take place in the Saturnian rings and that the ring particles would be covered by a layer of broken pieces. But in the context of the ring system, some new effect has to be taken into consideration. That is, as described before, the volatile water ice on the surface is constantly subject to erosion and redistribution effect as a result of solar ultraviolet radiation and meteoroid impact vaporization. As a result, most if not all of the ring particles would be covered by a layer of water ice even if they are originally of rocky composition. In other words, we expect that the ring particles - large or small - should have a two-layer structure even though they were not subject to any thermal differentiation effect as in the case of large icy satellites. In this manner, the rings can be painted white.

4. Summary

In the above we have discussed how impact gardening and ballistic transport of the volatile water across the ring plane could lead to the appearance of water ice composition of the Saturnian ring particles. In addition, we have also suggested that the absence of a thick ring around Jupiter could be the result of faster erosion than that of the Saturnian rings. This argument depends a lot on the volatility of the icy material in the rings. If a significant part of the ring material (embedded inside the ice matrices) is of rocky composition, the existence of a disk of rocky particles should remain since thermal evaporation and photosputtering would be very ineffective in eroding them. We are therefore returning to the same question asked by Charnoz *et al.* (2009) on the original composition of the Saturnian rings and that of the putative Jovian rings of much larger mass than found in the present system. One way out - as suggested by these authors - is perhaps to invoke tidal capture of the outer shells of thermally differentiated TNOs while leaving the inner core of rocky and metallic cores continued on their heliocentric orbits. A corollary of this scenario is that J15 Adrastea and J16 Metis, which define the outer edge of the main Jovian ring, would have relatively large density ($\rho \sim$ 3-4 g cm^{-3}) since they might represent the remnants of the inner core material of the captured TNOs.

Acknowledgment I thank the organizers of this IAU symposium, especially Professor Cesare Barbieri, for invitation and financial support. This work was partly supported by NSC Grant: NSC 96-2752-M-008-011-PAE and Ministry of Education under the Aim for Top University Program NCU.

References

Anderson, J. D., Johnson, T. V., Schubert, G., Asmar, S., Jacobson, R. A., Johnston, D., Lau, E. L., Lewis, G., Moore, W. B., Taylor, A., Thomas, P. C., & Weinwurn, G. 2005, *Science*, 308, 1291

Ayliffe, B. A. & Bate, M. R. 2009, *Mon. Not. R. Astron. Soc*, 397, 657

Burns, J. A., Simonelli, D. P., Showalter, M. R., Hamilton, D. P., Proco, C. C., Throop, H., & Esposito, L. W. 2004, *in* Jupiter, F. Bagenal, T. Dowling, & W. McKinnon (eds.), (Cambridge University Press), p.241

Dollfus, A. 1979a, *Icarus*, 37, 404

Dollfus, A. 1979b, *Icarus*, 40, 171

Elliot, J. L., Dunham, E. W., & Mink, D. J. 1977, *Nature*, 267, 328

Farmer, A. J. & Goldreich, P. 2007, *Icarus*, 188, 108

Hubbard, W. B., Brahic, A., Sicardy, B., Elicer, L. R., Roques, F., & Vilas, F. 1986, *Nature*, 319, 636

Ip, W.-H. 1983, *J. Geophys. Res.*, 88, 819

Ip, W.-H. 1984, *Icarus*, 60, 547

Ip, W.-H. 1988, *A&A*, 199, 340

Ip, W.-H. 2005, *Geophys. Res. Lett.*, 32, Issue 13, CiteID L 13204

Johnson, R. E. & Quickenden, T. I. 1997, *J. Geophys. Res.*, 102, E5, 10985

Johnson, R. E., Luhmann, J. G., Tokar, R. C., Bouhram, M., Berthelier, J., Sittler, E. C., Cooper, J. F., Hill, T. W., Michael, M., Liu, J., Crary, F. J., & Young, D. T. 2006, *Icarus*, 180, 393

Jurac, S. & Richardson, J. D. 2007, *Geophys. Res. Lett.*, 34, Issue 8, CitreID L08102

Manfroid, M. S., Haefner, R., & Bouchet, P. 1986, *A&A*, 157, L3

Morfill, G. E., Fechtig, H., Gruen, E., & Goertz, C. K. 1983, *Icarus*, 55, 439

Richardson, J. E. & Jurac, S. 2004, *Geophys. Res. Lett.*, 31, Issue 24, CiteIDL24803

Showalter, W. R., de Pater, I., Verbanac, G., Hamilton, D. P., & Burns, J. A 2008, *Icarus*, 195, 361

Van Helden, A. 1984a, *in* R. Greenberg & A. Brahic (Eds.), Plantary Rings, (University of Arizona Press), p. 12

Van Helden, A. 1984b, *in* T. Gehrels & M.S. Matthews (Eds.), Saturn, (University of Arizona Press), p. 23

Waite, J. H., Combi, M. R., Ip, W.-H., Cravens, T. E., McNutt, R. L., Kasprzak, W., Yelle, R., Luhmann, J., Niemann, H., Gell, D., Magee, B., Fletcher, G., Lunine, J., & Tseng, W.-L. 2006, *Science*, 311, 1419

Galileo's Medicean Moons: their impact on 400 years of discovery
Proceedings IAU Symposium No. 269, 2010
C. Barbieri, S. Chakrabarti, M. Coradini & M. Lazzarin, eds.
doi:10.1017/S1743921310007313

The Juno Mission

S. J. Bolton[1] and the Juno Science Team
[1]Southwest Research Institute, P.O. Drawer 28510
San Antonio, Texas 78228, United States
email: sbolton@swri.edu

Abstract. Juno is the next NASA New Frontiers mission which will launch in August 2011. The mission is a solar powered spacecraft scheduled to arrive at Jupiter in 2016 and be placed into polar orbit around Jupiter. The goal of the Juno mission is to explore the origin and evolution of the planet Jupiter. Juno's science themes include (1) origin, (2) interior structure, (3) atmospheric composition and dynamics, and (4) polar magnetosphere and aurora. A total of nine instruments on-board provide specific measurements designed to investigate Juno's science themes. The primary objective of investigating the origin of Jupiter includes 1) determine Jupiter's internal mass distribution by measuring gravity with Doppler tracking, 2) determine the nature of its internal dynamo by measuring its magnetic fields with a magnetometer, and 3) determine the deep composition (in particular the global water abundance) and dynamics of the sub-cloud atmosphere around Jupiter, by measuring its thermal microwave emission.

---

1. Juno Investigation of Jupiter: Goals and Objectives

Solar system formation models all begin with the collapse of a proto-solar nebula. Because Jupiter is mostly hydrogen and helium, it must have formed early, while the proto-solar nebula was still present. How this happened, however, is unclear. Models range from a proto-planetary core forming first all the way to a gravitational instability in the nebula triggering its collapse. Differences between these scenarios are profound. Even more importantly, the composition and role of icy planetesimals inplanetary formation hangs in the balance and with them, the origin of Earth and other terrestrial planets. The role of icy planetesimals, likely carriers of volatiles including the water and organics that are the fundamental building blocks of life and produced bio-molecules on early Earth, remains particularly crucial.

Juno measures water abundance and determines if Jupiter has a core, a crucial step in discovering the origin of this giant planet and our solar system. Juno will uncover vital chemical and physical clues to the nature of the nebula out of which the solar system formed and the manner and timing of giant planet formation. Next to the Sun, Jupiter is the largest object in the solar system. As such, it is both a record and a driver of the formation of the planets. By mapping the gravitation and magnetic fields, Juno investigates Jupiter's interior structure and measures the mass of its core. How deep Jupiter's zones, belts, and other features penetrate is one of the most outstanding fundamental questions in Jovian atmospheric dynamics. The mapping of variations in atmospheric composition, temperature, cloud opacity and dynamics at depth can help determine the global structure and dynamics of Jupiter's atmosphere below the cloud tops. Juno also investigates Jupiter's powerful magnetospheric dynamics create the brightest aurora in our solar system. How are the electrons and ions precipitated down into Jupiter's atmosphere to create the aurora? Juno directly measures the distributions of these charged particles, their associated fields, and the concurrent UV and IR emissions of Jupiter's polar magnetosphere. Jupiter's massive gravitational field shaped the dynamical environment in the terrestrial planet region, affecting the timing of the growth of Earth and its rocky neighbors as well as the delivery of water and organics to the surface of our planet. Jupiter is the archetype for extrasolar giant planets, now known to exist around a few percent or more of sunlike stars. An understanding of the formation of Jupiter from gas and icy planetesimals in the solar nebula therefore illuminates processes of planet formation throughout the universe.

Jupiter's core. The mass of Jupiter's core helps distinguish among competing scenarios for the planet's origin. We know that young stars lose their gaseous accretion disks rapidly, in 1-10 Myr (Strom *et al.* 1993). Because it is made mostly of hydrogen and helium, Jupiter had to form early, and hence prior to terrestrial planet formation. Being much more massive than the other planets, it probably grew more rapidly than any other planet. One set of models propose that a protoplanetary core (~10 or more Earth masses) was formed first by accretion in the cold outer part of the protosolar nebula in ~2 to 5 million years (Mizuno 1980, Lissauer 1993, Pollack *et al.* 1996, Wuchterl *et al.* 2000, Hersant *et al.* 2003). The collapse of the surrounding hydrogen and helium followed (Lissauer 1993; Pollack *et al.* 1996; Wuchterl *et al.* 2000), yielding a planet with a central dense core of at least 10 Earth masses (ME) and a hydrogen-helium envelope. An alternate model proposes an even faster process: a gravitational instability in the nebula triggers a collapse that forms the giant planets in about 0.1 Myr (Boss 1997, 2000). The simplest version of this model suggests a core mass of only zero to six ME. The early evolution of the solar system would have been very different, as would the number and nature of planetesimals captured by the giant planets, depending upon which formation mechanism is correct (Guillot & Gladman 2000). Analysis of the Juno gravity and microwave investigations constrains the core mass and the total mass of heavy elements sufficiently to resolve this planetary formation question.

Jupiter's water abundance. Given that oxygen is the third most abundant element in the universe and recognizing that icy planetesimals were the dominant carriers of heavy elements in the solar nebula, a measurement of Jupiter's global water abundance is pivotal in understanding giant planet formation and the delivery of volatiles throughout the solar system. Jupiter contains key evidence about the nature of the protoplanetary disk, or solar nebula, out of which the solar system formed. For example, if water is found to be enriched in similar proportion to nitrogen, carbon, sulfur, and the noble gases (~3 times solar)(Figure 1), a model producingplanetesimals from ice that condensed at less than

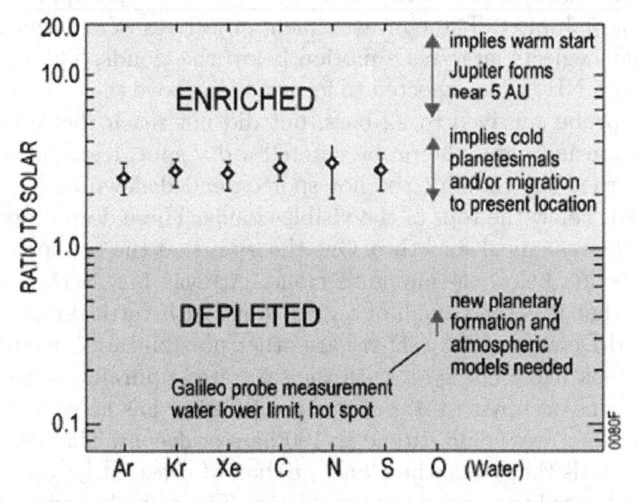

Figure 1. Juno's measurement of O discriminates among Jupiter's formation scenarios as shown in this figure. Abundances of Ar, Kr, Xe, C, and S are well determined on Jupiter at 3× Solar. O is not yet determined. Juno determines both the N and O abundances.

~30 K (cold) is favored (Owen *et al.* 1999, Mahaffy *et al.* 2000, Owen 2004). This could require inward migration of core-forming planetesimals (or Jupiter itself) from much larger distances or a model of the formation of the solar nebula, that includes the direct transfer of interstellar material in large agglomerations. If water is much more enriched than the noble gases, i.e., by 9 or more relative to solar, then the trapping of noble gases would be more characteristic of that expected through formation of planetesimals from ice grains that condensed at ~150 K near Jupiter's present location with subsequent cooling to 35 K (Gautier *et al.* 2001, Hersant *et al.* 2003). Determination of the mixing ratio of water to hydrogen and hence (through Galileo measurements) to the noble gases provides a direct test for the mode of origin of the icy planetesimals that enriched Jupiter. These planetesimals may have been the most abundant solid material in the early solar nebula and therefore may also be important to the delivery of volatiles to the inner planets (Owen and Encrenaz 2003). The Juno microwave radiometry investigation determines the water abundance in Jupiter with sufficient precision to resolve this question of planetary formation, and because it maps the water over all latitudes, it is not prone to the sampling bias by measurements at one or a few probe locations as Galileo was.

Juno produces five pole-to-pole latitudinal maps of microwave opacity as a function of altitude to depths greater than 100 bars (Figure 2). The independent swaths at different longitudes provide the ability to understand large scale features such as the Great Red Spot. The 0.1% precision of the radiometer allows us to measure small variations in radiance with respect to horizontal position and emission angle. With these measurements, we determine the global O/H and N/H ratios; we correlate the patterns of ammonia and water abundance below the clouds with the principal dynamical features at cloud-top level; we examine the deep roots of features like the Great Red Spot, the belts and zones, and potentially the 5- ?m hot spots. Context for these features is provided through coordinated Earth-based images and comparison with data from the E/PO imager, JunoCam.

Atmospheric Dynamics and the Galileo Probe. The depth of the major flow features is the most basic question of Jovian meteorology. The objectives of the Galileo probe mission were to measure composition, temperatures, winds, clouds, lightning, and radiative heating from the top of the ammonia cloud at 0.5 bars to below the base of the water cloud which was expected to lie at 5 bars. There it was supposed to sample the well-mixed interior of Jupiter. These measurement objectives were derived from an atmospheric model that neglects large-scale motion below the clouds. The three condensable gases H_2O, H_2S, and NH_3 were expected to form clouds above the 6, 2, and 0.8 bar levels, respectively. The probe survived to 22 bars, but did not reach the well-mixed interior. Based on remote sensing data, the probe entered a dry spot, a so-called 5μm hot spot. The surprise was that the roots of the hot spot extended down at least to the 22-bar level, about 150 km below the tops of the visible clouds. These deep roots are apparently part of a large-scale dynamical structure. One theory is that the hot spot is a giant downdraft extending 150 km below the ammonia cloud (Atreya *et al.* 1997, Owen *et al.* 1997); another theory is that it is the trough of a giant wave with vertical displacements of 150 km (Showman and Dowling, 2000). There are other possibilities. For instance, 99.9% of the planet might look like a hot spot, with the saturated updrafts concentrated in a few violent thunderstorms occupying 0.1% of the area. Jupiter has no solid or liquid surface, so the dynamical structures could extend to 100 bars or deeper. The dozen or more pairs of dark and light bands that circle the planet on lines of constant latitude are called belts and zones. The highspeed jets are on the boundaries. The prevailing view, based on clouds and chemical tracers, is that the belts are sites of downwelling, but the concentration of lightning in the belts (Gierasch *et al.* 2000) seems to contradict this view. Individual belts, zones, and ovals have persisted for over 100 years. This longevity is remarkable given that Jupiter is a fluid planet with no solid surface to provide stability. Deep roots and their large inertia may be the key to this longevity (Busse 1976, Ingersoll and Pollard, 1982).

The Polar Magnetosphere. A set of instruments measures the polar magnetosphere by determining:

- the electric currents along magnetic field lines;
- the electromagnetic emissions associated with aurora and electrostatic waves;
- the distribution of energetic particles and distribution of auroral and magnetospheric plasma;
- the ultraviolet auroral emissions.

Juno's polar orbit is ideal to answer the fundamental questions of how auroras are generated. Juno's instruments are designed to determine the physical processes occurring in the high latitude magnetosphere and allow us to relate them to auroral activity and to our understanding of the equatorial magnetosphere. At Earth, auroras are primarily driven by the energy of the solar wind. At Jupiter, the primary energy source is the rotation of the planet, but a second source is the motion of the galilean satellites across rotating jovian magnetic field lines. The solar wind also plays a role. These three sources are apparently reflected in the three types of auroras observed at Jupiter (Figure 3): the main ovals of emissions encircling the north and south magnetic poles; emissions emanating from the base of magnetic flux tubes connected to the Galilean satellites; erratic emissions poleward of the main ovals which will be observed by Juno on the two polar caps, also revealing interhemispheric symmetries and asymmetries. Despite their differences, the Jovian and terrestrial auroral displays are thought to be caused by similar processes: strong electric currents flowing along the magnetic field and electromagnetic fields accelerating the charged particles that bombard the upper atmosphere. The closure of these currents across the ionosphere and in the magnetosphere or solar wind transfers

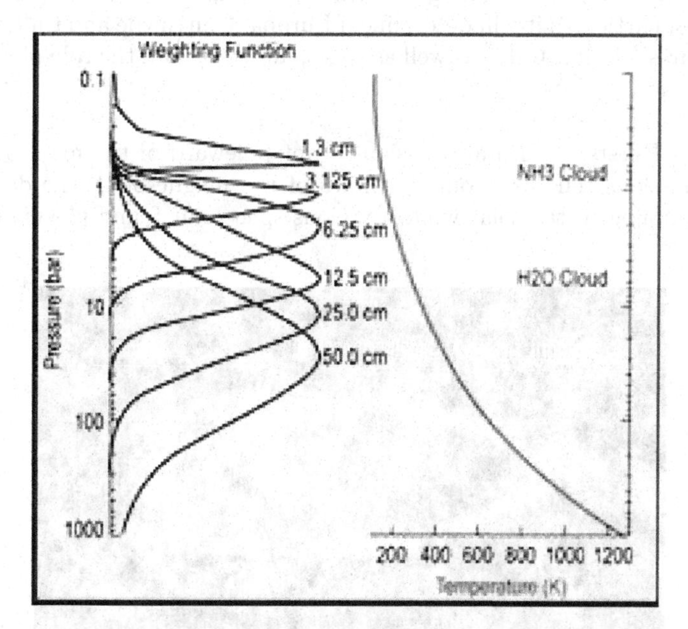

Figure 2. Contribution functions for the emission from Jupiter's atmosphere at nominal MWR frequencies. The ammonia cloud tops lie above the 1-bar pressure altitude. H_2O clouds are expected at higher pressure. The longest wavelength will penetrate Jupiter's atmosphere to depths below 100 bars.

momentum between the two electrically connected regions (Figure 4). Exploring how the auroral circuit provides magnetic coupling of Jupiter to its surrounding nebula, its satellites and the solar wind paves the way for a better understanding of astrophysical systems similarly dominated by rotation of a central magnetic object coupled to a surrounding plasma-such as a young star in a nascent planetary system.

Main Aurora. Jupiter is a rapid rotator with its volcanic moon Io as the main plasma source. Under the effect of centrifugal forces this plasma produces an equatorial disk that rotates with the giant planet, but the "rigidly" corotating plasma begins to slip beyond ~20 RJ. This breakdown of corotation results in a region of radially outward current in the plasma disk, which continues at either end along magnetic field lines to complete a circuit in the polar regions of Jupiter's ionosphere (e.g., Hill *et al.* 1983; see Figure 4). The ionosphere currents dissipate much energy (Joule heating), and the westward drag of the sub-corotating plasma disk on the ionosphere results in strong (several km/s) winds (cf., Cowley *et al.* 2003) in Jupiter's upper atmosphere. To carry the outward current (away from the ionosphere) magnetospheric electrons are accelerated into Jupiter's atmosphere, where they excite the main oval auroral emissions. The main oval morphology is remarkably stable, but the brightness varies substantially over days. Juno provides an unprecedented look at the poorly understood parts of the main oval circuit -the field-aligned currents, acceleration region, and electrojet winds-along with the auroral emissions.

Satellite footprint auroras. Localized auroral emissions are observed at the feet of the magnetic field lines that connect to the Galilean satellites. In addition, the Io interaction generates a wake extending up to halfway around Jupiter. Characterization of the electrodynamic coupling requires in situ measurement of particles and fields and remote sensing of auroras. Juno observes magnetospheric structures and the ionospheric response (currents, fields, particles): it will determine if Europa, Ganymede and Callisto have wake auroral structures (similar to Io) as well as spots, and ascertain the role played by Alfven waves.

Polar aurora. Bursts of auroral emissions erupt poleward of the main oval. Studying these polar emissions tells us about dynamics of the magnetotail, the dayside magnetopause and coupling to the solar wind. Previous spacecraft found plasma to be flowing

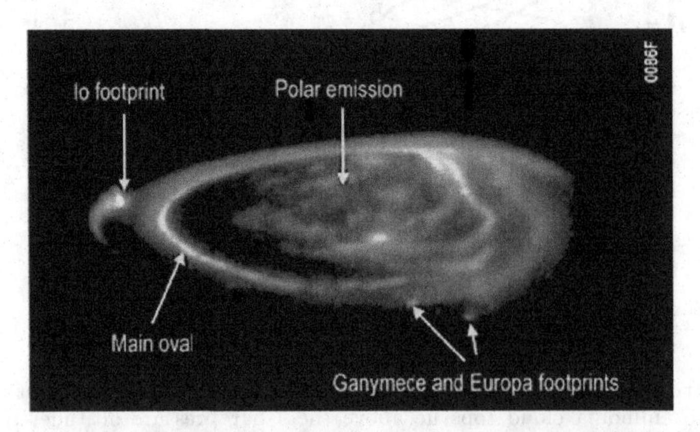

Figure 3. Three types of auroras revealed in this HST image of Jupiter's aurora. Each are signatures of momentum transfer processes.

downtail on the dawn side at distances of ~40 RJ. One model proposes that this jovian planetary wind was a signature of reconnection, with "plasmoids" being ejected down the tail. Analogous to substorms at Earth, we expect such processes to produce currents, particle acceleration and auroral emissions. Alternatively, these emissions may be related to reconnection at the dayside magnetopause. Juno distinguishes between these theories via simultaneous in situ measurements and remote sensing of the polar aurora.

Juno science objectives. The Juno science objectives are 1) to determine the O/H ratio (water abundance) and constrain core mass to decide among alternative theories of origin; 2) understand Jupiter's interior structure and dynamical properties by mapping its gravitational and magnetic fields; 3) map variations in atmospheric composition, temperature, cloud opacity and dynamics to depths greater than 100 bars; and 4) explore the three-dimensional structure of Jupiter's polar magnetosphere and auroras. Each objective will be addressed with a set of dedicated instruments on the Juno spacecraft.

2. Juno Science and Payload

Juno is a solar powered, spinning spacecraft that will be placed into an elliptical polar orbit around Jupiter. The launch is August of 2011 and, with gravitational assistance of an Earth flyby, the spacecraft will arrive at Jupiter after a 5 year journey. The payload consists of of nine instruments:

Gravity Science (JPL/ASI). The primary objective of the Gravity Science Experiment is to determine the internal structure of Jupiter by making detailed measurements of its complete gravity field from a polar orbit. The investigation is a Doppler radio science experiment that uses the telecommunications system. Mass distribution in the interior of Jupiter causes asymmetric variations in the gravity field of Jupiter, exerting a variable gravitational *pull* on the Juno spacecraft. These variations lead to tiny variations in the motion of the spacecraft around Jupiter, which are detected using the Doppler shift in the X and Ka band transponders used by the radio sub-system. A correction is usually applied for the effects resulting from the Earth's atmosphere during data processing.

Magnetometer (GSFC)-MAG. Investigations with MAG have three goals: mapping of the magnetic field, determining the dynamics of Jupiter's interior, and determination of the three-dimensional structure of the polar magnetosphere. To achieve these goals, the

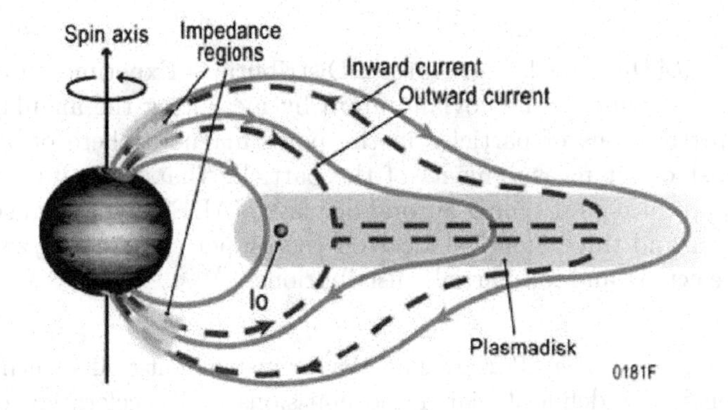

Figure 4. Juno measurements target each critical path in this closed circuit that transfers angular momentum from Jupiter to its nebula.

mission employs two Flux Gate Magnetometers and two Advanced Stellar Compasses (ASC) to provide accurate location and orientation information of the magnetometers on the Juno spacecraft for precise magnetic field mapping.

Microwave Radiometer (JPL)-MWR. The primary goal of the Juno Microwave Radiometer is to probe the deep atmosphere of Jupiter at radio wavelengths ranging from 1.3 cm to 50 cm using six octave-spaced radiometers to measure the planet's thermal emissions (Fig 2). The MWR objective is the determination of the water abundance in Jupiter's deep atmosphere. The MWR will obtain measurements of ammonia and water in the Jupiter atmosphere, which are the principle absorbers in the microwave region, by scanning Jupiter along the orbital track as the spacecraft spins. These observations will allow scientists to determine the global (well mixed) water abundance on Jupiter which represents the oxygen abundance. The Juno MWR avoids the synchrotron emission from Jupiter's magnetosphere observing from above the poles and beneath the Jovian radiation belts. MWR achieves high accuracy to measure water abundance in the deep atmosphere by using *relative limb darkening*, a parameter that depends on the emission angle of the radiation. The vertical profile of water abundance is obtained by using multiple frequencies, much like the retrieval of temperature profiles on earth with multi-spectral infrared measurements from orbiting weather satellites. The MWR uses six antennae mounted on the spacecraft body, which sweep across the planet as the spacecraft spins to measure the radiation at six different wavelengths along the orbital track. Successive orbits will map the planet longitudinally. The six different wavelengths observed by the MWR, combined with the emission angle dependence will provide a good idea of the atmospheric temperature profile (see Figure 2). The latitudinal dependence of the temperature profile and depth will enable inference of the circulation of Jupiter's deep atmosphere to a much greater depth than that obtained by the Galileo probe.

Energetic-particles (APL)-JEDI. JEDI will measure the energy and angular distribution of Hydrogen, Helium, Oxygen, Sulfur and other ions in the polar magnetosphere of Jupiter using the time of light versus energy technique. JEDI consists of three separate sensing heads using time of flight (TOF) versus energy to sort incoming ions into mass species and energy, and uses foiled and un-foiled measurements to discriminate electrons from ions.

Plasma (SwRI)-JADE. The Jovian Auroral Distributions Experiment (JADE) will resolve the plasma structure of the Jovian aurora by measuring the angular, energy and compositional distributions of particles in the polar magnetosphere of Jupiter. JADE will make the first direct measurements of the particles that precipitate into Jupiter's atmosphere and produce its stunning auroral displays. JADE comprises a single head ion mass spectrometer and three identical electron energy per charge analyzers to measure the full auroral electron and ion particle distributions.

Plasma Waves (U of Iowa)-Waves. The Waves instrument will identify the regions of auroral currents that define Jovian radio emissions and acceleration of the auroral particles by measuring the radio and plasma spectra in the auroral region. The Waves instrument uses an electric dipole antenna and a magnetic search coil to measure electromagnetic waves and to discriminate electrostatic and magnetostatic waves.

Ultraviolet (SwRI)-UVS. UVS records the wavelength, position and arrival time of detected ultraviolet photons during the time when the spectrograph slit views Jupiter during each turn of the spacecraft. Using a 1024 256 micro channel plate (MCP) detector, it will provide spectral images of the UV auroral emissions in the polar magnetosphere which allows to relate the these auroral measurements with JADE.

Visible Camera (Malin)-JunoCam. JCM is a camera which will provide the first 3-color images of Jupiter as the Juno spacecraft approaches the poles for context, public engagement and E/PO. It uses four filters mounted directly on the detector to obtain the first close up color images of the poles. The Juno mission plans to invite students to work alongside the science team to capture these images once the Juno spacecraft is in orbit around Jupiter.

Infrared Imager/Spectrometer (ASI)-JIRAM The Jupiter InfraRed Auroral Mapper JIRAM is the first Italian instrument of this kind to be sent to Jupiter. The primary goal of JIRAM is to probe the upper layers of Jupiter's atmosphere down to pressures of 5-7 bars at infrared wavelengths in the 2-5 μ interval using an imager and a spectrometer. By means of its high contrast imaging and spectroscopy, JIRAM will study the dynamics and chemistry of auroral regions and their link to Jupiter's magnetic field and magnetosphere.

3. Summary

Juno's investigation of Jupiter will provide fundamental information on the history of volatiles in early solar system. The distribution and state of these volatiles will shed light on how the planets received their share of volatiles such as water and carbon which eventually led not only to the formation of Earth, but life itself. Understanding the origin of our own solar system by investigating Jupiter's formation may provide information for finding Earth like planets accompanying the recent observed giant planets outside our solar system.

Acknowledgement

This work was funded by the National Aeronautics and Space Administration (NASA).

References

Janssen, M. A., Hofstadter, M. D., Gulkis, S., Ingersoll, A. P., Allison, M., Bolton, S. J., & Kamp, L. W. 2005, Microwave Remote Sensing of Jupiter's Atmosphere from an Orbiting Spacecraft *Icarus*, 173, 447-453

Strom, S. E., Edwards, S., & Skrutskie, M. F. 1993, *Evolutionary time scales for circumstellar disks associated with intermediate- and solar-type stars*, Protostars and Planets III, E. H. Levy and J. I. Lunine, Eds., University of Arizona Press, 837

Mizuno, H. 1980, *Formation of the giant planets. Prog. Theor. Phys.*, 64, 544-557

Lissauer, J. J. 1993, *Planet formation. Ann Rev. Astr. Astrophys.* 31, 129

Pollack, J. B., Hubickyj, O., Bodenheimer, P., Lissauer, J. J., Podolak, M., & Greenzweig, Y. 1996, Formation of the giant planets by concurrent accretion of solids and gases, *Icarus* 124, 62.

Hersant, F., Gautier, D., & Lunine, J. I. 2003, Enrichment in volatiles in the giant planets of the solar system, *Planetary and Space Sci.*, in press.

Wuchterl, G., Guillot, T., & Lissauer, J. J. 2000, *Giant planet formation. Protostars and Planets IV*, V. Mannings, A. P. Boss, and S. S. Russel, Eds., University of Arizona Press, 1081-1109

Boss, A. P. 1997, Giant planet formation by gravitational instability, *Science* 276, 1836, 2000: Possible rapid gas-giant planet formation in the solar nebula and other protoplanetary

Guillot, T. & Gladman, B. 2000, *Late planetesimal delivery and the composition of giant planets. Proc. Disks, Planetesimals, and Planets Conf.*, F. Garzon *et al.*, Eds., ASP Conf. Ser., 475-485

Owen, T. C. & Encrenaz, T. 2003, Element abundances and isotope ratios in the giant planets and Titan, *Sp. Sci. Rev.*, 106, 121-138

Owen, T. C., Atreya, S. K., Mahaffy, P., Niemann, H. B., & Wong, M. H. 1997, On the origin of Jupiter's atmosphere and the volatiles on the Medicean stars. Three Galileos: The Man, the Spacecraft, the Telescope, C. Barbieri *et al.*, Eds., Kluwer Academic Publishers, 289-297

Niemann, H. B., Atreya, S. K., Carignan, G. R., Donahue, T. M., Haberman, J. A., Harpold, D. N., Hartle, R. E., Hunten, D. M., Kasprzak, W. T., Mahaffy, P. R., Owen, T. C., & Way, S. H. 1998, The composition of the Jovian atmosphere as determined by the Galileo probe mass spectrometer. *J. Geophys. Res.*, 103, 22831-22846

Gautier, D., Hersant, F., Mousis, O., & Lunine, J. I. 2001, Enrichments in volatiles in Jupiter: A new interpretation of the Galileo measurements., *Astrophys. J. Lett.*, 550, L227-L230 (Erratum 559, L183)

Owen, T. C. & Encrenaz, T. 2003, Element abundances and isotope ratios in the giant planets and Titan, *Sp. Sci. Rev.*, 106, 121-138

Atreya, S. K., Wong, M. H., Owen, T. C., Niemann, H. B., Mahaffy, P. R. 1997, *Chemistry and clouds of the atmosphere of Jupiter: A Galileo perspective. Three Galileos: The Man, the Spacecraft, the Telescope*, C. Barbieri *et al.*, Eds., Kluwer Academic Publishers, 249 260

Showman, A. P. & Dowling, T. E. 2000, Nonlinear simulations of Jupiter's 5-micron hot spots., *Science*, 289, 737-1740

Gierasch, P. J. 2004, Stability of jets on Jupiter and Saturn, Icarus, 167(1), 212-219. Gierasch, P. J., A. P. Ingersoll, D. Banfield, S. P. Ewald, P. Helfenstein, A. Simon-Miller, A. Vasavada, H. H. Breneman, D. A. Senske, and the Galileo imaging team, 2000: Observation of moist convection in Jupiter's atmosphere, *Nature*, 403, 628-629

Busse, F. H. 1976, Simple model of convection in Jovian atmosphere, *Icarus*, 29(2), 255-260

Ingersoll, A. P. & Pollard, F. 1982, Motions in the interiors and atmospheres of Jupiter and Saturn: Scale analysis, anelastic equations, barotropic stability criterion, *Icarus*, 52, 62-80

Hill, T. W., Dessler, A. J. & Goertz, C. K. 1983, *Magnetospheric models. Physics of the Jovian Magnetosphere*, A. J. Dessler, Ed., Cambridge University Press, 353-394.

Cowley, S. W. H., Bunce, E. J., Stallard, T. S., *et al.* 2003, Jupiter's polar ionospheric flows: Theoretical interpretation, *Geophys. Res. Lett.*, 30(5), art. no. 1220

Galileo's Medicean Moons: their impact on 400 years of discovery
Proceedings IAU Symposium No. 269, 2010
C. Barbieri, S. Chakrabarti, M. Coradini & M. Lazzarin, eds.

doi:10.1017/S1743921310007325

Seeking Europa's Ocean

Robert T. Pappalardo
Jet Propulsion Laboratory
California Institute of Technology
email: robert.pappalardo@jpl.nasa.gov

Abstract. Galileo spacecraft data suggest that a global ocean exists beneath the frozen ice surface Jupiter's moon Europa. Since the early 1970s, planetary scientists have used theoretical and observational arguments to deliberate the existence of an ocean within Europa and other large icy satellites. Galileo magnetometry data indicates an induced magnetic field at Europa, implying a salt water ocean. A paucity of large craters argues for a surface on average only ~40-90 Myr old. Two multi-ring structures suggest that impacts punched through an ice shell ~20 km thick. Europa's ocean and surface are inherently linked through tidal deformation of the floating ice shell, and tidal flexing and nonsynchronous rotation generate stresses that fracture and deform the surface to create ridges and bands. Dark spots, domes, and chaos terrain are probably related to tidally driven ice convection along with partial melting within the ice shell. Europa's geological activity and probable mantle contact permit the chemical ingredients necessary for life to be present within the satellite's ocean. Astonishing geology and high astrobiological potential make Europa a top priority for future spacecraft exploration, with a primary goal of assessing its habitability.

Keywords. Europa, habitability, ocean, convection, ice

Four hundred years ago, the discovery of the four Galilean moons of Jupiter forever changed our sense of place in the Universe (Figure 1). No longer could humans claim to be at its center, and the Copernican Revolution was underway. Today one of those moons—Europa—has the potential to again revolutionize science and our sense of place in the Universe, if we were to find life there.

Beneath the icy surface of Europa may be the solar system's most promising location for extant life beyond Earth. As one of the most geophysically and astrobiologically fascinating bodies in our solar system, Europa's exploration is key in advancing our understanding of habitable zones in our solar system, and other solar systems. If Europa contains a global ocean within, its volume is expected to be 2-3 times the volume all of Earth's oceans. Here I review some of the history, evidence, and implications of a subsurface ocean within Europa.

1. Oceans Come and Go

In the 1970s, it was realized that internal heat might be sufficient to melt oceans within large icy moons such as Europa, at least early in solar system history (Consolmagno and Lewis; 1976; Fanale *et al.*, 1977). However, soon afterward, Reynolds and Cassen (1979) realized that the icy shell which caps an internal ocean would thicken as the satellite lost heat over geological time, eventually reaching a critical thickness at which it would transition from conductive to convective heat loss. When convection occurs, warm ice at the base of a floating icy shell would move buoyantly upward, and cold ice near the top of the icy shell sinks downward. They estimated that solid-state convection of ice should cause a satellite to lose heat very rapidly, freezing an internal ocean in about 100 million years. Even if Europa and other large icy satellites once had oceans, Reynolds and Cassen (1979) reasoned that they should be frozen solid today.

Later calculations indicated that tidal heating might change the story, for Europa at least. Cassen *et al.* (1982) and Squyres *et al.* (1983) argued that if Europa's icy shell is sufficiently warm and dissipative, then the icy shell might be so thin that tidal heating would not initiate. Thus, it seemed that Europa might generate enough internal heating as it is squeezed by Jupiter's gravity that it might maintain an ocean beneath its icy surface layer. The most detailed analyses were performed by Ojakangas and Stevenson (1989), who showed that Europa might possess a conductive icy shell that averages ~20-25 km thick above a liquid water ocean.

Voyager images of Europa were limited to ~2 km/pixel at best (Figure 2). They revealed tantalizing hints of an icy surface consisting of bright plains and darker mottled terrain, criss-crossed by mysterious ridges and wider bright and dark bands (Lucchitta and Soderblom, 1982). The bands of the bright plains showed evidence for lateral motions of the ice, with translations of ~20 km, with new material (presumed to be solid-state ice, but plausibly liquid water) filling the gaps (Schenk and McKinnon, 1989; Pappalardo and Sullivan, 1994). The origins of ridges and mottled terrain remained mysterious, but prescient inference by Lucchitta and Soderblom (1982) suggested that vertical tectonism and intrusion dominates Europa's geology. The age of Europa's surface remained fully unknown: only a handful of certain craters were observed, suggesting a very young surface; however, if large impact craters flattened through relaxation (Squyres *et al.*, 1983)

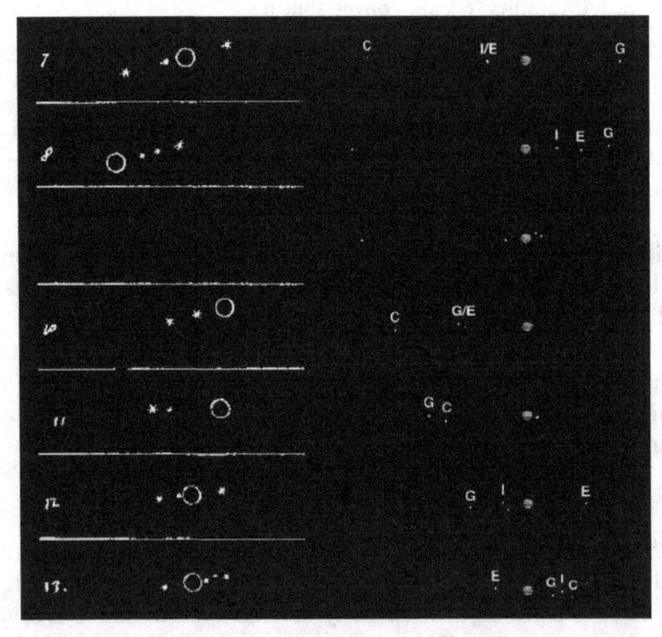

Figure 1. Galileo's sketches of the Medicean Stars, from January 7-13, 1610 (with the exception of Jan. 9, which was cloudy), compared to a simulated view of the Galilean satellites on the same nights from the Jet Propulsion Laboratory's Solar System Simulator (space.jpl.nasa.gov) viewed from Earth at about 1 hr past the time of sunset in Padua, Italy. Labels indicate the positions of the satellites that Galileo identified: Io (I), Europa (E), Ganymede (G), and Callisto (C). On some nights, two satellites apparently were too close to each other or to Jupiter for Galileo to distinguish them as separate objects. Note that Europa and Io blended together into one point in the January 7 discovery sketch.

and the small pits observed in the mottled terrain represented numerous small impact craters (Malin and Pieri, 1986) then the surface might be quite old.

2. Europa's Ocean

These uncertainties and predictions set the stage for the Galileo mission, which included 12 close fly-bys of Europa during the years 1996-2000 (Alexander *et al.*, 2009). The results opened our eyes to the intricacies of Europa and the likelihood of an ocean within.

Doppler gravity data from the Galileo flybys indicates that Europa's interior is layered (Figure 3, inset), with an iron core and rocky mantle, topped by an H_2O-rich layer about 100 km thick (Schubert *et al.*, 2009). However, gravity data alone cannot distinguish whether any of the H_2O layer consists of liquid water.

It was the Galileo magnetometer measurements which provided most direct evidence for liquid water within Europa today, through indication of an induced magnetic field (Khurana *et al.*, 2009). Jupiter's magnetic field is tilted by about 10° compared to the equatorial plane in which Europa and the other Galilean satellites orbit. As Jupiter rotates with its rapid 10 hour period, its magnetic field rotates along with it. But because of the tilt, Jupiter's moons alternately find themselves above then below the magnetic equator of Jupiter. This means that the moons feel Jupiter's magnetic field alternating in polarity, cycling at the synodic period of the satellite relative to Jupiter as it rotates, i.e. 11.2 hr for Europa.

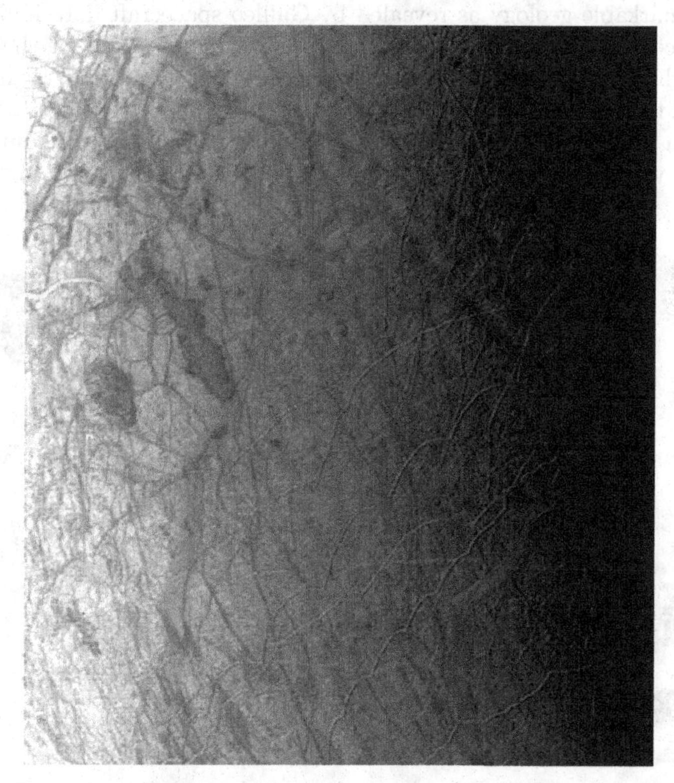

Figure 2. A portion of one of the best Voyager images of Europa at ~2 km/pixel, showing a bewildering array of spots, ridges, bands, pits, and domes. Cycloidal ridges are prominent in the lower right portion of the scene.

During Galileo's nominal mission, firm evidence arose for an induced field and thus a conducting layer at shallow depth within Callisto (Kivelson *et al.*, 1999). The only reasonable conductive material is salt water, maintained by the shallow thermal gradient of the satellite, probably in combination with salt and/or ammonia antifreeze (Spohn and Schubert, 2003). An induced field was hinted at for Europa, but confirmation awaited the last close fly-by of Europa in 2000, when Galileo for the first time encountered Europa at close range while the moon was beneath Jupiter's magnetic equator (Kievelson *et al.*, 2000). The data confirmed an induced magnetic field, indicating a subsurface ocean within the outer 30 km or so of the satellite (Zimmer *et al.*, 2000). Ultimately an ocean was inferred within Ganymede as well, though that moon's intrinsic magnetic field makes it difficult to confidently tease out this result (Kivelson *et al.*, 2002).

Unlike its larger sibling satellites Ganymede and Callisto, the relatively thin H_2O layer at Europa means that pressures never get high enough for high-density ices to form beneath its ocean. This means that Europa's salty ocean is not sandwiched between ice layers as are the probable oceans of Ganymede and Callisto; instead Europa's ocean is expected to be in direct contact with a rocky mantle below (Schubert *et al.*, 2009). A rocky seafloor means that the chemical nutrients that might support life can be supplied directly from the rocky mantle, making Europa astrobiologically attractive, as discussed below.

3. Europa's Youthful Surface

Europa's remarkable geology as revealed by Galileo spacecraft data indirectly tells of a subsurface ocean beneath (Figure 3). Most remarkable of all about the surface is its confirmed youth. There are only 23 known impact features larger than 10 km in diameter on the ~75% of the surface sufficiently imaged to identify them (Schenk and Turtle, 2004). Combined with models of the flux of recent impactors at Europa, this implies that the surface is very young, with an average age of about 40-90 Myr, with an uncertainty of

Figure 3. Schematic view of Europa's icy shell, illustrating selected geological features and their interpretation. Insets illustrate the likely deep interior structure of Europa, and the context of its icy shell and ocean floor.

about a factor of 2 (Zahnle *et al.*, 2003; Bierhaus *et al.*, 2009). Much of Europa's surface activity has transpired in the 1% of Solar System history since dinosaurs populated the Earth. Such a geologically recent surface argues that if a subsurface ocean was involved in the formation of Europa's surface features, that ocean should still be present today.

Europa's impact structures hint of a warm icy lithosphere and an ocean below (Figure 4). Europa's largest impact craters, ~26 km diameter Pwyll and 23 km diameter Manannán, are more shallow than similarly sized craters on the terrestrial planets or on most other icy satellites, suggesting that they impacted a warm icy lithosphere (Schenk and Turtle, 2009). Two features termed "multi-ring structures" are not morphologically craters but are certainly the sites of large impacts: they show numerous concentric rings, display disaggregated centers, and have radial chains of secondary craters (Moore *et al.*, 1998, 2001). These multi-ring structures are believed to have penetrated through an ice shell about 20 km in thickness into liquid water below, resulting in their unusual morphologies (Schenk and Turtle, 2004).

Europa's troughs and ridges trace patterns that indirectly betray the existence of an ocean below. Most satellites are in synchronous rotation, having despun early in Solar System history such that they always show the same face to their parent planets as they orbit. However, a global ocean can permit the floating icy shell to slip relative to the satellite interior, undergoing "non-synchronous rotation," taking ~10^7-10^9 yr for the icy shell to slip once around over the interior (Greenberg and Weidenshilling, 1984; Ojakangas and Stevenson, 1989). Whether significant stresses should result depends on the rotation rate relative to the relaxation rate (thus viscosity) of the icy shell (Wahr *et al.*, 2009). Several researchers have compared the predicted pattern and distribution of non-synchronous rotation stresses to Europa's global-scale lineaments, concluding that this process has likely occurred through at least some portion of Europa's geologically recent history (Kattenhorn and Hurford, 2009; Selvans *et al.*, 2010). There have also been

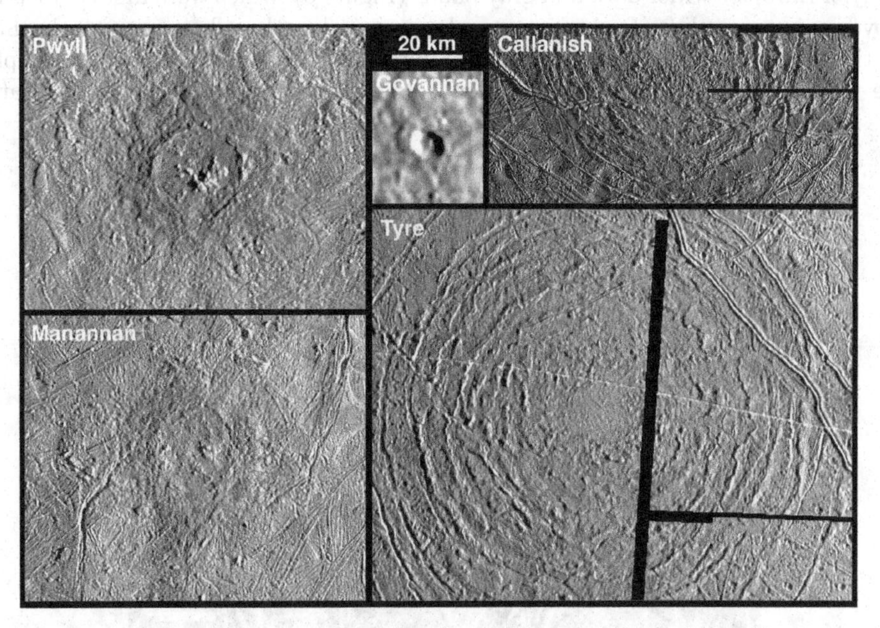

Figure 4. Europa's largest impact craters Pwyll and Manannán, and the multi-ring structures Callanish and Tyre, which probably formed by impacts that penetrated to Europa's ocean. The 11 km diameter crater Govannan has a more conventional morphology.

suggestions that true polar wander of Europa's icy shell has occurred, flipping Europa's icy shell through time (e.g. Schenk *et al.*, 2008).

Because Europa's orbit is somewhat elliptical, with an eccentricity of 0.0094, it experiences a changing gravitational pull from its parent planet as it orbits Jupiter each Europa day (eurosol) of 3.55 days. This radial tide should cause Europa's icy shell to flex by about 30 m as it orbits, with slight variation in the predicted tidal amplitude depending on the thickness and rigidity of the ice shell (Moore and Schubert, 2000). A librational tide is also induced as Europa travels in its elliptical orbit. This tidal flexing means that the stresses at all non-equatorial points on the surface continually rotate direction and change magnitude as Europa orbits (Greenberg *et al.*, 2002). Some of Europa's fractures and ridges trace out cycloidal patterns (Figure 2), displaying a series of arcs connected at sharp cusps; this unique pattern is probably related to propagation of cracks in the rotating and fluctuating stress environment (Hoppa *et al.*, 1999). Indirectly, these cycloidal features betray a hidden ocean, because the stress required to form them would be too small if Europa has no ocean, as then the ice shell would flex by <1 m (Hoppa *et al.*, 1999; Moore and Schubert, 2000).

Europa's ubiquitous ridges typically have a distinctive double ridge morphology, with a valley in between (Figure 5). Formation of these ridges is probably related to the rise of water or warm ice along fractures (Kattenhorn and Hurford, 2009; Prockter and Patterson, 2009). A geophysically plausible model for their formation suggests that ridges are produced as the opposing sides of faults grind back-and-forth against each other as Europa orbits, generating heat akin to rubbing your hands together. This would warm the ice to either side sufficiently to induce buoyant uplift to form a ridge and/or would allow compressive stress to act on these weak plains, and might even generate enough heat to create melt along ridge axes (Nimmo and Gaidos, 2002). Subsequent shearing of these weak tabular zones might permit the formation of wider and more complexly lineated ridges (Aydin, 2006).

Many of Europa's wider dark and gray bands (Figure 6), which were first characterized in Voyager images, show fine striations along their lengths when seen at the higher resolutions that Galileo images provided (Sullivan *et al.*, 1998). These may be places where the icy shell has spread apart along older ridges, exposing warm icy material

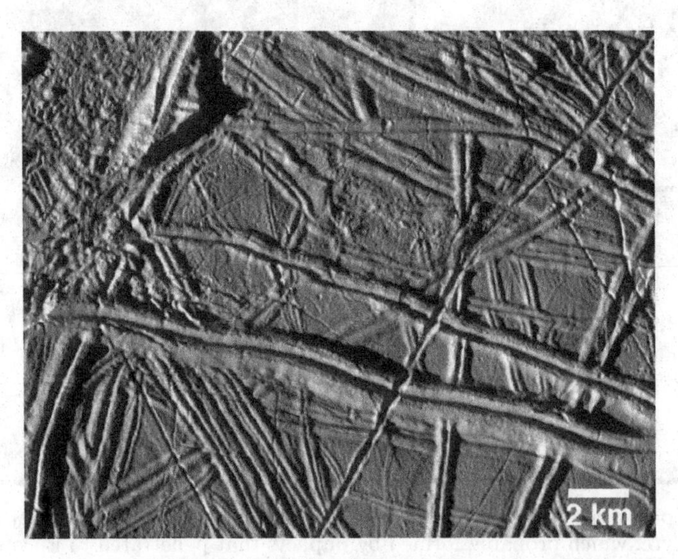

Figure 5. Typical double ridges on Europa's surface.

from below, which is then fractured as extension continues (Prockter and Patterson, 2009). This model suggests that bands are icy analogs to the spreading centers that exist on Earth's rocky ocean floors, with new lithospheric ice created along an axial trough and shunted laterally then commonly faulted (Prockter *et al.*, 2002). Unlike Earth's plate tectonics, however, there are no subduction zones identified on Europa, and only rare examples of folding have been identified (Prockter and Pappalardo, 2000), so the compensation of band extension remains an unresolved issue.

Wherever Europa's surface is somehow breached by geological activity to expose subsurface material, that material is relatively dark and a ruddy ochre in color, in contrast to the bright icy older surface. Multispectral characterization suggests that the colored materials contain hydrated sulfate salts and hydrated sulfuric acid (Dalton *et al.*, 2007; Carlson *et al.*, 2009). Salts may originate in the ocean (McCord *et al.*, 1999; Zolotov and Kargel, 2009), while sulfuric acid hydrate may be a product of radiation-driven chemistry, radiolytically generated as high-energy charged particles of Jupiter's magnetosphere slam into Europa's surface. None of these materials is colored, but these same materials may gain their color from irradiation of sulfur-containing materials and growth of ochre-colored sulfur chains (Carlson *et al.*, 2009).

Europa's face is freckled by pits, domes, and dark spots collectively known as "lenticulae," Latin for "freckles" (Carr *et al.*, 1998; Pappalardo *et al.*, 1998). Individual lenticulae are commonly about 5-10 km across, and some pepper the surface in clusters. They are probably the surface expression of upwelling warm ice diapirs, implying that Europa's icy shell is convecting (Barr and Showman, 2004), as Reynolds and Cassen (1979) had

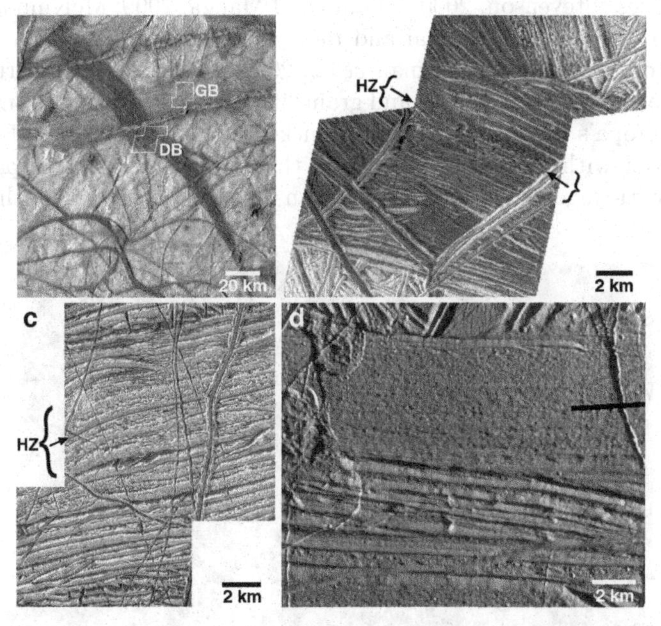

Figure 6. Examples of bands on Europa. (a) A dark wedge-shaped dark band (DB) cross-cuts a brighter (older) gray band (GB); outlines mark locations of the two subsequent views. (b) At high-resolution, the dark band shows multiple subparallel ridges and troughs that flank a central trough (arrows) and hummocky zone (braces, HZ). (c) At high resolution, the gray band shows similar characteristics of a central trough (arrow) and hummocky zone (braces, HZ). (d) A band seen at high resolution and high solar incidence angle displays subparallel ridge and troughs with characteristics of normal fault blocks. These characteristics have been compared to spreading centers on Earth.

first predicted. This implies that diapirs of warm ice apparently risen up from the base of the floating icy shell, and some have impinged on the cold surface layer to warp it upward into domes, while others have pierced the cold surface to flow out onto the surface, though piercing the coldest near-surface ice implies additional processes such as compositional buoyancy of the diapirs (Pappalardo and Barr, 2004). Pits may be locations where warm diapirs have triggered melting within the ice shell, resulting in downwarping of the surface as melt drains downward in the ice shell below (Singer *et al.*, 2010).

Europa shows large disrupted regions aptly termed "chaos," where the surface is broken into raft-like plates that have rotated and tilted in a sea of rubbly debris (Carr *et al.*, 1998). Some argue these are places where the icy shell has completely melted, perhaps above a warm plume of water below (Greenberg *et al.*, 1999); others infer that these are places where ice convection has been especially intense, disrupting the cold icy surface above, and partially melting near-surface ice to allow fragmented blocks to slide and turn (Collins *et al.*, 2000). Either way, chaos tells of a warm and active icy shell, with at least partial melting of the ice (Collins and Nimmo, 2009).

Reynolds and Cassen (1979) predicted that convection should rapidly freeze an ocean, but the Galileo magnetometer data and the youthfulness of Europa's surface imply that Europa's ocean is still there. In recent decades, significant advances have been made in understanding of convection, tidal heating, and the properties of ice. Thus, it now seems that ice convection might not freeze an underlying ocean after all, but instead ice convection and an ocean can co-exist through time (Spohn and Schubert, 2003; Barr and Showman, 2004).

The thickness of Europa's icy shell is uncertain, but geological and geophysical modeling provides clues (Stevenson, 2000; Nimmo and Manga, 2009; McKinnon *et al.*, 2009). Consideration of the amount of tidal and radiogenic heat from Europa predict an icy shell about 20 to 30 km thick (Hussman *et al.*, 2002; Spohn and Schubert, 2003). Work using the most pertinent ice rheology and grain size data suggests that convection should initiate when Europa's ice shell is 15-25 km thick (McKinnon, 1999), so Europa's lenticulae are consistent with an icy shell > 15 km thick. Most telling, Europa's two largest bulls-eye-like multi-ringed structures indicate an icy shell about 20 km thick when they

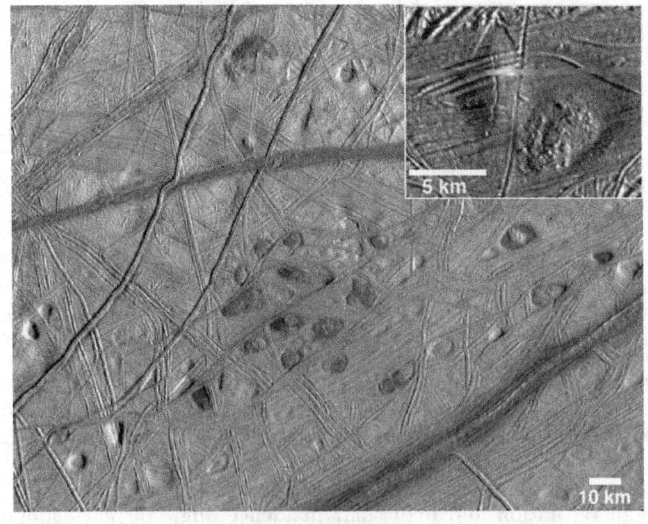

Figure 7. Pits, spots, and domes on Europa, commonly termed "lenticulae" in a regional view, and a high resolution view (inset). Illumination is from the east (right) in both images.

formed in geologically recent times (Schenk, 2002). Some researchers have argued that Europa's ice shell might be only a few kilometers thick (Greenberg *et al.*, 2002); however, such a model is inconsistent with thermal, convection, and impact modeling. Instead, an icy shell ~20 km thick is consistent with all of the geological evidence, while providing sufficient tidal amplitude to permit the formation of Europa's ubiquitous tectonic features.

4. Europa and the Search for Life

On Earth, everywhere there is water, there is life, so it is reasonable that the search for life in our solar system focuses on the search for water. Europa's probable ocean suggests is an important destination in the search for life in our Solar System. But life needs more than just water to exist, it also needs the chemical elements from which organic molecules can be built and a source of energy to power metabolism.

In the lightless depths of Europa, photosynthesis is an unlikely metabolic mechanism. Instead, it is chemosynthetic life which might plausibly exist, if the chemical disequilibrium to power metabolic reactions exists in Europa's ocean (Hand *et al.*, 2009). In fact, Europa's intense geological activity and relatively thin icy shell combine to give its ocean advantages for habitability that other icy satellite oceans in our Solar System do not have. Geological breaching of Europa's icy shell combined with partial melting and convective transport could permit the ocean and surface to exchange water and ice, providing important advantages to life and the search for life.

The chemical energy for life could enter Europa's ocean from above and from below. Europa's surface is bathed in intense radiation from high-energy charged particles trapped in Jupiter's magnetosphere. These particles slam into Europa's ice-rich surface, transforming some of its H_2O into molecules of oxygen (O_2), peroxide (H_2O_2), and other oxidized compounds (Johnson *et al.*, 2004; Paranicas *et al.*, 2004). If these oxidants can be transported to the subsurface ocean by geological processes, they would make an ideal chemical fuel for life.

Meanwhile, if Europa's rocky mantle is hot, then reduced chemical nutrients could pour into the base of the ocean from below, potentially at black smokers analogous to those

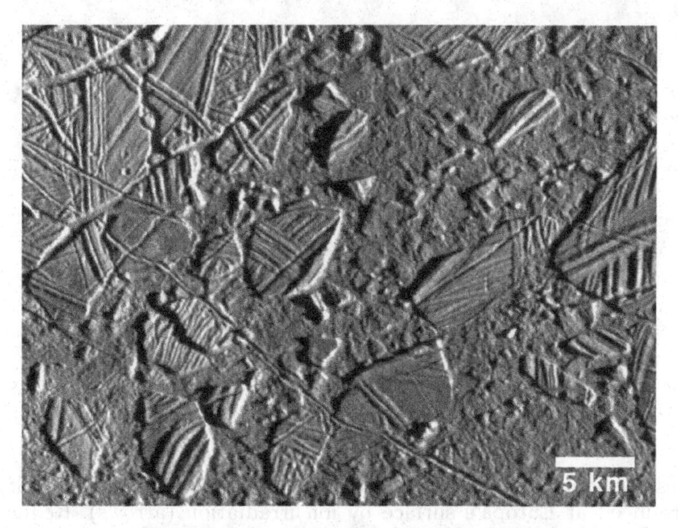

Figure 8. Chaos terrain on Europa, where city-sized icy blocks have rotated and translated in a matrix of disrupted and crumbled blocks.

comprising hydrothermal vents that power communities of organisms on the terrestrial ocean floor (McCollom, 1999). Whether Europa's mantle is significantly tidally heated is currently unknown, and depends on the complex thermal-orbital evolution of Europa and the other Galilean satellites (Moore and Hussman, 2009).

If Europa's icy shell is only several kilometers thick and can completely melt through in places, then oxidants could be directly delivered to the ocean (Greenberg *et al.*, 2000). However, in the more likely scenario where Europa's icy shell is ~20 km thick, then convective activity and partial melting could transport surface oxidants to the ocean. Partial melting in chaotic terrain, in lenticulae, and along ridge axes may allow melt to sink downward through the cold near-surface ice and to the convective zone below on a time scale of ~10-100 yr (Barr *et al.*, 2002). In turn, convection can circulate material from the top to bottom of the convective zone on a time scale of ~10^5 yr (Pappalardo and Barr, 2004). Through this indirect route, oxidants may be supplied to Europa's ocean below. In addition, some oxidants can be generated within the ice shell and ocean

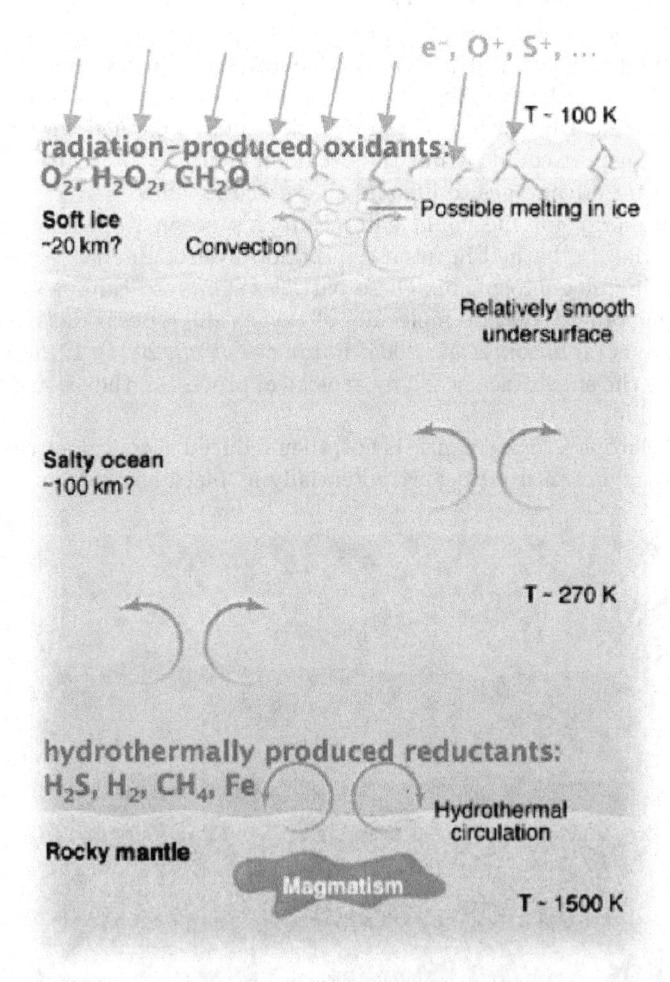

Figure 9. Representation of Europa's ice shell, ocean, and rocky mantle (after Stevenson, 2000). Oxidants are produced at Europa's surface by ion irradiation (arrows). Reductants might be produced at the ocean floor if the mantle rock is hot, promoting hydrothermal circulation. Such oxidants and reductants could serve to fuel life in Europa's ocean.

by radioactive decay (Hand *et al.*, 2007). Understanding transport mechanisms within Europa's icy shell is key to understanding the habitability of Europa's ocean.

The search for life at Europa will begin slowly, and requires patience. The Jupiter Europa Orbiter spacecraft presently on the drawing boards as part of the international Europa Jupiter System Mission is the next required step in evaluating the potential habitability of Europa (Greeley *et al.*, 2009). Such an orbiting spacecraft would be able to characterize Europa's ocean, including its thickness and salinity, using the techniques of gravity and altimetry to measure Europa's tidal deformation, and magnetometry to much better characterize the satellite's induced magnetic field signature. Ice-penetrating radar could search for pockets of shallow water within the icy shell such as in chaos regions and along ridges, and could be correlated to geological indicators to understand surface-ice-ocean material exchange processes; a radar signal through sufficiently cold ice might even be able to directly detect the ocean (Blankenship *et al.*, 2009). Compositional measurements, imaging, and charged particle readings could combine to indicate how Europa's surface features are created and evolve, and whether the surface chemistry is indicative of ocean chemistry, and whether that chemistry is conducive to life. The Jupiter Europa Orbiter could launch as soon as 2020, reach Jupiter by 2025, and enter Europa orbit in 2028.

In the more distant future, humans may land a spacecraft on Europa. Life cannot survive on Europa's surface because of the intense radiation by Jupiter's magnetospheric charged particles, but signs of life might be sought just tens of centimeters below the surface, where ice would protect any organic materials that might be dredged from the ice-ocean interface to near the surface. Signs of life might be detectable with a landed spacecraft which would use sophisticated chemical analyses and a simple microscope to search (Hand *et al.*, 2009).

Europa is an astrobiologically and geophysically fascinating world, arguably offering the best prospect for present-day life in our Solar System beyond Earth. Europa's warm, salty ocean is in direct contact with a mantle that could be geologically active, supplying chemical nutrients from below, while its icy shell is thin and its surface is youthful, so recent or current activity could churn nutrients downward into the ocean. If life exists at Europa, its signs could be exposed at or near the surface.

More than four hundred years after its discovery, Europa may some day again bring about a revolution in scientific thought, and—if life is found there—may again change our sense of place in the Universe.

Acknowledgements This work was carried out at the Jet Propulsion Laboratory, California Institute of Technology, under a contract with the National Aeronautics and Space Administration.

References

Alexander, C., Carlson, R., Consolmagno, G., Greeley, R., & Morrison, D. 2009, The exploration history of Europa *Europa*, R. T. Pappalardo *et al.*, eds., pp. 3–26. Univ. of Arizona Press, Tucson

Aydin, A. 2006, Failure modes of the lineaments on Jupiter's moon, Europa: Implications for the evolution of its icy crust, *J. Struct. Geol.*, 28, 2222–2236

Barr, A. C., Nimmo, F., Pappalardo, R. T., & Gaidos, E. 2002, Shear heating and solid-state convection: Implications for astrobiology, *Lunar Planet. Sci. Conf. XXXIII*, abstract #1545

Barr, A. C. & Showman, A. P. 2009, Heat transfer in Europa's icy shell, *Europa*, R. T. Pappalardo *et al.*, eds., pp. 405–430. Univ. of Arizona Press, Tucson

Blankenship, D. D., Young, D. A., Moore, W. B., & Moore, J. C. 2009, Radar sounding of Europa's subsurface properties and processes: The view from Earth, *Europa*, R. T. Pappalardo *et al.*, eds., pp. 631–654. Univ. of Arizona Press, Tucson

Bierhaus, E. B., Zahnle, K., & Chapman, C. R. 2009, Europa's crater distributions and surface ages, *Europa*, R. T. Pappalardo *et al.*, eds., pp. 161–180. Univ. of Arizona Press, Tucson

Carlson, R. W., Calvin, W. M., Dalton, J. B., Hansen, G. B., Hudson, R. L., Johnson, R. E., McCord, T. B., & Moore, M. H. 2009, Europa's surface composition, *Europa*, R. T. Pappalardo *et al.*, eds., pp. 283–328. Univ. of Arizona Press, Tucson

Carr, M. H., Belton, M. J. S., Chapman, C. R., Davies, M. E., Geissler, P., Greenberg, R., McEwen, A. S., Greeley, R., Sullivan, R., Head, J. W., Pappalardo, R. T., Klaasen, K. P., Johnson, T. V., Moore, J., Neukum, G., Schubert, G., Burns, J. A., Thomas, P., & Veverka, J. 1998, Evidence for a subsurface ocean on Europa, *Nature*, 391, 363–365

Cassen, P. M., Peale, S. J., & Reynolds, R. T. 1982, Structure and thermal evolution of the Galilean satellites, *Satellites of Jupiter*, D. Morrison, ed., pp. 93–1128, Univ. of Ariz. Press, Tucson

Collins, G. & Nimmo, F. 2009, Chaos on Europa, *Europa*, R. T. Pappalardo *et al.*, eds., pp. 259–282. Univ. of Arizona Press, Tucson

Collins G. C., Head, J. W., Pappalardo, R. T., & Spaun, N. A. 2000, Evaluation of models for the formation of chaotic terrain on Europa, *J. Geophys. Res.*, 105, 1709–1716

Consolmagno, G. J. & Lewis, J. S. 1976, Structural and thermal models of icy Galilean satellites, *Jupiter*, (T. A. Gehrels, ed.), pp. 1035–1051. Univ. of Arizona, Tucson

Dalton, J. B. 2007, Linear mixture modeling of Europa's nonice material using cryogenic laboratory spectroscopy, *Geophys. Res. Lett.34*, L21205, doi: 10.1029/2007GL031497

Fanale F. P., Johnson, T. V., & Matson, D. L., 1977, Io's surface and the histories of the Galilean satellites, *Planetary Satellites*, (J. A. Burns, ed.), pp. 379–405. Univ. of Arizona, Tucson

Greeley, R., Pappalardo, R. T., Prockter, L. M., & Hendrix, A. 2009, Future exploration of Europa, *Europa*, (J. A. Burns, ed.), (R. T. Pappalardo *et al.*, eds.), pp. 655–695, Univ. of Arizona, Tucson

Greenberg, R. & Weidenschilling, S. J. 1984, How fast do Galilean satellites spin?, *Icarus*, 58, 186–196

Greenberg R., Hoppa, G. V., Tufts, B. R., Geissler, P., Riley, J., & Kadel, S. 1999, Chaos on Europa, *Icarus*, 141, 263–286

Greenberg, R., Geissler, P., Tufts, B. R., & Hoppa, G. V. 2000, Habitability of Europa's crust: The role of tidal-tectonic processes, *J. Geophys. Res.*, 105, 17, 551–17, 562

Greenberg, R., Geissler, P., Hoppa, G., & Tufts, B. R. 2002, Tidal tectonic processes and their implications for the character of Europa's icy crust, *Rev. Geophysics 40*, 1004, doi:10.1029/2000RG000096

Hand, K. P., Carlson, R. W., & Chyba, C. F. 2007, Energy, chemical disequilibrium, and geological constraints on Europa, *Astrobiology*, 7, 1–18

Hand, K. P., Chyba, C. F., Priscu, J. C., Carlson, R. W., & Nealson, K. H. 2009, Astrobiology and the potential for life on Europa, *Europa*, R. T. Pappalardo *et al.*, eds., pp. 589–630. Univ. of Arizona Press, Tucson

Hoppa, G. V., Tufts, B. R., Greenberg, R., & Geissler, P. E. 1999, Formation of cycloidal features on Europa, *Science*, 285, 1899–1902

Hussman, H., Spohn, T., & Wieczerkowski, K. 2002, Thermal equilibrium states of Europa's ice shell: Implications for internal ocean thickness and surface heat flow, *Icarus*, 156, 143–151

Johnson, R. E., Burger, M. H., Cassidy, T. A., Leblanc, F., Marconi, M., & Smyth, W. H. 2009, Composition and detection of Europa's sputter-induced atmosphere, *Europa*, R. T. Pappalardo *et al.*, eds., pp. 507–528. Univ. of Arizona Press, Tucson

Kattenhorn, S. A. & Hurford, T. 2009, Tectonics of Europa, *Europa*, R. T. Pappalardo *et al.*, eds., pp. 199–236. Univ. of Arizona Press, Tucson

Khurana, K. K., Kivelson, M. G., Hand, K. P., & Russell, C. T. 2009, Electromagnetic induction from Europa's ocean and the deep interior, *Europa*, R. T. Pappalardo *et al.*, eds., pp. 571–587. Univ. of Arizona Press, Tucson

Kivelson, M. G., Khurana, K. K., Stevenson, D. J., Bennett, L., Joy, S., Russell, C. T., Walker, R. J., Zimmer, C., & Polansky, C. 1999, Europa and Callisto: Induced or intrinsic fields in a periodically varying plasma environment, *J. Geophys. Res.*, 104 4609–4625

Kivelson, M. G., Khurana, K. K., Russell, C. T., Volwerk, M., Walker, R. J., & Zimmer, C. 2000, Galileo magnetometer measurements strengthen the case for a subsurface ocean at Europa, *Science*, 289, 1340

Kivelson, M. G., Khurana, K. K., & Volwerk, M. 2002, The permanent and inductive magnetic moments of Ganymede, *Icarus*, 157, 507–522

Lucchitta, B. K. & Soderblom, L. A. 1982, The geology of Europa, *The Satellites of Jupiter*, (D. Morrison, ed.), pp. 521–555. Univ. of Arizona, Tucson, Arizona

Malin, M. C. & Pieri, D. C. 1986, Europa, *Satellites*, (J. A. Burns and M. S. Matthews, Eds.), pp. 689–716. Univ. of Arizona Press, Tucson

McCollom, T. M. 1999, Methanogenesis as a potential source of chemical energy for primary biomass production by autotrophic organisms in hydrothermal systems on Europa, *J. Geophys. Res*, 104, 30729-30742

McCord, T. B., Hansen, G. B., Matson, D. L., Johnson, T. V., Crowley, J. K., Fanale, F. P., Carlson, R. W., Smythe, W. D., Martin, P. D., Hibbitts, C. A., Granahan, J. C., Ocampo, A., & the NMS Team 1999, Hydrated salt minerals on Europa's surface from the Galileo NIMS investigation, *J. Geophys. Res*, 104, 11827–11851

McKinnon, W. B. 1999, Convective instability in Europa's floating ice shell, *Geophys. Res. Lett*, 26, 951–954

McKinnon, W. B., Pappalardo, R. T., & Khurana, K. K. 2009, Europa: Perspectives on an ocean world, *Europa*, (R. T. Pappalardo *et al.*, eds.), pp. 697–709, Univ. of Arizona, Tucson

Moore, W. B. & Schubert, G. 2000, The tidal response of Europa, *Icarus*, 147, 317–319

Moore, W. B. & Hussman, H. 2009, Thermal evolution of Europa's silicate interior, *Europa*, R. T. Pappalardo *et al.*, eds., pp. 369–380. Univ. of Arizona Press, Tucson

Moore J. M., Asphaug, E., Sullivan, R. J., Klemaszewski, J. E., Bender, K. C., Greeley, R., Geissler, P. E., McEwen, A. S., Turtle, E. P., Phillips, C. B., Tufts, B. R., Head, J. W., Pappalardo, R. T., Jones, K. B., Chapman, C. R., Belton, M. J. S., Kirk, R. L., & Morrison, D. 1998, Large impact features on Europa: Results of the Galileo nominal mission, *Icarus*, 135, 127–145

Moore J. M., Asphaug, E. B., Michael, J. S., Bierhaus, B., Breneman, H. H., Brooks, S. M., Chapman, C. R., Chuang, F. C., Collins, G. C., Giese, B., Greeley, R., Head, J. W., Kadel, S., Klaasen, K. P., Klemaszewski, J. E., Magee, K. P., Moreau, J., Morrison, D., Neukum, G., Pappalardo, R. T., Phillips, C. B., Schenk, P. M., Senske, D. A., Sullivan, R. J., Turtle, E. P., & Williams, K. K. 2001, Impact features on Europa: Results of the Galileo Europa Mission (GEM)., *Icarus*, 151, 93–111

Nimmo, F. & Gaidos, E. 2002, Strike-slip motion and double ridge formation on Europa, *J. Geophys. Res.*, 107, 1–8

Nimmo, F. & Manga, M. 2009, Geodynamics of Europa's icy shell, *Europa*, (R. T. Pappalardo *et al.*, eds.), pp. 381–404, Univ. of Arizona, Tucson

Ojakangas, G. W. & Stevenson, D. J. 1989, Thermal state of an ice shell on Europa, *Icarus*, 81, 220–241

Pappalardo, R. T. & Barr, A. C. 2004, Origin of domes on Europa: The role of thermally induced compositional buoyancy, *Geophys. Res. Lett.*, 31, L01701, doi:10.1029/2003GL019202

Pappalardo, R. T. & Sullivan, R. J. 1996, Evidence for separation across a gray band on Europa, *Icarus*, 123, 557–567

Pappalardo, R. T., Head, J. W., Greeley, R., Sullivan, R. J., Pilcher, C., Schubert, G., Moore, W., Carr, M. H, Moore, J. M., Belton, M. J. S., & Goldsby, D. L. 1998, Geological evidence for solid-state convection in Europa's ice shell, *Nature*, 391, 365–368

Paranicas, C., Cooper, J. F., Garrett, H. B., Johnson, R. E., & Sturner, S. J. 2009, Europa's radiation environment and its effects on the surface, *Europa*, R. T. Pappalardo *et al.*, eds., pp. 529–544. Univ. of Arizona Press, Tucson

Prockter, L. M. & Pappalardo, R. T. 2000, Folds on Europa: Implications for crustal cycling and accommodation of extension, *Science*, 289, 941–943

Prockter, L. M. & Patterson, G. W. 2009, Morphology and evolution of Europa's ridges and bands, *Europa*, (R. T. Pappalardo *et al.*, eds.), pp. 237–258, Univ. of Arizona, Tucson

Prockter L. M., Head, J., Pappalardo, R., Sullivan R., Clifton, A. E., Giese, B., Wagner, R., & Neukum, G. 2002, Morphology of europan bands at high resolution: A mid-ocean ridge-type rift mechanism, *J. Geophys. Res.*, 107, 1–26

Reynolds, R. T. & Cassen, P. 1979, On the internal structure of the major satellites of the outer planets, *Geophys. Res. Lett.*, 6, 121–124

Schenk, P. M. 2002, Thickness constraints on the icy shells of the Galilean satellites from a comparison of crater shapes, *Nature*, 417, 419–421

Schenk, P. & McKinnon, W. B. 1989, Fault offsets and lateral crustal movement on Europa: Evidence for a mobile ice shell, *Icarus*, 79, 75–100

Schenk, P. M. & Turtle, E. P. 2009, Europa's impact craters: Probes of the icy shell, *Europa*, R. T. Pappalardo *et al.*, eds., pp. 181–198. Univ. of Arizona Press, Tucson

Schenk P., Matsuyama, I., & Nimmo, F. 2008, True polar wander on Europa from global-scale small-circle depressions, *Nature*, 453, 368–371

Schubert, G., Sohl, F., & Hussman, H. 2009, Thermal evolution of Europa's silicate interior, *Europa*, R. T. Pappalardo *et al.*, eds., pp. 353–368. Univ. of Arizona Press, Tucson

Selvans, Z. A., Wolf, A. S., & Pappalardo, R. T. 2010, A critical comparison of Europa's lineaments to non-synchronous rotation stresses, *J. Geophys. Res.*

Singer, K. N., McKinnon, W. B., & Schenk, P. M. 2010, Pits, spots, uplifts, and small chaos regions on Europa: Evidence for diapiric upwelling from morphology and morphometry, *Lunar Planet. Sci. Conf. 41*, Abstract #2195

Spohn, T. & Schubert, G. 2003, Oceans in the icy Galilean satellites of Jupiter?, *Icarus*, 161, 456–467

Squyres S. W., Reynolds, R. T., Cassen, P., & Peale, S. J. 1983, Liquid water and active resurfacing of Europa, *Nature*, 301, 225–226

Stevenson, D. 2000, Europa's Ocean: The case strengthens, *Science*, 289, 1305–1307

Sullivan, R. & 12 colleagues 1998, Episodic plate separation and fracture infill on the surface of Europa, *Nature*, 391, 371–373

Wahr, J., Selvans, Z. A., Mullen, M. E., Barr, A. C., Collins, G. C., Selvans, M. M., & Pappalardo, R. T. 2009, Modeling stresses on satellites due to nonsynchronous rotation and orbital eccentricity using gravitational potential theory, *Icarus*, 200, 188–206

Zimmer C., Khurana, K. K., & Kivelson, M. G. 2000, Subsurface oceans on Europa and Callisto: Constraints from Galileo magnetometer observations, *Icarus*, 147, 329–347

Zahnle, K., Schenk, P., Levison, H. F., & Dones, L. 2003, Cratering rates in the outer solar system, *Icarus*, 163, 263–289

Zolotov, M. Yu & Kargel, J. S. 2009, On the chemical composition of Europa's icy shell, ocean, and underlying rocks, *Europa*, R. T. Pappalardo *et al.*, eds., pp. 431–458. Univ. of Arizona Press, Tucson

Galileo's Medicean Moons: their impact on 400 years of discovery
Proceedings IAU Symposium No. 269, 2010

C. Barbieri, S. Chakrabarti, M. Coradini & M. Lazzarin, eds.
doi:10.1017/S1743921310007337

Europa Lander Mission: A Challenge to Find Traces of Alien Life

Lev Zelenyi[1], **Oleg Korablev**[2], **Elena Vorobyova**[2,1], **Maxim Martynov**[3], **Efraim L. Akim**[4], **Alexander Zakahrov**[1]

[1]Space Research Institute (IKI), Moscow, Russia, Fax: +7 495 3331248

[2]IKI

[3]Lavochkin Assosiation, Russia

[4]Keldysh Institute of Applied Mathematics, Russia

email: korab@iki.rssi.ru

Abstract. An international effort dedicated to science exploration of Jupiter system planned by ESA and NASA in the beginning of next decade includes in-depth science investigation of Europa. In parallel to EJSM (Europa-Jupiter System Mission) Russian Space Agency and the academy of Science plan Laplace-Europa Lander mission, which will include the small telecommunication and science orbiter and the surface element: Europa Lander. In-situ methods on the lander provide the only direct possibility to assess environmental conditions, and to perform the search for signatures of life. A critical advantage of such in situ analysis is the possibility to enhance concentration and detection limits and to provide ground truth for orbital measurements. The science mission of the lander is biological, geophysical, chemical, and environmental characterizations of the Europa surface. Remote investigations from the orbit around Europa would not be sufficient to address fully the astrobiology, geodesy, and geology goals. The science objectives of the planned mission, the synergy between the Europa Lander and EJSM mission elements, and a brief description of the Laplace-Europa Lander mission are presented.

Keywords. Europa; Jupiter; Galilean satellites; habitability; space research.

---

1. Introduction

The investigation of giant planets and their satellites is important for understanding of the formation and evolution of our Solar system, measuring distribution of chemical elements, study of internal structure and atmospheres of giant planets and their satellites, assessing the habitability of these distant worlds. It has been an important part of NASA research program starting from Pioneer and Voyager missions in the beginning of 1970th. Jupiter system arouses the most interest of the scientific community after pioneering Voyager and Galileo missions. Along with the planet and its extraordinary magnetosphere, the Galilean satellites of Jupiter, and Europa in particular, attract special attention.

To some extent Europa is a similar to the Moon. The radius of Europa is a little smaller, but the average density (about 3 g cm^{-3}) is close to that of the Moon (3.3 g cm^{-3}), suggesting that Europa is mostly a mineral body. Like the Moon, Europa and other Galilean satellites are nearly in synchronous rotation when the rotation period almost equals the orbital period and the satellite faces the planet with always the same side.

Water ice on the surface of Europa has been discovered by means of ground-based astronomy long before Voyager spacecrafts arrived to Jupiter (Kuiper, 1957, Moroz, 1965, Johnson and McCord 1971, Pilcher *et al.*, 1972). Near-infrared spectra of Europa and Ganimede demonstrate a broad feature near 1.5μm associated with H_2O ice band. The dayside surface temperature was estimated at 120-135K. First images of Europa's surface have been obtained in 1979 during Voyager 1 and 2 flybys. The most part of surface looks relatively smooth and bright; few equatorial areas being darker and spotted. The observed density of impact craters is far lower than on other Solar system bodies. The main morphological feature of Europa surface is a network of dark linear details, extending over thousands of kilometers. The smoothness of the surface indicates that the ice remained relatively warm, soft and mobile during an important part of the geological history of the satellite. Many observations indicated that there might be a global liquid ocean beneath the icy crust, which could be theoretically a habitat for some form of life (Reynolds *et al.*, 1983).

In 1995 the Galileo spacecraft entered into the orbit of Jupiter's satellite. Since 1997 the mission was focused on Europa. High resolution (up to several meters) images of Europa have been obtained (Greeley *et al.*, 1998a). The width of many linear features appears to be only about one kilometer. It was proposed that the prominent linear, often double and multiple ridges have been formed as a result of extrusion of viscous material. The Earth analog of such process could be the formation of middle oceanic ridges as a result of repeated magma flows. The main source of energy in the tectonic process on Europa is a tidal interaction with Jupiter and other satellites. The resonance keeps the satellite from perfectly circular orbit, inducing a forced eccentricity that results in tidal flexing and internal dissipation of energy. The formation of the bands and lineations of the shell could be explained by global stress fields due orbital eccentricity, orbital recession, and nonsynchronous rotation. The displacement of the tidal bulge due to nonsynchronous rotation is the best explanation of ridges and cycloidal features assuming relatively thin ice crust (McEwen 1986, Greeley *et al.*, 1998b, Hoppa *et al.*, 1999).

Impact craters rates have shown the age of most areas does not exceed 100 million years (Zahnle *et al.*, 2003, 2008). The young surface requires recent geologic activity, which is more likely in the presence of liquid water. The appearance of chaotic terrains, rafts, and many other geological evidences presume recent release of water on the surface. However, a search for on-going geologic activity on Europa surface over the 20-year period between Voyage and Galileo observations has shown no detectable changes (Phillips *et al.*, 2000).

Gravity measurements by the Galileo show that Europa is an internally differentiated rocky body with the thickness of the outer water shell in the range of about 80 to 170 kilometers (Anderson *et al.*, 1998). These data cannot distinguish between water in liquid and ice phases. Current models of the internal structure of differentiated Europa predict a Fe-FeS core with radius of order 500–600 km depending on model, mineral mantle about 1000 km and water crust below 200 km (80–160 km) (Kuskov and Kronrod, 2005). The contact of the putative liquid ocean and the mineral floor is a key point for a hypothesized habitat of the Europan ocean.

The exploration of outer planets continues to play an important role in the program of leading space agencies. Currently JUNO, a New Frontiers mission to study the atmosphere, magnetosphere and gravitational field of Jupiter is being prepared by NASA for launch in August 2011 (Matousek 2007). After analyzing several concepts of Europa missions of different cost/complexity levels since 1996 (Langmaier 2008; Gershman, 1999, Gershman, 2003), NASA concluded that multiple and single flyby missions have significantly less science value when compared to a dedicated moon orbiter. A first dedicated mission at Europa consisting of a single lander was considered risky and limited to only one site. Participating in the successful Cassini-Huygens programme ESA actively joined the program of outer planets exploration. In 2007 two prominent developments, ESA Cosmic Vision Class L proposal and NASA Flagship mission proposal were merged into a joint space mission to the Jupiter system: Europa Jupiter System Mission (EJSM, http://opfm.jpl.nasa.gov/europajupitersystemmissionejsm/) with an approximate launch date in 2020–2022 (Blanc *et al.*, 2009, Clark *et al.*, 2009). Recently this project was selected among two outer planet missions for further development (Phase A) with final selection planned for 2012–2013. The project includes two spacecrafts for exploration of Jupiter, Ganimede and Europa. In the framework of EJSM, NASA plans Jupiter Europa Orbiter (JEO) for investigation Jupiter, Io and focused on Europa and its habitability. It is designed to follow-up on the major discoveries of the Galileo and Voyager missions at Europa, and in particular to characterize its ocean. JEO will be built to withstand the intense radiation at Europa orbit, and would consist of an orbiter with 11 science instruments designed for extensive mapping of Europa. On the way to Europa, JEO will tour the Jovian system and make routine and frequent observations of Jupiter, its satellites and its environment. ESA plans Jupiter Ganymede Orbiter (JGO) for investigation of Jupiter, Ganimede and Callisto. It would consist of a solar powered orbiter with 10 science instruments designed for remote sensing of Jupiter and the inner satellites and for extensive mapping of Ganymede. Prior to Ganymede arrival, JGO will intensely investigate Callisto from a resonant orbit, and make extensive observations of the Jupiter system to complement those of JEO. The Jupiter Magnetospheric Orbiter (JMO) proposed by the Japanese Aerospace Exploration Agency (JAXA) may be included into EJSM project to characterize plasma environment of the Jupiter system. JMO would undertake detailed in-situ studies of Jupiter's magnetosphere affording the opportunity for "3-point" investigations of the Jupiter system together with JGO and JEO.

Traditionally the Soviet, and later the Russian space program was limited to the inner part of Solar system (Moon, Mars and the most successfully Venus). Still, some preliminary studies of scientific projects to Jupiter system were carried out in the USSR in the 80th. Starting from 2007 Russian Academy of Sciences and Roscosmos considers the possibility to include in EJSM a Europa surface element, a soft lander, independently launched by Russia [Blanc *et al.*, 2009]. The scientific objectives of this lander will be to search for signatures of possible present and extinct life, and *in situ* studies of Europa internal structure, surface and environment. After preliminary assessment of the project in 2008-2009 it has been decided to complement the mission with an orbiter of Europa, to support independent data relay. The Orbiter observations will be also used for the selection of the landing site in real time, and conduct supportive science experiments. This mission nominated Laplace – Europa Lander will continue with assessment study in 2010-2012.

Cooperating with the EJSM mission, the Russian Europa Lander will: 1) complement and support remote sensing by JEO and JGO investigations with in-situ measurements to provide ground truth for the remote sensing data; 2) complement and validate remote sensing data by JEO and JGO in the Jupiter system by remote observations from Russian Europa Orbiter with different instruments; 3) provide a multipoint sounding of Jupiter plasma environment together with JGO, JEO and possibly with Japanese JMO. In turn, Europa Lander will benefit from high-resolution mapping and other data collected by JEO for landing site selection. Also, the possibility of data relay from Europa Lander via JGO is being considered. The launch of Europa Lander should follow the launch of JEO, and the mapping information from JEO should arrive during the cruise/approach phase of Europa Lander mission.

2. Europan Ocean

The existence of a global subsurface ocean on Europa is supported by many above observations. Already the analysis of Voyager imaging suggested that orientations of the linear features cannot be satisfactorily explained by stress fields due to the non-synchronous rotation. In the meanwhile, stress directions inferred for many bands, are consistent with stresses resulting from sliding of the crust of Europa as a single unit (Schenk and McKinnon 1989). Such a reorientation, denominated as polar wander, may result from spatial variations in the thickness of the shell (Ojakangas and Stevenson 1989). Such polar wander probably also explains symmetry patterns in the distribution of chaotic terrain, pits, and uplift features (Greenberg *et al.*, 2003; Schenck *et al.*, 2008].

The polar wander is a good evidence of a global ocean in the past. The most convincing evidence of the contemporary global ocean comes from Galileo magnetic field measurements (Kivelson *et al.*, 2000; Zimmer *et al.*, 2000). The induced magnetic field of Europa variable in direction and strength in function of Europa position within Jupiter's strong magnetic field was predicted by Colburn and Reynolds (1985). The data cannot be explained by localized pockets of salty water, and require a nearly complete spherical shell of salty water, with likely low salinity, lower than that of the Earth Ocean (Schilling *et al.*, 2004; Hand and Chyba 2007). The best estimate of the ice thickness from conductivity models is 4 km.

Estimates of Europa's ice-shell thickness range from several hundred meters to $\geqslant 30$ km. All of them are model dependent, and suffer from unknown parameters and insufficiently precise observations. Two major lines of evidence based on tidal deformations due to orbit eccentricity and asynchronous rotation, see Billings and Kattenhorn (2005) for a review. Estimates based on tidal flexing (e.g., Williams and Greeley, 1998, Billings and Kattenhorn 2005) give a thin shell of brittle ice, under which many authors suppose a thick convective low-viscosity layer. Buoyancy models (e.g., Carr *et al.*, 1998, Williams and Greeley, 1998) basing on somewhat limited stereo and photoclinometric topography of the satellite, and estimation based on the resistance of the ice shell to tidal stress leading to formation of cycloids (Hoppa *et al.*, 1999; Greenberg *et al.*, 2000) indicate the total depth of the ice of about few kms. Crater depth and morphology analysis (Greeley *et al.*, 1998b; Schenk, 2002] provide estimates of the ice shell varying from few to 18–20 km. The extreme position in this debate, a thick shell of a few tens of kilometers likely convective in its lower part (e.g., Pappalardo *et al.*, 1999) is supported by thermodynamic models (Hussmann *et al.*, 2002; Tobie *et al.*, 2003). The uncertainty in thickness translates directly into an uncertainty in the heat transfer mechanism: if the shell is thick, the rigid surface could be underlain by a layer of convecting water ice (e.g., Tobie *et al.*, 2003; Showman and Han, 2004), whereas a thin shell would instead transport the heat by conduction, and can account for melting of water through the surface (Greenberg *et al.*, 1999).

In spite of numerous observational evidences the existence of the global ocean on Europa remains a subject of debate (Carr *et al.*, 1998; Pappalardo *et al.*, 1999; Stevenson 2000; Schenk *et al.*, 2008). Galileo observations and measurements cannot give a definitive constraint about the thickness of the ice shell, provided the ocean is there. The icy crust is likely variable in thickness, inhomogeneous in composition, and possibly evolving in time on geologically short time scales (Zahnle *et al.*, 2003; Figueredo and Greeley 2004). Likely thin or locally thinned ice shell allows ocean water to reach the surface, possibly periodically on a relatively short time scales.

The thickness of the shell above the ocean is of fundamental importance both for the habitability of the ocean (because it is a quantitative indication of energy available from different sources), and for the ability to discover a signature of life on the surface or shallow subsurface of Europa.

3. Habitability of Europa: Alien Life?

If the subsurface ocean exists on Europa, it is characterized by supply of biologically important elements from the floor, and energy sources, from tidal effects, radioactivity, and geochemical mechanisms. Europan subsurface ocean environment may be similar to that of the deep ocean hydrothermal vents on Earth where remarkable life forms have been detected. There is a fundamental difference among finding life at the depth on the Earth (with biomass may be comparable to that on the surface, see e.g., Whitman *et al.*, 1998) or even the finding the life on Mars and the possibility to find life on Europa. The existence of "deep" life on Earth is not a proof that life has originated at depth. Similarly, interplanetary transfer of microorganisms could be a possible way of insemination of Mars or other bodies in the solar system. Situation is very different on Europa. According (Pierazzo and Chyba, 2001) about 10^{12} kg of carbon ($\sim$ 0.1% of the Earth biomass) could accumulate on Europa since its formation. However, there is no atmosphere on Europa, and interplanetary dust particles or small meteorites hit the surface at very high velocities leading to their total destruction. Therefore, the mechanism, which might deliver the prebiotic organics to early Earth (Mukhin *et al.*, 1989; Chyba and Sagan, 1992), or possibly inseminated Mars (Mileikowsky *et al.*, 2000), would hardly work on Europa.

It means that if life is found on Europa, it is (1) life originated in the depth of subsurface ocean independently of the free energy from the Sun, and (2) life originated independently of the life on the Earth, likely, an alien or endemic biology. Such a discovery would mean that the traditional view of planetary habitability, widely discussed in relation with exasolar planets, and based on distance to the star and the surface temperatures should be broadened (Chyba and Philips, 2002).

The analysis of Galileo mission data enables to develop new astrobiological concepts that speculate upon extant life on Europa (Reynolds *et al.*, 1983; Sieger *et al.*, 1998; Kargel *et al.*, 2000; Chyba 2000). Several basic features of Europan evolution has been postulated by Kargel *et al.* (2000): (1) initial carbonaceous chondrite rock; (2) global primordial low-temperature aqueous differentiation and formation of an impure primordial hydrous crust; (3) brine evolution and intracrustal differentiation; (4) degassing of Europa's mantle and gas venting; (5) hydrothermal processes, and (6) chemical alteration of the surface. These assumptions would result in a more realistic ocean/asthenosphere/lithosphere model and in crust/ocean enriched by magnesium and sodium sulfates, substances discovered on the surface of Europa by Galileo spectroscopy. A wide range of possible ocean/crust compositions is the reason to construct different scenarios for the origin and evolution of putative ecosystems. The main energy sources, which could provide the support for the extant life at Europa are the tidal heating and the convection (Reynolds *et al.*, 1983), radiolytic generation of oxidants (Chyba 2000; Sieger *et al.*, 1998), chemoautotrophy (Kargel *et al.*, 2000; Sieger *et al.*, 1998), and possibly ionotrophy (Irwin and Schulze-Makuch, 2003). Microbial metabolites, dead biomass and also primordial organics must serve as a nutrient and energy sources for the attendant heterotrophic communities. From these considerations one may develop models of complicated and diverse ecosystems, protected from harsh surface conditions, and evolving successfully in the planetary interior. These potential habitats are based on aquatic biochemistry in salty ocean, ice, and on the seafloor.

Although the conditions at Europa and at the Earth differ significantly, extreme terrestrial ecosystems have been always considered and extensively analysed as useful models for astrobiology. There are two basic reasons for these considerations:

First, terrestrial ecosystems serve as (the only to date) basis for definition of life. Commonly, the Earth-centric view point implies the Earth-centric perspective for spreading of biological life. But the utimate goal of astrobiological search for life is to find an isolated extraterrestrial biosphere. One should agree with Conrad and Nealson (2001) who have claimed that we must define life in universal measurable terms and should have a clear idea of the end-member states for this search— what does life, or the absence of life look like at multiple spatial and temporal scales. The chemical composition, structures, energy consumption etc., have been proposed as examples of such universal characteristics of life. The totality of above characteristics has to be analysed in order to choose appropriate criteria, testifying life in putative Europan ecosystems.

Second, the considerable optimism regarding the habitability of Europa is to large extent based on the presence of strong tolerance of microbial and biochemical activity in Earth environments, which could be evaluated as possible Europan analogues. Prebiotic processes under hydrothermal conditions may have been important in the origin of terrestrial life (e.g., Mukhin, 1976; Cody *et al.*, 2000). There has been a lot of research on the ability of microorganisms to grow in extreme environments with respect to temperature, salinity, acidity, radiation, and pressure (Kargel *et al.*, 2000; Marion *et al.*, 2003). Polar investigations confirm that cold environments served as a refuge for life during long-term glaciations. Cold ecotopes are biochemically active at subzero temperatures, and represent a number of features characteristic for adaptation of living microbial populations to low temperatures (Abyzov, 1993; Vorobyova *et al.*, 1996, 1997; Priscu *et al.*, 1999; Priscu, 2002; Castello and Rogers, 2005; Gilichinsky *et al.*, 2007). In all cases active life at subzero temperatures occurs in presence of liquid water, maintained due to highly concentrated brines, or at water-mineral interface where the freezing temperature increases.

Several extreme ecosystems on the Earth have been proposed as potential Europa analogs for biomarkers' investigation, selection of life criteria, and validation of experimental approaches. These include lake and sea-ice biotopes, deep-ice cores, permafrost, and deep sea hydrothermal vents. The most intriguing ones are the lake Vostok in Antarctica, and chemosynthetic communities in the Gulf of Mexico at $\sim$ 540-m depth (Sassen *et al.*, 1999). In the Golf of Mexico oxidation of methane hydrates drives biochemical processes which support complex chemosynthetic communities and consumers as well. It is supposed that similar hydrothermal vents exist below the most recently resurfaced areas of Europa ice crust. These sites are the most promising for the search of extant life.

The particular feature of a putative Europan or other Jovian satellites' biospheres in comparision with Mars is that they could not be completely dormant (anabiotic). In contrast to Mars, Jovian satellites seem not suffered great cataclysms during their evolution, environmental conditions have not change drastically, and putative life developed through specific adaptations. If biosphere has emerged on Europa it should be active now.

Whether the life could have originated in the absence of sunlight? If so, the scientific rewards for successful search for life on Europa could hardly be greater. The Lander provide a unique possibility of finding traces of biotic or pre-biotic materials from the recent ocean outflows on the surface, which explains high science merit of such mission.

4. Science Goals of Europa Lander Mission and Possible Measurement Strategies

Scientific questions to be addressed by instruments on the surface of Europa and even by deep submarine expeditions were extensively discussed in the refereed literature (Chyba and Philips, 2002; Cooper *et al.*, 2002; Gershman *et al.*, 2003; Zimmerman *et al.*, 2005; Langmaier and Elliott, 2008 and the references therein). We repeat the most obvious science goals, effectively addressed by surface science.

The main scientific questions of Europa exploration can be formulated as follows:

- Is there a liquid water beneath the ice crust of Europa?
- Does the global ocean really exist on Europa?
- What is the depth and distribution of the icy crust?
- Are environmental conditions on Europa suitable for life?
- Are there traces of extinct life on Europa?
- Are there evidences of life on Europa at present time?

Clearly, remote studies from the orbit around Europa would not be sufficient to fully characterize surface environment and to address astrobiology goals. Measurements on the surface are important for geodesy and geology, to study the Europa ocean and to characterize locally the ice crust. Laplace-Europa Lander project will tackle the fundamental questions of internal structure, surface environment, and habitability of Europa. The main objectives of the mission will be to softly land on the chosen location, to collect multiple samples from one or multiple nearby sites to provide access to the shallow subsurface (to reach unaltered by radiation material). The landing site will be selected using the most accurate remote imagery to land on geologically youngest area, and to probe the material as recently exposed to the surface as possible. The following science and measurement objectives are being considered:

- To corroborate the theory of the liquid ocean, to characterize of the thickness and stiffness of the icy crust, to study the internal structure by means of different geophysical measurements; to characterize the seismicity of Europa, and to measure the magnetic field on the surface
- To conduct a detailed study of surface material, characterize physical (electrical and heat conductivity, stiffness, etc.) and chemical (pH, redox potential) parameters, analyze the composition of ice and admixtures, including isotopic ratios by means of gas chromatography with mass spectrometry (GCMS) and other methods;
- To characterize environment with particular attention to its capability to support life, and to search the traces of extinct or extant life in the surface and shallow subsurface (organic components, anions, cations, salinity, elements relevant to primary biological productivity, e.g. N, O, P, S, Mg, potential metabolism products), isotopic composition ($^{13}C/^{12}C$, $^{15}N/^{14}N$, etc.) at high sensitivity, by means of GSMS, Raman spectroscopy, other methods;
- To conduct observations and measurements in regional, local and micro scales, to study morphology and mineralogy of the surface and to validate remote orbital observations;
- To perform local measurements of radiation conditions, secondary ions, exosphere of the satellite and volatiles near the surface (CH_4, NH_3, CO_2, etc.)

A list of potential experiments to be considered on the surface of Europa, and their relevance to three major classes of science goals: Conditions, Composition, and Habitability is presented in Table 1. This list comprises many experiments with duplicated science goals, and should not be considered as a model payload, but rather as a long list of potential candidates to the model payload. Most of these methods contribute to the

Table 1. A list of possible instruments on the surface of Europa from a lander (adapted from Korablev *et al.*, 2010).

Instrument	Conditions	Composition	Habitability	Prototype	Mass (estimated)
Seismometer	•		○	OPTIMISM/Mars 96	495g +electronics
Gravimeter	•		○	GRAS/Phobos 11	250g
Tiltometer	•		○	Huygens	(300g)
Magnetometer	•	○	○	MMO Bepi Colombo	770g
TV camera set	•	○	○	CIVA/Rosetta; Phobos 11	1200g
Optical microscope	•	•	○	Beagle-2; Phobos 11	300g
IR spectroscopy	•	•	○	No direct prototype; technique well established	(2000g)
IR close-up spectrometer	•	•	•	CIVA/Rosetta MicrOmega/ExoMars	(1000g)
GCMS	○	•	•	GAP/Phobos 11; COSAC/Rosetta	(5000g)
Wet chemistry set		•	•	Urey/ExoMars[1]	2000g
Immuno-arrays		○	•	SOLID/ExoMars[1]	(1000g)
ATR spectroscopy	○	•	•	MIMA/ExoMars[1] for FTS analyzer	(2000g)
Raman spectroscopy	○	•	•	RAMAN-LIBS/ExoMars[1]	1100g[2]
LIBS	○	•	○	RAMAN-LIBS/ExoMars[1]	1100g[2]
Laser-ablation MS	○	•	○	LASMA/Phobos 11	1000g
XRF	○	•	○	APXS/Rosetta	640g
XRS	○	•	•	No prototype	(2000g)
XRD[3]	•	○		XRD/ExoMars	1200g
Various sensors	•	○	○	MUPUS/Rosetta	2350g
Radiation dose	•		○	RADOM/Chandrayaan-1	100g

Notes:
• =direct support; ○ =indirect support
[1] Experiments considered for ExoMars, but presently excluded from science package
[2] Target mass for both Raman and LIBS
[3] XRD experiment makes sense with a sample of pristine ice material from depth, non-altered, or little altered by radiation and meteorites.

assessment of the habitability of Europa, and many chemical analysis experiments have a high potential for biochemical detection of life (see Table 1). Specific life-contrasting tests might include isotopic ratios (GCMS and TDLAS, but likely concentration needed); chirality (difficult detection by UV methods, Raman, could also be assessed by GCMS); wet chemistry set, and immuno-arrays. A strategy to assess the feasibility of these experiments and a proper balancing between instruments proposed for a direct search of life and instruments for standard/advanced in-situ chemical and physical characterization of Europa is to be developed.

The most trustworthy experiments to be put on the surface of Europa identified so far are:

- seismometer, to estimate the thickness of the ice,
- a set of sensors for physical characterization
- chemical analytic package with high exobiology potential, based on GCMS
- IR spectrometer to link orbital and surface measurements
- a set of cameras, and microscopes

Three method to sample the surface, which are potentially compatible with the resources allocation of the lander are: the robotic arm/grinder, a drill to reach the depth of several tens of cm and to deliver the sample into the lander, and a melting probe with a mass of ~ 5 kg (1 kg instrument) to reach the depth of ~ 3 meters.

The science goals of the orbital element of the Laplace-Europa Lander mission are still to be considered. It is reasonable to duplicate some key investigations of NASA JEO, e.g., high-resolution imaging for landing site selection, lidar to characterize the figure of the satellite, near-IR mapping spectroscopy for surface composition, possibly a long-wave penetrating radar to map the thickness of the ice crust. A number of complementary measurements will be considered, to characterize in situ the ion and neutral composition of Europa environment, to measure the radiation dose, and to perform remote studies of Jupiter and other satellites.

5. Mission concept

A number of missions to explore Europa have been studied during the past decade. A comparison of mission architectures is presented in NASA report by Langmaier and Elliot (2008). This study has concluded that the most of the science at lowest mission risk can be achieved with a powerful science Europa orbiter. Such a Flagship-class mission, Europa Explorer (Clark *et al.*, 2007) has become the basis of NASA share of the EJSM. However, previous studies of large Jupiter system missions included a number of surface elements (Greeley and Johnson, 2004).

An analysis of mass allocation for various landing strategies shows (Balint, 2004) that a classical soft lander is the most advantageous in terms of mass. A number of hard penetrator solutions are being considered for Europa (Gowen *et al.*, 2010). However, the high-velocity penetrator concept for an airless body is yet to be demonstrated. Russian Laplace-Europa Lander mission includes a soft lander (total mass of 1210 kg), and a small telecommunication and science orbiter (395 kg) (Zelenyi *et al.*, 2010a). Comparing with many previous landers with mass below $\sim$ 400 kg, Laplace-Europa Lander is relatively large and may afford a considerable science package.

Two variants of the interplanetary cruise are possible. First, a conventional scheme, including chemical propulsion and a series of gravitational maneuvers around Venus and the Earth has been considered. Alternatively, electric propulsion during the heliocentric cruise coupled with a single gravitational maneuver near the Earth can be employed. In either case the mass of filled transport modules (propulsion system) for such class of mission takes up to 80–85% from total mass of the spacecraft.

To enter into the orbit around Europa in the vicinity of Jupiter a series of gravitational maneuvers near Galilean satellites is required in order to save propellant mass. Extremely strong radiating belts of Jupiter have to be taken into account. Lengthy approach to the planet is unacceptable due to enormous cumulated radiation dose, destroying the subsystems of the spacecraft. The trajectory in the vicinity of Jupiter should be chosen in order to minimize the stay within Europan orbit, and to exclude whenever possible entering within the Io orbit. Estimations of charged particle fluxes and radiation doses under various shielding in different parts of the trajectory were made using different empirical models at each stage of the computations (Podzolko *et al.*, 2009). The chosen sequence of gravitational maneuvers is shorter than two years. This stage is completed with the insertion into a circular polar orbit around Europa.

The calculations have shown, that even a heavy-class launch vehicle Proton with upper stage booster Breeze-M does not allow sufficient mass for both Orbiter and Lander to be delivered to Europa using chemical propulsion, even using multiple gravitational maneuvers. It is feasible, however when using the electric propulsion during the interplanetary cruise associated with gravitational maneuver near the Earth. In either case, a heavy-class launch vehicle is required to carry out the spacecraft (SC) to the escape trajectory.

This scenario, using the electric propulsion and one gravitational maneuver near the Earth is accepted as the basis for the further analysis. The ballistic scheme of mission consists of following basic stages:

- Transfer to low earth orbit (200 km) by Proton launcher
- Acceleration to interplanetary trajectory by upper stage Breeze; jettisoning of Breeze;
- Earth-Earth cruise using electric propulsion;
- Gravitational maneuver at the Earth;
- Earth-Jupiter cruise using electric propulsion; jettisoning of electric propulsion module;
- Breaking in the sphere of Jupiter attraction and insertion into initial high-apogee orbit;
- Increasing the pericenter altitude to Ganymede orbit;
- Multiple rendezvous with Ganymede and Callisto to reduce the relative speed of approach to Europa;
- Insertion into a circular orbit around Europa with an altitude of 100 km; jettisoning of the braking propulsion unit;
- Orbital flight; separation of the landing module;
- Deceleration of the landing module, and landing.

Once the spacecraft arrives to Jupiter system, it approaches Jupiter (10^5 km) and enters into an initial orbit. During the first approach the orbit is inclined at 40° with respect to the equator plane to minimize harmful influence of radiation belts of Jupiter. This inclination is compensated by an impulse in apocenter, and the spacecraft enters in a transfer orbit with a pericenter of 9×10^5 km, and an apocenter of 2×10^7 km in the plane of Galilean satellites. During further flight at resonance orbit a series of rendezvous with the satellites is conducted, allowing to reduce propellant required for the insertion into the orbit round Europa. This stage will last ~ 23 months (see Figure 1).

From this orbit remote studies of the surface will be conducted, and the landing site meeting certain topography conditions will be chosen. The landing module will be separated to perform an active soft landing onto the surface. The orbital module remains on the orbit and serves as a relay for the lander.

The landing is performed in two stages. First, the velocity of the landing module is reduced, than measurements of the altitude are performed, and the inertial speed of descent is defined. Second, the lander is rendered along the local vertical, and the vertical speed is reduced from about 3 km/s to nearly zero at about 20 m above the surface.

6. Mission design

The mass breakdown of the Europa mission is given in Table 2. The spacecraft (see Figure 1) consists of four following basic elements:

- The electric propulsion transport module;
- The braking module.
- The orbiter;
- The lander.

The electrorocket transport module provides acceleration and deceleration during the heliocentric part of the flight. It is also responsible for the attitude control and stabilization of the spacecraft during this period. The traction is provided by means of eight plasma engines SPD-140 with a thrust of ~0.17 N each, and a specific impulse of 28000 m/s. The xenon is used as the propellant. The engines are integrated in four blocks, each of them includes two SPD-140 engines and a 1-axis deployment/attitude control mechanism. The velocity vector is controlled aligning simultaneously the orientation of the four engine blocks, and the solar panels. Measures are taken to prevent contamination of the solar panels, and the spacecraft itself with the plasma exhaust. The electric propulsion transport module with the solar panels is jettisoned from the spacecraft before the arrival to the Jupiter system.

Braking propulsion system module provides corrections during the cruise, and the braking impulses near Jupiter and the forming the orbit around Europa. The propulsion system module consists of two propellant tanks, the main cruise engine, four thruster units; it includes also valves and constructional elements. Four identical spherical tanks with a capillary intake are used; two are intended for the oxidizer, and two for the fuel. The cruise propulsion system consists of four engines with the general traction of ~1600 N and a specific impulse of 3000 m/s. The thruster units provide operating moments for attitude control and stabilization, and also velocity impulses for small trajectory corrections. Each block consists of four engines with traction of 50 and one engine with traction of 10 .

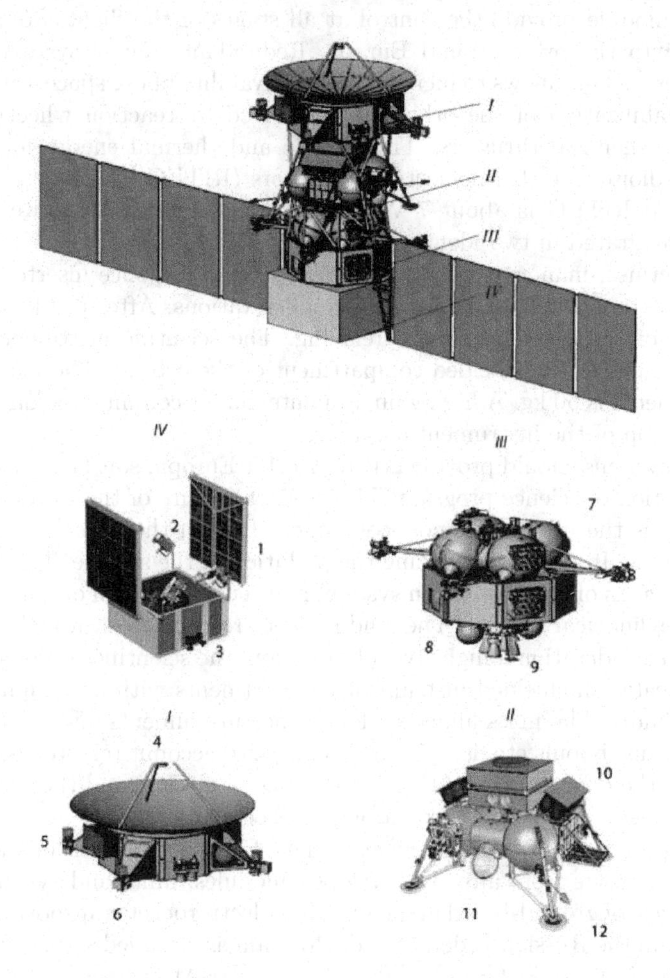

Figure 1. Europa Lander mission elements : I.— Orbiter: 4— High-gain antenna, 5— attitude control thrusters, 6— nuclear power units; II.— Landing module: 10— nuclear power units, 11— science payload compartment; 12— landing system; III.— Braking propulsion system: 7— propellant tanks, 8— structure, 9— main engine; IV.— Electroroket propulsion system: 1— solar panels, 2— electrojet engines, 3— attitude control thrusters. Xenon tanks are inside.

Table 2. Estimated mass budget of the Europa mission employing electrorocket trust during interplanetary cruise. Upper stage Breeze booster is not included.

Unit	Mass, kg
Orbital module	395
Landing module	550
Braking propulsion system module	385
Electrorocket propulsion system module	860
Intermediate structure	70
SC without propellant	**2260**
EPS module propellant (xenon)	1435
Propulsion system module propellant	2005
Landing module propellant	660
SC with propellant	**6360**

The orbital module is the principal structural element of the spacecraft. The systems of the orbital module provide the control at all stages of the flight, from cruise trough the insertion into the orbit around Europa. Redundant control system of the lander remains in reserve. This allows to increase the survivability of the spacecraft. The attitude control and stabilization of the orbiter is supported by reaction wheels, and a set of one-component chemical thrusters. The electric and thermal energy for the orbiter is provided by radioisotope thermoelectric generators (RITEG). Electric power generated by one standard RITEG is about 7 W. To provide for necessary power of 210W, the RITEGs are integrated in two identical blocks (15 RITEGs in each).

During the cruise, maneuvers in the Jupiter system, and once inserted into the orbit around Europa the orbiter serves for science observations. After the lander is released, the main goal of the orbiter turns to relaying. The scientific instruments and service systems are located in the shielded compartment of the orbiter. The mass allocated for science instruments is 50 kg. A high-gain 4-m antenna for communication with the Earth is fixed on the top of the instrument container.

The lander systems should provide ΔV to deorbit Europa, soft landing on the surface, and the realization of science program. The distinct feature of the Europa lander (w.r.t. Lunar landers) is the use of the monocomponent fuel on the base of hydrazine for the main brake engine. It allows minimizing the pollution of the surface. The lander includes a block of fuel tanks of the propulsion system, realized in the form of four spherical tanks connected by cylindrical spacers. The landing feet are fixed to each fuel tank.

Taking into consideration a high level of radiation, the scientific equipment and service systems are located in shielded instrument compartments with aluminum alloy walls of > 10 mm thickness. The mass allocated for science instruments and related subsystems (sampling systems, booms etc. is ~ 50 kg. Taking into account remoteness from the Sun, the lander like the orbiter uses RITEG as a source of electric and thermal energy. Data exchange with the Earth is provided through the orbiter.

The heritage from previous developments. The Europa's Lander mission will extensively use the heritage from already developed modules, units and systems which have passed a full cycle of ground-based and flight. The electrorocket transport module is being developed within the Russian Federal Space Program as a unified service framework. The first flight opportunity will be in the frame of Electro-M geostationary meteo-satellite (2014). It will also be used for Interheliozond mission for solar studies, also targeting the launch date in 2014.

Most of the heritage comes from developments within the frame of Phobos Sapmle Return Mission (Phobos-Soil) in preparation for the launch in 2011 (Zelenyi *et al.*, 2010b). The orbital module and the propulsion system are already developed and are the part of Phobos-Soil. Recently they have passed a full cycle of qualification. The same type of orbital module will be used for the Luna-Globe (2013) project and is planned as a basis of Mars-Net (2016) and Venera-D (2018) orbiters.

Russia has a long experience in automated landing on the Moon and planets. A number of successful landings on the Moon have been realized from 70th to 80th years. Algorithms of landing are developed and will be renewed during another lunar mission Luna-Resource, which includes landing.

Nevertheless, a number of the critical technologies has to be developed at the earliest stages of the project. A particular attention will be paid to electronic components tolerant to extremely high radiation doze. Radioisotope thermoelectric generators which developed now in the frame of Luna-Resource program, have to be qualified for application on Europa. A lot of ground experiments is required to test the technology of low-temperature of ice drilling.

References

Abyzov, S. S. 1993, In *Antarctic Microbiology* (Ed. E. Friedmann), Wiley & Sons, New York, USA, pp. 265–295.

Anderson, J. D., Schubert, G., Jacobson, R. A., Lau, E. L., Moore, W. B., & Sjogren, W. L. 1998, *Science*, 281, 2019–2022.

Balint, T. 2004, *JPL report D30050*, 60 p. NASA JPL/California Institute of Technology, Pasadena, CA, September 22

Billings, S. E. & Kattenhorn, S. A. 2005, *Icarus* 177, 397–412

Blanc, M., Alibert, Y., Andre, N., *et al.* 2009, *Experimental Astronomy*, 23, 849–892

Castello, J. D. & Rogers., S. O. 2005, *Life in ancient ice*, Princeton Univ. Press. 307 p

Chyba, C. & Sagan, C. 1992, *Nature*, 355, 125–132

Chyba, C. F. 2000, *Nature*, 403, 381–382

Chyba, C. F. & Phillips, C. B. 2002, *Europa as an Abode of Life. Origins of Life and Evolution of the Biosphere*, 32, 47–68

Clark, K., Greeley, R., Pappalardo, R., & Jones, C. 2007, *Europa Explorer Mission Study: Final Report, JPL report D41283.*

Clark, K., Stankov, A., Pappalardo, R. T., Greeley, R., Blanc, M., Lebreton, J.-P., & van Houten, T. 2009, *The Europa Jupiter System mission, 40th Lunar and Planetary Science Conference*, March 23–27, The Woodlands, Texas, id.2338

Cody, G. D. *et al.* 2000, *Science*, 289, 1337–1340.

Colburn, D. S. & Reynolds, R. T. 1985, *Icarus*, 63, 39–44

Conrad, P. G. & Nealson, K. H. 2001, *Astrobiology*, 1, 15–24

Cooper, J. F., Philips, C. B., Green, J. R., Wu, X., Carlson, R. W., Tamppari, L. K., Terrile, R. J., Johnson, R. E., Eraker, J. H., & Markis, N. C. Europa exploration: science and mission priorities, The future of solar system exploration, 2003–2013, *ASP conference series*, v.272, M. V. Sykes, ed., 2002.

Figueredo, P. H. & Greeley R. 2004, *Icarus*, 167, 287–312

Hand, K. P. & Chyba, C. F. 2007, *Icarus*, 189, 424–438

Hoppa, G. V., Tufts, B. R., Greenberg, R., & Geissler, P. E. 1999, Science, 285, 1899–1902

Hussmann, H., Spohn, T., & Wieczerkowski, K. 2002, *Icarus*, 156 (1), 143–151.

Gershman, R. 1999, Conceptual design of a Europa Lander mission, *IEEE Aerospace conference*, v.1, pp. 421–431

Gershman, R., Nilsen, E., & Oberto, R. 2003, *Acta Astronautica*, 52, 253–258

Gilichinsky, D. A., Wilson, G. S., Friedmann, E. I., *et al.* 2007, *Astrobiology*, 7, 275–311

Gowen, R. A., Smith, A., Fortes, A. D., *et al.* 2010, *Adv. Space Res.*, in press.

Greeley, R., Sullivan, R., Klemaszewski, J. *et al.* 1998a, *Icarus*, 135, 4–24

Greeley, R., Sullivan, R., Coon, M. D., Geissler, P. E., Tufts, B. R., Head, J. W., Pappalardo, R. T., & Moore, J. M. 1998b, *Icarus*, 135, 25–40

Greeley, R. & Johnson, T. 2004, *Report of the NASA Science Definition Team for the Jupiter Icy Moons (JIMO)*, Technical Report, National Aeronautics and Space Administration, February 13

Greenberg, R., Hoppa, G. V., Tufts, B. R., Geissler, P., Riley, J., & Kadel, S. 1999, *Icarus*, 141, 263–286

Greenberg, R., Leake, M. A., Hoppa, G. V., & Tufts, B. R. 2003, *Icarus*, 161(1), 102–126

Irwin, L. N. &. Schulze-Makuch, D. 2003, *Astrobiology*, 3, 813–820

Johnson, T. V. & McCord, T. B. 1971, *Astroph. J.*, 169, 589

Kargel, J. S., Kaye, J., Head, J. W. III, Marion, G. M., Sassen, R., Crowley, J., Prieto, O., Grant, S. A., & Hogenboom, D. 2000, *Icarus*, 148, 226–265

Kivelson, M. G., Khurana, K. K., Russell, C. T., Volwerk, M., Walker, R. J., & Zimmer, C. 2000, *Science*, 289, Is. 5483, 1340–1343

Korablev, O., Gerasimov, M., Dalton, J. B., Lebreton, J.-P., Hand, K., & Webster, C. 2010, Submitted to *Ads. Space Res.*

Kuiper, G. P. 1957, *Astron. J.*, 62, 295–295

Kuskov, O. L. & Kronrod, V. A. 2005, *Icarus*, 177, 550–569

Langmaier, J. & Elliott J. 2008, *JPL publication* 08–1, January

Leith, A. & McKinnon, W. 1996, *Icarus*, 120, 387–398

Marion, G. M., Fritsen, C. H., *et al.* 2003, *Astrobiology*, 3, 785–811

Matousek, S. 2007, *Acta Astronautica*, 61, 932–939

McCord T. B., G. B. Hansen, F. P. Fanale, *et al.* 1998, *Science*, 280, 1242–1245

McEwen, A. S. 1986, *Nature*, 321, 49–51

Mileikowsky, C., Cucinotta, F. A., Wilson, J. W., *et al.* 2000, *Planet. Space Sci.*, 48, 1107–1115

Moroz, V. I. 1965, *Astronomicheskii Zhurnal*, 42, 1287

Mukhin, L. M. 1976, *Origins of Life* 7, 355–368

Mukhin, L. M., Gerasimov, M. V., & Safonova, E. N. 1989, *Nature*, 340, 46–48

Nansen, O. L. 1975, *Icarus*, 26, 24–29

Ojakangas, G. & Stevenson, D. 1989, *Icarus*, 81, 242–270

Pappalardo, R. T., Belton, M. J. S., Breneman, H. H., *et al.* 1999, *J. Geophys. Res.*, 104, E10, 24015–24056

Pappalardo, R. T., Clark, K., Greeley, R., *et al.* 2007, Europa Explorer: A Mission to Explore Europa and Investigate Its Habitability, DPS meeting #39, #28.02; *Bull. American Astronomical Soc.*, 39, 465

Phillips, C. B., McEwen, A. S., Hoppa, G. V. *et al.* 2000, *J. Geophys. Res.*, 105, 22579–22598

Pierazzo, E., Chyba & C. F. 2002, *Icarus*, 157, 120–127

Pilcher, C. B., Ridgway, S. T., & McCord, T. B. 1972, *Science*, 178, 1087–1089

Podzolko, M. V., Getselev, I. V., Gubar', Yu. I., & Veselovsky, I. S. 2009, *Solar System Research*, 43, 116–120

Pollack, J. B., Witteborn, F. C., Edwin, F. E., Strecker, D. W., Baldwin, J. B., & Bunch, B. E. 1978, *Icarus*, 36, 271–303

Prieto-Ballesteros, O., F. Gomez *et al.* 2010, *Adv. Space Res.*, submitted

Priscu, J. C. *Perennially ice covered Antarctic lakes: oases for life and models for other icy worlds. Astrobiology in Russia workshop, March 25–29, 2002*, St. Petersburg, Russia, p. 116–130

Priscu, J. C., Adams, E. E., Lyons, W. B., Voytek, M. A., Mogk, D. W., Brown, R. L., McKay, C. P., Takacs, C. D., Welch, K. A., Wolf, C. F., Kirshtein, J. D., & Avci, R. 1999, *Science*, 286, 2141–2144

Reynolds, R. T., Squyres, S. W., Colburn, D. S., & McKay, C. P. 1983, *Icarus*, 56, 246–254

Russell, C. 2003, The Cassini-Huygens Mission, Kluwer

Sassen, R., S. Joye, S. T. Sweet, D. A. DeFreitas, A. V. Milkov, & I. R. MacDonald 1999, *Organic Geochem.*, 30, 485–497

Schenk, P. M. 2002, *Nature*, 417, 419–421

Schenk, P. M., McKinnon W. B. 1989, *Icarus*, 79, 75–100

Schenk, P., Matsuyama, I., & Nimmo, F. 2008, *Nature*, 453, 368–371

Schilling, N., Khurana, K., & Kivelson, M. G. 2004, *J. Geophys. Res.* 109, doi:10.1029/2003JE002166

Sieger, M. T., Simpson, W. C., & Orlando, T. M. 1998, *Nature*, 394, 554–556

Showman, A., & Han, L. 2004, *J. Geophys. Res.* 109 (E1), CiteID E01010

Stevenson, D. J. 2000, *Science*, 289, 1305–1307

Tobie, G., Choblet, G., & Sotin, C. 2003, *J. Geophys. Res.* 108 (E11), 10–1, doi:10.1029/2003JE002099

Vorobyova E., Soina V., Gorlenko M., *et al.* 1997, *FEMS Microbiol. Rev.*, 20, 277–290

Vorobyova, E. A., Soina, V. S., & Mulyukin, A. L. 1996, *Adv.Space.Res.*, 15, 103–108

Williams, K. K., & Greeley, R. 1998, *Geophysical Res. Lett.*, 25, 4273–4276

Whitman, W. B., Coleman, D. C., & Wiebe, W. J. 1998, *Proc. Natl. Acad. Sci. U.S.A.*, 95, 6578–6583

Zahnle, K., Schenk, P., Levison, H., & Dones, L. 2003, *Icarus*, 163(2), 263–289

Zahnle, K., Alvarellos, J. L., Dobrovolskis, A., & Hamill, P. 2008, *Icarus*, 194(2), 660–674

Zelenyi, L. M., Zakharov, A. V., Polischuk, G. M., & Martynov, M. B. 2010a, *Solar System Res.*, 44, 15–25

Zelenyi, L., Korablev, O., Martynov, M., *et al.* 2010b, Submitted to *Adv. Space Res.*

Zimmer, C., Khurana, K. K., & Kivelson, M. G. 2000, *Icarus*, 147(2), 329–347

Zimmerman W. F., James Shirley, Robert Carlson, Tom Rivellini, Mike Evans, Europa Small Lander Design Concepts, American Geophysical Union, Fall Meeting 2005, abstract #P54A-08, 2005.

Galileo's Medicean Moons: their impact on 400 years of discovery
Proceedings IAU Symposium No. 269, 2010
C. Barbieri, S. Chakrabarti, M. Coradini & M. Lazzarin, eds.
doi:10.1017/S1743921310007349

Atmospheric moons Galileo would have loved

Sushil K. Atreya

Department of Atmospheric, Oceanic, and Space Sciences,
University of Michigan, Ann Arbor, MI 48109-2143, USA
email: atreya@umich.edu

Abstract. In the spirit of the symposium and the theme of the session of this presentation, "Our solar system after Galileo, the grand vision," I review briefly a relatively recently discovered phenomenon in the solar system – existence of atmospheres on certain moons, including Io, one of the four moons Galileo discovered four centuries ago. The origin of such atmospheres is discussed, and comparisons are made between various *gassy* moons.

Keywords. Solar system: general, planets and satellites: individual (Enceladus, Ganymede, Io, Jupiter, Moon, Saturn, Neptune, Titan, Triton, Uranus), Sun: abundances

---

1. Historical Perspective

The night of January 8, 1610 was Galileo Galilei's eureka! moment. On this night, Galileo realized that the objects he had observed in close proximity to Jupiter the previous night with his "occhiale" (telescope) were not fixed stars, but were actually the moons of Jupiter, for they had moved and their configuration around Jupiter had changed substantially. With observations during the course of next seven days, Galileo confirmed without any doubt the presence of four moons orbiting Jupiter. Galileo named them the "Medicean Moons," after the Medicis of Florence, and now known as the Galilean Moons. On March 13, 1610, two months after the observations, Galileo published his findings in *Sidereus Nuncius* that is generally translated as Starry Messenger, or sometimes Starry "Message", apparently referring to Galileo's response to Father Orazio Grassi who had accused Galileo of pretending to be a herald from heaven after seeing the title of the book. With the observations of the moons of Jupiter, Galileo had placed the Copernican hypothesis of heliocentric system on a firm footing. It marked arguably the most monumental turning point for humanity.

With his telescope Galileo went on to reveal a multitude of other mysteries of the heavens, including but not limited to the "seas" and craters on the moon, full planetary phases of Venus and the sunspots not being satellites of the Sun, but no celestial body baffled him so much as Saturn. When Galileo first observed Saturn in 1610, it appeared to him as if the planet was surrounded by two large "moons," giving an appearance of cup handles (Figure 1), unlike anything he had seen at Jupiter. Two years later, he found the moons had disappeared! In 1616, they had taken on the shape of "arms" around Saturn (Figure 1)! With his curiosity, Galileo's frustrations grew, but his twenty-magnification telescope lacked the quality for resolving the objects around distant Saturn. Four decades later in 1655, Christiaan Huygens, a Dutch optical physicist and a great admirer of Galileo, finally resolved the mystery using a far better quality telescope – what Galileo had observed around Saturn between 1610 and 1616 were disk-shaped rings (Huygens would discover their true shape in 1659). In the same year, 1655, Huygens also discovered a large moon of Saturn, Titan.

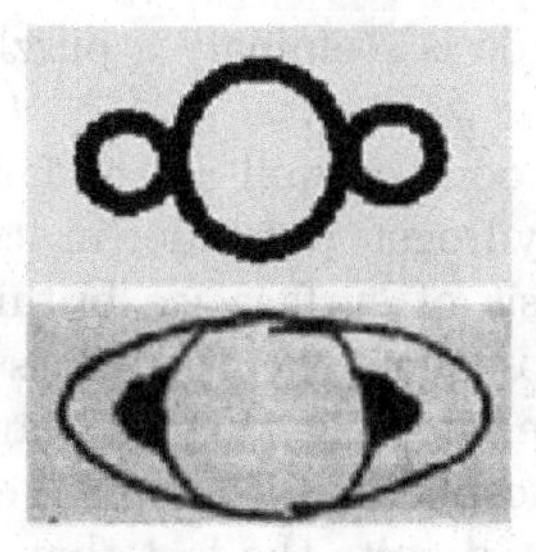

Figure 1. Galileo's observations of Saturn, showing two large "moons" in 1610 (top) and "arms" in 1616 (bottom).

2. Gases on Io, and lack of them on the other Galilean moons

Although Galileo's attempts to discover moons around Saturn were unsuccessful and must have been disappointing to him, it is not too outrageous to imagine he must have been convinced of their existence, especially after discovering the moons of Jupiter, and his many attempts at unraveling the mysterious changing nature of the objects around Saturn. He must have wondered of the make-up of the moons of Jupiter and Saturn. Even more important than Galileo's successes and failures at demystifying the solar system is the revolution he had started, by pointing a new way of looking at the solar system, the universe, our place in the cosmos. So, it was only a matter of time before more mysteries of the cosmos would be revealed by more and more innovative means. In this paper, I will attempt to summarize our understanding of one such mystery – the presence of gases around just a handful of the more than one hundred known moons in the solar system. Only Jupiter's moon, Io, Saturn's moon, Titan, Neptune's moon, Triton, have bound atmospheres, whereas Saturn's moon, Enceladus, shows plumes of volatiles (Table 1). In addition, traces of molecules are found near the surfaces of Europa and Ganymede. Since many chapters in this book are devoted to the Galilean moons, here only a brief discussion of their atmospheric aspects is presented.

Unbeknown to him then, Io, one of the moons that Galileo discovered around Jupiter on January 7, 1610, turns out to be the most volcanically active object in the solar system, with over 400 active volcanoes, spewing ashes and gases as high as 500 km above Io's surface (Figure 2), as revealed first by the Voyager spacecraft nearly four centuries after Galileo's epic discovery of Io. Io's prolific volcanism results from the tidal heat generated within the moon's interior due to its orbital eccentricity. Io's 2:1 mean orbital motion resonance with Europa (Io orbits Jupiter twice for every orbit of Europa) and 4:1 resonance with Ganymede's mean orbital motion maintain a slight eccentricity of 0.0041 in Io's orbit, without which the geologic activity of Io would be minimal. The same orbital resonance causes a slight eccentricity in Europa's orbit also, resulting in tidal heating in the interior of this moon. As a consequence of this heating, models predict possible presence of a salty water ocean beneath Europa's icy crust. Observations by the Galileo spacecraft from 1995 to 2003 revealed an extraordinarily young and smooth surface with few impact craters (Figure 2), which appears to be the result of flows of warm ice from the interior through cracks and fissures in Europa's surface. On the other hand, impact craters are common on Ganymede and Callisto (Figure 2), as these moons experience progressively smaller tidal heating than the two inner Galilean moons Europa and Io. The smaller tidal heating is also one reason why Callisto, Ganymede and Europa have little volatiles above their surfaces. Fueled by incessant volcanism, a tenuous atmosphere of sulfur dioxide with a pressure of approximately a billionth of the Earth's atmosphere (Table 1), or equivalent to that between 110 km and 160 km above Earth's surface, is formed on Io. However, Ganymede is seemingly a puzzle, especially when compared to Saturn's moon Titan.

The largest of the Galilean moons as well as the largest moon in the solar system, Ganymede, has only traces of hydrogen peroxide (in ice), oxygen and ozone molecules (Table 1), produced by radiolysis of surface ice by high energy charged particles in Ganymede's environment and subsequent chemistry. It is then quite striking that Saturn's moon Titan, which is roughly the same size (Figure 2) and mass as Ganymede (Table 1), is endowed with a massive atmosphere, with even greater pressure than that at the Earth's surface (Table 1). This is despite the fact that Titan was born in a colder part of the solar nebula than Ganymede. In fact, that may have helped Titan. The presence of an atmosphere on the relatively small bodies such as moons depends on a number of interrelated factors, particularly the acquisition, retention, and then the accumulation of the volatiles. The warmer planetesimals that formed Ganymede may not have trapped sufficient quantities of the relevant volatiles to begin with unlike the Titan-forming planetesimals, since the trapping efficiency of certain volatiles (see below) increases at lower

Figure 2. Images of the four Galilean moons alongside Jupiter, together with Saturn's moon, Titan (lower right). Io (top) is the innermost of the four Galilean moons, followed by Europa, Ganymede and Callisto. The image of Io, showing volcanic eruption, was taken by the New Horizons spacecraft during its flyby of Jupiter in 2007, while the images of the other Galilean moons are from the Galileo spacecraft between 1995 and 2003. The image of Titan was taken by the Cassini Visual and Infrared Mapping Spectrometer in 2005.

Table 1. Key physical, orbital and environmental characteristics of *atmospheric* moons in the solar system.

Characteristic[a]	**Moon**	**Io**	**Ganymede**	**Titan**	**Enceladus**	**Triton**
R_{planet}	60.27 R_E	5.9 R_J	14.99 R_J	20.25 R_S	3.95 R_S	14.33 R_N
M $[10^{22} kg]$	7.35	8.94	14.82	13.5	0.0108	2.14
R_e $[km]$	1738	1815	2631	2575	252	1352
ρ $[kg/m^3]$	3340	3570	1936	1880	1608	2064
g $[m/s^2]$	1.622	1.796	1.43	1.35	0.11	0.78
T_s [days]	27.32	1.769	7.16	15.95	1.37	5.877
i [degrees]	5.9	0.04	0.20	0.33	0.02	157
e	0.05	0.0041	0.001	0.029	0.005	0.00
Albedo	0.12	0.63	0.4	0.2	0.99 (bond) 1.37 (geometric)	0.76
v_e $[km/s]$	2.38	2.56	2.75	2.64	0.237 ($v_e < v_{T\,(N_2)}$!)[c]	1.50
Surface T $[K]$	40 – 396	130[b]	110	94	114 – 157	38
Surface P $[mb]$	negligible	$(3 - 40) \times 10^{-6}$	negligible	1500	plume (unbounded)	0.016
Atmosphere	traces of H, He, Na, K, H_2O	SO_2	traces of O_2, O_3, H, H_2O_2 in ice	N_2, CH_4	H_2O, CO_2, CO, CH_4, NH_3, N_2	N_2, CH_4

[a] R_{planet} - mean radial distance between the moon and its parent body (E: Earth, J: Jupiter, S: Saturn, N: Neptune); M - mass; R_e - mean equatorial radius; ρ - mean density; g - equatorial surface gravity; T_s - sidereal period; i - orbital inclination; e - orbital eccentricity; v_e - equatorial escape velocity.
[b] Io's mean surface temperature. Temperature of some volcanoes, such as the lava eruptions near Pillan Patera can reach as high as 2000 K, much hotter than the hottest basaltic eruptions on Earth, presumably indicating silicate volcanism (McEwen *et al.* 1998).
[c] $v_{T\,(N_2)}$ - mean thermal velocity of nitrogen ranges from 0.26 to 0.31 km/s for temperatures of 114-157 K above Enceladus' tiger stripes, indicating that even heavy molecules such as nitrogen can overcome Enceladus' gravity and escape, as $v_{T\,(N_2)} > v_e$, implying that Enceladus cannot maintain a permanent atmosphere in the absence of continuous plumes.

temperatures. Although both Ganymede and Titan had nearly equal probability of retaining the gases, especially the heavier ones such as nitrogen, against escape, whatever little volatiles Ganymede did acquire, were subjected to an irreversible fatal fate, unlike Titan. Unlike Saturn, Jupiter's magnetospheric charged particle environment is intense, severe, and extensive. As a consequence, neither any significant processing to form stable atmospheric gases from the initial inventory of volatiles, nor their accumulation over time could occur at Ganymede, because of charged particle stripping of the atmospheric gases Ganymede might have mustered to form early on. What is then so unique about Ganymede's twin Titan, which succeeded in having a large atmosphere? We will attempt to gain an insight into its possible origin, evolution and fate.

3. Gases on the Saturnian moon Titan, and plumes on Enceladus

Molecular nitrogen (N_2) and methane (CH_4) comprise the bulk of Titan's atmosphere whose pressure is 1500 mb (Table 1). Nitrogen is approximately 95% by volume, and methane is roughly 5%. The atmosphere also has trace quantities of heavy hydrocarbons (acetylene, ethane, propane, benzene, etc.), nitriles (such as hydrogen cyanide), and the heavy noble gases – primordial and radiogenic argon (^{36}Ar and ^{40}Ar) – at parts per million by volume levels or less. In addition, Titan has an extensive haze extending from the lower atmosphere to the ionosphere above 900 km. The heavy hydrocarbon and nitrile molecules are formed by the methane-nitrogen photochemistry in the neutral atmosphere and the ionosphere above it. The same molecules go on to form the haze and soot on Titan (Wilson & Atreya 2004), similar to the photochemical smog in urban and industrial areas on Earth. However, unlike Earth, haze is key to maintaining the very atmosphere of

Table 2. Current protosolar elemental abundance[a], together with elemental enrichment factors at Jupiter and Saturn (updated from Atreya & Wong 2005).

Elements	Sun (protosolar)	Jupiter/Sun	Saturn/Sun
He/H	0.09705	0.807 ± 0.02	$0.567 - 0.824$[d]
Ne/H[b]	2.10×10^{-4}	0.059 ± 0.004	?
Ar/H[c]	3.62×10^{-6}	2.51 ± 0.50	?
Kr/H	2.14×10^{-9}	2.03 ± 0.38	?
Xe/H	2.10×10^{-10}	2.11 ± 0.40	?
C/H	2.75×10^{-4}	4.30 ± 1.03	9.3 ± 1.8[e]
N/H	6.76×10^{-5}	4.90 ± 1.87	$2 - 4$
O/H	5.13×10^{-4}	0.48 ± 0.17 (hotspot)	?
S/H	1.55×10^{-5}	2.88 ± 0.69	?
P/H	2.57×10^{-7}	1.21	$5 - 10$

[a] The protosolar elemental abundances are calculated from the present-day solar photospheric values of Grevesse *et al.* (2005; Table 1), after accounting for the effects of diffusion at the bottom of the convective zone on the chemical composition of the photosphere, together with the effects of gravitational settling and radiative accelerations, as discussed in Grevesse *et al.* (2005). The Grevesse *et al.* (2005) values represent an improvement over the earlier conventional standard compiled by Anders & Grevesse (1989). The new solar values result from the use of 3D hydrodynamic model of the solar atmosphere, non-LTE effects, and improved atomic and molecular data. The Jupiter results are those measured by the Galileo Probe Mass Spectrometer (GPMS).
[b] Grevesse *et al.* (2005) derived neon using oxygen as proxy. Direct Ne data from X-ray stars gives 2.7 times greater solar Ne (Drake & Testa 2005), so that Grevesse *et al.* (2005) solar Ne/H has been raised by 2.7, i.e. from 7.76 (-5) to 2.1 (-4). Thus, the Jupier/Sun Ne = 0.059 ± 0.004.
[c] Similar to Ne, Grevesse *et al.* (2005) derived solar argon using oxygen as proxy, which is inappropriate. Therefore, Ar has been changed back to meteoritic (or solar photospheric) value of 3.62×10^{-6} (Anders & Grevesse 1989) in this table.
[d] Based on reanalysis of Voyager data by Conrath & Gautier (2000). Subject to revision following analysis of Cassini/CIRS (Composite Infrared Spectrometer) data.
[e] Cassini/CIRS measurement (Flasar *et al.* 2005).

Titan, as discussed later. First, we discuss the origin of the main components of Titan's atmosphere, nitrogen and methane.

Titan's nitrogen: origin, evolution. Nitrogen on a planet or satellite can be primordial or secondary. If primordial, it was delivered as N_2 by the planetesimals that formed the object. If it is secondary, it formed from another chemical compound during or subsequent to the accretionary process. Neptune's moon Triton, and Pluto are an example of the former, as both Triton and Pluto formed in the part of the solar nebula where the temperatures were cold enough to trap N_2 directly into the icy planetesimals, i.e. below approximately 35 K. That was not the case for Titan, as the temperatures in the subnebula of Saturn where Titan was formed were much too warm to trap N_2. This claim was bolstered by the very small amount of primordial argon (^{36}Ar) detected in Titan's atmosphere by the Gas Chromatograph Mass Spectrometer (GCMS) on the Huygens probe of the Cassini-Huygens Mission (Niemann *et al.* 2005). Had nitrogen been delivered as N_2, Titan should have nearly a million times more ^{36}Ar than detected (Niemann *et al.* 2005; Atreya *et al.* 2009), as both argon and nitrogen require nearly similar low temperatures for trapping in ice. More specifically, the measured Ar/N_2 was found to be about 2×10^{-7} in Titan's present day atmosphere, instead of the solar $Ar/N_2 = 0.11$ (Table 2) it would have had if nitrogen was delivered directly as N_2 (the Ar/N_2 ratio would have been smaller by a factor of 3-5 in Titan's past compared to that measured today because Titan has lost a large amount of nitrogen via escape during its geologic history). Thus, similar to the nitrogen in the atmosphere of Earth and the other terrestrial planets, Titan's nitrogen atmosphere must be secondary in origin, forming from a compound, most likely ammonia (NH_3). Ammonia is also the principal reservoir of nitrogen at Saturn and Jupiter where it is found to be greatly enriched relative to the abundance based on the solar nitrogen elemental abundance (Table 2).

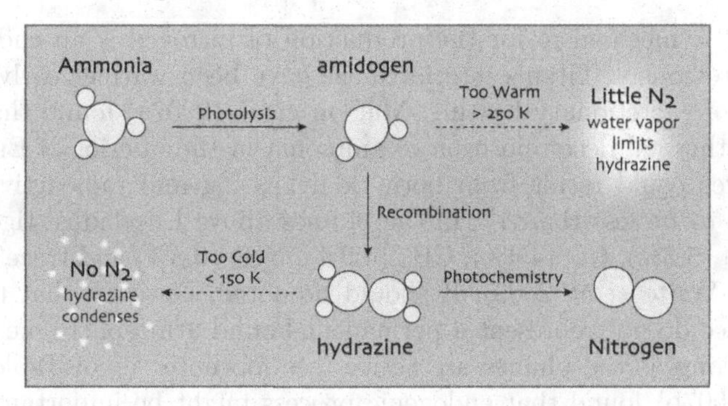

Figure 3. Pathways for production of nitrogen from photochemistry of ammonia.

Nitrogen was detected on Titan first by the Voyager flyby spacecraft in 1979, using the radio occultation technique. Even prior to that various authors had suggested possible presence of nitrogen based on Titan's formation models (review in Atreya *et al.* 2009). The models suggested ammonia as the source. Atreya *et al.* (1978) calculated a surface pressure of about 20 bars of N_2 on *primordial* Titan from their pre-Voyager photochemical model of ammonia (Figure 3). Voyager measured a surface pressure of 1.5 bar, which was subsequently confirmed by *in situ* measurements carried out by the GCMS on the Huygens probe. Titan's present nitrogen is believed to be a relic of a much larger primordial atmosphere in Titan's past as nitrogen has been escaping from the moon over geologic time. It is estimated that Titan started out with 3-5 times more nitrogen, hence a pressure of 5-8 bars of N_2 in its primordial past (Niemann *et al.* 2005), or within a factor of two of Atreya *et al.*'s predictions. However, photochemical production of nitrogen from ammonia could take place only when Titan was much warmer than today, as the vapor pressure of ammonia is negligible at Titan's current temperature. Models show that Titan was indeed much warmer during its accretionary heating phase. During that time its surface temperature is estimated to have been as high as 400 K, allowing copious ammonia and water vapor to be present in the atmosphere, together with methane. Atreya *et al.* (1978) and Atreya (1986) show that the most suitable window for photochemical production of nitrogen is between 150 K and 250 K.

For temperatures greater than 250 K, the saturation vapor pressure of water is appreciable so that reactions between water vapor and intermediate molecules take place that prevent subsequent reactions to continue to completion of nitrogen production (Figure 3). On the other hand, at temperatures below 150 K, although water vapor does not interfere with the ammonia-to-nitrogen chemistry because the H_2O vapor pressure is much too low, the key intermediate molecule of the ammonia photochemistry, hydrazine (N_2H_4), condenses, which practically chokes its decomposition by the solar ultraviolet photons, hence the failure to produce nitrogen from it (Figure 3). Using a radiative transfer model of Titan's primordial atmosphere (Adams 2006) and the photochemical chemical scheme of Atreya *et al.* (1978) and Atreya (1986), it is found that it would take approximately 30 million years to form a 10 bar atmosphere of nitrogen in Titan's past (Adams 2006; Atreya *et al.* 2009). This is well within the 100 million years or so Titan spent in the above suitable range of temperatures according to evolutionary models of Titan. Another important consideration in the estimate of the time scale for nitrogen production is the solar flux, which was much greater before the Sun reached its main sequence phase at 50 million years, as much as 1000 times greater in the ultraviolet range where the photodissociation of ammonia takes place. This allowed for a more efficient and vigorous production of nitrogen, and was factored into above estimate of time scale.

Another possible mechanism for the production of nitrogen is an endogenic process. As mentioned previously, Titan's interior must have been warmer early in its geologic history because of accretionary heating. Matson *et al.* (2007a) found that nitrogen can be formed from thermal decomposition of ammonia in the interior of Enceladus, where high temperatures could result from both tidal heating and radioactive decay (nitrogen is estimated to be less than 1% in the plumes above Enceladus' tiger stripes, with H_2O (90%), CO_2 (5.3%), CO (4.4%), CH_4 (0.9%), NH_3 (0.8%) and trace molecules comprising the rest, Waite *et al.* 2009). It should be noted, however, that the gases above Enceladus' surface do not represent a permanent bound atmosphere, i.e. they would be present only as long as the plumes are active (see footnote "c" of Table 1 for details). Matson *et al.* (2007b) found that endogenic process might be important at Titan also. We estimate as much as 15-20% of Titan's nitrogen could have been produced by such a process. Since the nitrogen produced in Titan's interior would most likely be stored as clathrate hydrate, it may be released from time to time when the clathrates are destabilized. This would lead to very different nitrogen isotope abundances on Titan than measured by the Huygens GCMS, raising concerns about the validity of the endogenic process (Atreya *et al.* 2009). Nevertheless, it is potentially an attractive mechanism for producing nitrogen supplementary to that from the inevitable photochemical process, but requires further work. Another idea is that comets supplied ammonia or nitrogen bearing organic molecules from which nitrogen was formed on Titan. However, the nitrogen isotope abundances on Titan do not bear any resemblance to the values in the comets for which the nitrogen isotope abundances are available (Atreya *et al.* 2009). The hydrogen isotope ratios (D/H) in Titan's hydrogen and methane molecules are also different, being about one half that in the water vapor of a handful of comets (only four) where it has been measured. However, this isotope ratio may not be a good indicator of the source of Titan's ammonia, since the comparison should really be made with the D/H in Titan's surface ice, but such measurements are presently lacking. Moreover, the D/H measurement needs to be carried out in many more and diverse comets than the current four from which it is available. Even when the appropriate data for Titan and comets become available, a word of caution about overinterpretation of the data is in order – isotope fractionation occurs on processing of the surface material by cosmic rays, photochemistry or reevaporation-condensation. Hence a true comparison of primordial isotopes may still be lacking. In any event, the cometary contribution to Titan is an idea that also needs to be developed further.

Titan's methane: significance, origin, evolution, cycle. The existence of a nitrogen atmosphere on Titan is critically dependent on the existence of methane gas in the atmosphere. Methane provides the much needed "greenhouse" warming to prevent nitrogen from condensing out of the atmosphere. In the stratosphere, it amounts to as much as 100 K, due to the absorption of the solar infrared flux by haze. Haze results from polymerization and condensation of the photochemical products of methane, particularly, polyynes, polycyclic aromatic hydrocarbons and nitriles in the neutral upper atmosphere (Wilson & Atreya 2004, 2009) and heavy organics in the ionosphere (Vuitton *et al.* 2009). In the troposphere, opacity induced by collisions of methane-nitrogen, nitrogen-hydrogen (from methane) and nitrogen-nitrogen molecules results in a net temperature increase of 12 K. Without above heating effects initiated by methane, Titan's atmosphere would shrink substantially due to the condensation of nitrogen.

The origin of methane on Titan is less certain than nitrogen. Methane may have been delivered directly as methane to Titan, or it could have formed on Titan from other carbon bearing entities. The lack of detection of krypton and xenon (Kr & Xe) by

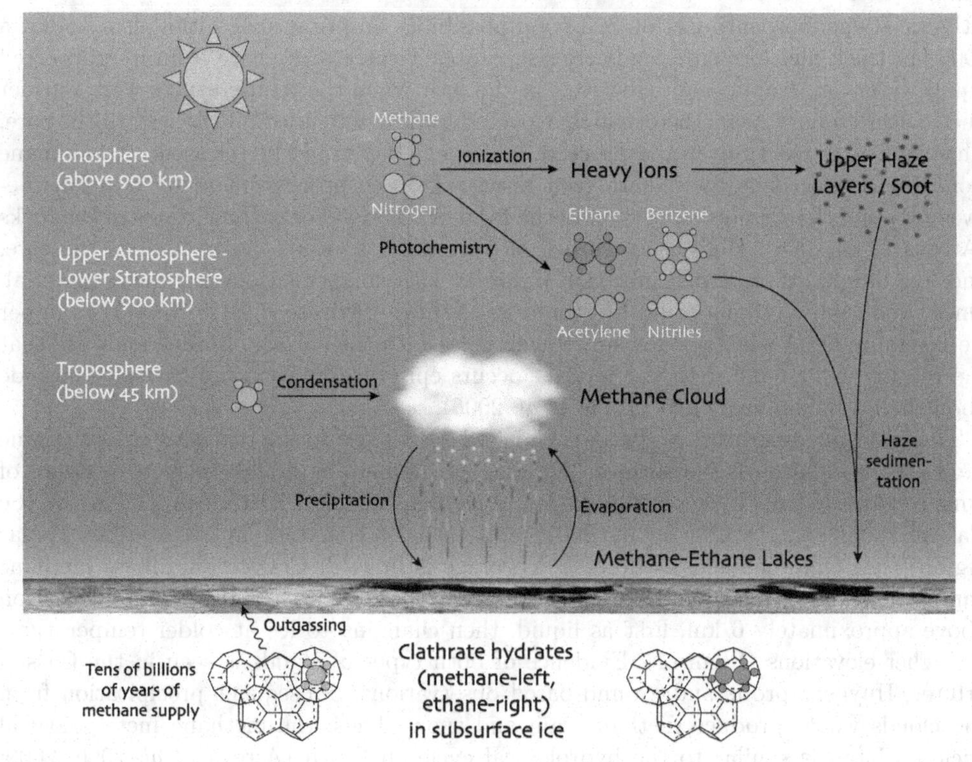

Figure 4. The cycle of methane on Titan. Parts of this figure were inspired by figures in the authors previous publications, including a figure on pp. 48-49 of Atreya (2007a), Figure 1 of Atreya (2007b) and Figure 3 of Lunine & Atreya (2008).

the Huygens GCMS seems to provide a clue. Had methane arrived at Titan directly as methane, these heavy noble gases would have been found in far greater quantities relative to N_2 (see Table 2, column 2) than the *upper limit* mole fraction of 10 ppbv obtained from the analysis of the Huygens GCMS data since the planetesimals would have trapped these heavy noble gases along with methane. Arguments have been made that the lack of detection of Kr and Xe may not be indicative of the lack of their existence on Titan. It is suggested that Kr & Xe may be sequestered as clathrates in the ice below Titan's purported water-ammonia ocean, 50-100 km beneath the surface. If so, it is somewhat puzzling that they are not released to the atmosphere from time to time, considering that radiogenic argon (^{40}Ar) is released from the rocks in Titan's core some 2000 km below the surface and detected in the atmosphere. Suggestions have been made also that Kr and Xe may be trapped in aerosols. If that were the case, reevaporation of these heavy noble gases would occur as the aerosols descend to the surface, which is 20 K warmer than the tropopause. Release of Xe and Kr from the surface would also take place upon impacts that occur from time to time. Since the present data pose a dilemma and the models are less than satisfactory, *in situ* measurements at the surface and below will be required in the future to fully resolve the important question of the origin of methane on Titan.

If methane did form on Titan, it would have done so in a process known as serpentinization (Atreya *et al.* 2006). Such a process produces methane in terrestrial oceans, at high temperatures (350-500 °C) in Black Smokers, and at relatively mild ones (30-90 °C) in Lost City, some 15 km from the spreading center on sea floor near the Mid-

Atlantic Ridge. Serpentinization is a good possibility on primordial Titan also. Today, a 2000-km thick high-pressure ice barrier separates Titan's rocky core from its purported liquid water-ammonia ocean. However, in the past when the temperatures were warmer due to accretionary heat, the ocean is expected to have extended all the way to the core. The water-rock reactions would proceed, hydrogen (H_2) would be released, and methane would be produced by metal-catalyzed Fischer-Tropsch process involving reactions between H_2 and the primordial carbon in the form of CO_2, CO or carbon grains in the rocks (Atreya *et al.* 2006). High temperature serpentinization would occur close to the core, and low temperature serpentinization above it with smaller suspended rocky-material. Once produced, methane could be stored as clathrate hydrate ($CH_4 \cdot 5.75H_2O$) in upper ice layers in Titan's interior for subsequent release to the surface. Models indicate that the release of methane from the interior occurs episodically, with the last such episode about half a billion years ago (Tobie *et al.* 2006).

The surface temperature of Titan is 94 K, which is close to the triple point of methane (and ethane). Once on the surface, methane can remain as liquid. In fact, evidence of numerous wide spread lakes has been found at high northern latitudes ($>70\,^{\circ}$N) by the Cassini radar (e.g., Stofan *et al.* 2007) and to a lesser extent in the southern polar region (e.g., Ontario Lacus) where the coverage is limited. Convection allows methane vapor above the surface liquid to rise in the troposphere where it undergoes condensation above approximately 6 km, first as liquid, then changing to ice at colder temperatures at higher elevations (Figure 4). Evidence of both types of clouds is seen in the Cassini orbiter, Huygens probe and ground-based observations. Subsequent precipitation from the clouds would produce methane rain and snow. Thus, the methane meteorological cycle of Titan is similar to the hydrological cycle on Earth (Atreya *et al.* 2006, 2009; Lunine & Atreya 2008). For all practical purposes it is a closed cycle, without loss of methane, as any methane removed to the interior from the surface through cracks and fissures is expected to be a temporary local loss as it is likely to be eventually released back out. However, permanent destruction of methane occurs in the stratosphere and the ionosphere.

Irreversible destruction of methane takes place by photochemical conversion of methane to ethane, acetylene, benzene, etc. (Figure 4) in approximately 30 million years (Wilson & Atreya 2004, 2009), which is a relatively short time in the geologic history of Titan. Thus methane must be released from its storage in Titan's interior from time to time, if Titan were to maintain a relatively stable atmosphere. Despite the photochemical destruction of methane, calculations indicate that Titan's interior may hold tens of billions of years of supply of methane if it was destroyed at today's rate (Atreya *et al.* 2009). There is no energy crisis on Titan! The principal product of methane photolysis is ethane. As much as 40% of the methane is expected to be converted to ethane over geologic time (Wilson & Atreya 2009). Ethane condenses in the upper troposphere of Titan, and ethane clouds have indeed been detected. However, evidence of seas or oceans of ethane predicted in the 1980's is not found. Episodic release of methane could explain why, at least partially. Instead of kilometer deep global ethane oceans, a maximum of only a few meters deep ethane liquid could have formed on Titan's surface since the last episodic release of methane (Wilson & Atreya 2009). Some of that ethane liquid would become a component of the methane lakes, and is indeed detected in Ontario Lacus (Brown *et al.* 2008), and some may be sequestered as clathrate in subsurface ice.

Prebiotic molecules. It has been speculated that organics might be present on the Galilean moon Europa, if it accreted from carbonaceous-chondrite type material, or organic molecules formed on Europa if the carbon-containing crust is in contact with purported subsurface ocean (Zolotov & Kargel 2009). Organic molecules have not been detected on Europa, however, either because of their low concentration, oxidation, or destruction by energetic charged particles of Jupiter's magnetosphere. On the other hand, numerous organic molecules have been detected at Titan, and there is even a possibility of the presence of complex prebiotic molecules. The trace hydrocarbons and nitriles produced in gas phase in the upper atmosphere and ionosphere of Titan either go on to form haze or condense in the colder lower stratosphere and at the tropopause (45 km, 74 K). A consequence of this phase and chemical change is that very low abundance of these trace molecules is present in gas phase for detection and quantification in the atmosphere. The remote sensing data from the Cassini orbiter and *in situ* data from the Huygens probe show that is indeed the case. On the other hand, the same molecules gradually sediment out of Titan's atmosphere, and deposit on to the surface as ices and haze. In time their concentration would build up. The surface thus holds the secrets of the atmosphere and the ionosphere, but the constituents are in far greater concentrations. A slow cooking of the more recent surface deposits by the flux of cosmic rays reaching the ground could result in the formation of complex organic molecules, perhaps even prebiotic molecules. There is even a likelihood that prebiotic molecules were already present in Titan's surface, but without any connection to the atmosphere. In its primordial accretionary period, Titan was most likely above the freezing point of water allowing for the possibility of liquid water. Ammonia and methane must have been abundant in addition to water, and solar ultraviolet flux was substantially greater. In other words, the conditions seem suitable for the formation of prebiotic molecules, as in the Miller-Urey experiment. For these reasons, and the very real likelihood that Titan's interior communicates with the surface, the surface is key to Titan's mysteries (Atreya 2007b).

4. Conclusion

Galileo Galilei's paradigm shifting discoveries four centuries ago paved the way for the new discoveries in the solar system, including finding gaseous environments around certain moons. Galileo would be thrilled to learn how far we have come in such a short time, and how rapidly we are moving in finding new solar systems, as he himself did when he discovered a "miniature solar system" on January 7, 1610 from Padova!

Acknowledgement

Graphics and formatting help provided by Minjin Baek is much appreciated.

References

Adams, E. Y. 2006, *Ph.D. Thesis*, University of Michigan, Ann Arbor, USA

Anders, E. & Grevesse, N. 1989, *Geochim. Cosmochim. Acta*, 53, 197

Atreya, S. K., Donahue, T. M., & Kuhn, W. R. 1978, *Science*, 201, 611

Atreya, S. K. 1986, *Atmospheres and ionospheres of the outer planets and their satellites* (Berlin, Heidelberg and New York: Springer-Verlag)

Atreya, S. K. 2007a, *Scientific American*, 296(5), 42

Atreya, S. K. 2007b, *Science*, 316, 843

Atreya, S. K. & Wong, A. S. 2005, *Space Sci. Revs*, 116, 121

Atreya, S. K., Adams, E. Y., Niemann, H. B., Demick-Montelara, J. E., Owen, T. C., Fulchignoni, M., Ferri, F., & Wilson, E. H. 2006, *Planet. Space Sci.*, 54, 1177

Atreya, S. K., Lorenz, R. D., & Waite, J. H. 2009, in: R. H. Brown, J.-P. Lebreton & J. H. Waite (eds.), *Titan from Cassini-Huygens* (Heidelberg, London and New York: Springer Dordrecht), p. 177

Brown, R. H., Soderblom, L. A., Soderblom, J. M., Clark, R. N., Jaumann, R., Barnes, J. W., Sotin, C., Buratti, B., Baines, K. H., & Nicholson, P. D. 2008, *Nature*, 454, 607

Conrath, B. J. & Gautier, D. 2000, *Icarus*, 144, 124

Drake, J. J. & Testa, P. 2005, *Nature*, 436, 525

Flasar, F. M. & the CIRS Team 2005, *Science*, 308, 975

Grevesse, N., Asplund, M., & Sauval, J. 2005, in: G. Alecian, O. Richard & S. Vauclair (eds.), *Element stratification in stars, 40 years of atomic diffusion*, EAS Publications Series, Vol. 17, p. 21

Lunine, J. I. & Atreya, S. K. 2008, *Nat. Geosci.*, 1, 159

Matson, D. L., Castillo, J. C., Lunine, J. I., & Johnson, T. V. 2007a, *Icarus*, 187, 569

Matson, D. L., Atreya, S. K., Castillo-Rogez, J., Johnson, T. V., Adams, E. Y., & Lunine, J. I. 2007b, *Eos Trans. AGU*, 88(52), Fall Meet. Suppl., Abstract P21D-04

McEwen, A. S., Keszthelyi, L., Spencer, J. R., Schubert, G., Matson, D. L., Lopes-Gautier, R., Klaasen, K. P., Johnson, T. V., Head, J. W., Geissler, P., Fagents, S., Davies, A. G., Carr, M. H., Breneman, H. H., Belton, M. J. S., 1998, *Science*, 281, 87

Niemann, H. B. & the GCMS Team 2005, *Nature*, 438, 779

Stofan, E. R., Elachi, C., Lunine, J. I., Lorenz, R. D., Stiles, B., Mitchell, K. L., Ostro, S., Soderblom, L., Wood, C., Zebker, H., Wall, S., Janssen, M., Kirk, R., Lopes, R., Paganelli, F., Radebaugh, J., Wye, L., Anderson, Y., Allison, M., Boehmer, R., Callahan, P., Encrenaz, P., Flamini, E., Franscescetti, G., Gim, Y., Hamilton, G., Hensley, S., Johnson, W. T. K., Kelleher, K., Muhleman, D., Paillou, P., Picardi, G., Posa, F., Roth, L., Seu, R., Shaffer, S., Vetrella, S., West, R. 2007, *Nature*, 445, 61

Tobie, G., Junine, J. I., & Sotin, C. 2006, *Nature*, 440, 61

Vuitton, V., Lavvas, P., Yelle, R. V., Galand, M., Wellbrock, A., Lewis, G. R., Coates, A. J., & Wahlaund, J.-E. 2009, *Planet. Space Sci.*, 57, 1558

Waite, J. H., Lewis, W. S., Margee, B. A., Lunine, J. I., McKinnon, W. B., Glein, C. R., Mousis, O., Young, D. T., Brockwell, T., Westlake, J., Nguyen, M.-J., Teolis, B. D., Niemann, H. B., McNutt, R. L., Perry, M., & Ip, W.-H. 2009, *Nature*, 460, 487

Wilson, E. H. & Atreya, S. K. 2004, *J. Geophys. Res.*, 109, E06002

Wilson, E. H. & Atreya, S. K. 2009, *J. Phys. Chem.*, 113, 11221

Zolotov, M. Y. & Kargel, J. S. 2009, in: R. T. Pappalardo, W. B. McKinnon & K. Khurana (eds.), *Europa* (University of Arizona Press), p. 449

Galileo's Medicean Moons: their impact on 400 years of discovery
Proceedings IAU Symposium No. 269, 2010
C. Barbieri, S. Chakrabarti, M. Coradini & M. Lazzarin, eds.

doi:10.1017/S1743921310007350

The Study of Mercury

Louise M. Prockter and Peter D. Bedini

Johns Hopkins University Applied Physics Laboratory,
11100 Johns Hopkins Road
Laurel, MD 20723, U.S.A.
email: Louise.Prockter@jhuapl.edu

Abstract. When the MErcury Surface, Space ENvironment, GEochemistry, and Ranging (MESSENGER) spacecraft enters orbit about Mercury in March 2011 it will begin a new phase in an age-old scientific study of the innermost planet. Despite being visible to the unaided eye, Mercury's proximity to the Sun makes it extremely difficult to observe from Earth. Nonetheless, over the centuries man has pursued a quest to understand the elusive planet, and has teased out information about its motions in the sky, its relation to the other planets, and its physical characteristics. A great leap was made in our understanding of Mercury when the Mariner 10 spacecraft flew past it three times in the mid-1970s, providing a rich set of close-up observations. Now, three decades later, The MESSENGER spacecraft has also visited the planet three times, and is poised to add significantly to the study with a year-long orbital observation campaign.

Keywords. History and philosophy of astronomy; Planets and satellites, general; Celestial mechanics, Space vehicles.

As one of the "wandering stars", Mercury has been observed since antiquity and references to it can be found in the lore of ancient civilizations around the world. The Chinese associated Mercury with the direction north and the element water. In Hindu mythology the planet was called Budha, a god who presided over Budhavara, or Wednesday. The Norse associated Mercury with their god Odin, who also presided over the middle day of the week (Woden's Day). Because Mercury is so close to the Sun, it travels quickly relative to other celestial bodies. It can be seen only during the days surrounding its greatest elongations, when it is visible just before sunrise or after sunset. It is these characteristics of swiftness and elusiveness that prompted a number of cultures to associate Mercury with their messenger gods, such as the Babylonian god Nabu, the Egyptian god Thoth, and the Greek god Hermes. The Maya represented Mercury as one or more owls, which served as messengers to the underworld. The English name for the planet comes from the Roman messenger god Mercurius, the Roman equivalent of Hermes.

The first known references to Mercury in writing are found in Mesopotamia in the 7th century BCE. The cuneiform tablets known as the MUL.APIN were most likely written by Assyrian astronomers and describe observations taken between 1300 and 1000 BCE (Figure 1) (Schaefer, 2007). At the same time that these tablets were being created, the Maya also were charting the motion of the planet. Records of detailed observations are found in the Dresden Codex (Makemson, 1957). Observations of Mercury figured prominently in the early efforts to develop a geometric model of the heavens. One of the first attempts at this was around 370 BCE by the Greek astronomer Eudoxus, who compiled an extensive star catalog, which included the visible planets. To explain the measurements represented in his catalog, Eudoxus' planetary model incorporated more than two dozen spheres. Mercury, along with each of the other four visible planets, was assigned four spheres: one to describe daily motion, one for motion through the zodiac, and two to represent retrograde motion (Heath, 1921).

Figure 1. Cuneiform MUL.APIN tablet, found in Mesopotamia in the 7th century BCE, thought to contain the first written references to Mercury.

In the second century AD Ptolemy published his great scientific treatise, *Almagest*, in which he describes a simpler geocentric model of the planets that was accepted as correct for centuries afterward. In this work he compiled all known observations of early Babylon and Hellenistic Egypt. He included seven observations of Mercury, the earliest of which was made on 15 November 265 BCE (Jones, 2006). Throughout the Middle Ages astronomers continued to observe the planets and to record their movements. Pre-optical instruments such as astrolabes and quadrants grew in sophistication and produced very accurate measurements. The German astronomer Bernard Walther is the first astronomer known to have used a clock for indicating the time of astronomical observations, and did so to record Mercury's appearance in the sky on 16 January 1484 (Beaver, 1970). In the following century, Copernicus used unpublished observations by Walther in developing his heliocentric planetary model.

In the first years of the seventeenth century, Johannes Kepler, using observations made by the Danish astronomer Tycho Brahe, calculated that on 29 May 1607 Mercury would pass directly between Earth and the Sun. Using a lens-less pinhole camera, he tracked "a little daub, quite black, approximately like a parched flea" passing across the bright image of the Sun. Convinced at first that he had witnessed a transit of Mercury, he published his results before realizing that he had indeed not observed Mercury after all (Caspar, 1993). As with similar observations thought to be of Mercury transits made in 807 AD and in 1278 AD, the dark spots moving on the Sun's disk were likely sunspots.

In 1608 the telescope was invented in Holland and it immediately revolutionized astronomy. Within one year of its introduction it was being used for astronomical studies throughout Europe, including those by Galileo Galilei in Padua, Italy, Thomas Harriot in London, England, and Simon Marius in Ansbach, Germany. Although the instrument brought man closer to the heavens than ever before, the early lenses were still too crude to reveal any of Mercury's secrets. As the state of the art of telescope manufacture developed, so did the number and quality of the observations.

With the completion of better planetary position tables in 1627 (the Rudolphine tables), Kepler once again predicted a transit of Mercury, this time to occur on

7 November 1631. Although it occurred slightly earlier than expected, the transit was observed and documented by the French astronomer, Pierre Gassendi. Unlike Kepler, who two decades earlier had mistaken a sunspot for Mercury, when a spot first appeared on the Sun's image, Gassendi believed it to be too small to be Mercury, and concluded that it was instead a sunspot (Van Helden, 1976). Only after further observation did the astronomer realize that the dark object moved too quickly to be a sunspot, and that he indeed was recording the transit of Mercury. The transit took 5 hours and had occurred about 4.75 hours before the predicted time. Although observed by others, Gassendi is the only astronomer to publish the results of his observation. Unfortunately, Kepler died the year before his calculation was proven true. Nevertheless, the ability to predict the passage of Mercury in front of the Sun was of great importance in confirming his planetary model, which was heliocentric, like Copernicus', but which described the orbits of planets as ellipses rather than circles.

Although confirmation of the transit of Mercury was consistent with the heliocentric model, an important observation remained to be made. If Venus and Mercury orbited the Sun inside of Earth's orbit, then they should exhibit repeatable phases. Galileo published his observations of the phases of Venus in 1610, but it wasn't until 1639 that advances in the telescope allowed Mercury's phases to be confirmed. This was accomplished by the Italian Jesuit, Johannes Baptista Zupus, using telescopes made by Francesco Fontana. The results of these observations were published by Fontana in 1646 (Figure 2). The application of the new telescope to the study of the heavens led to huge advances in the understanding of our cosmos, as new worlds presented themselves for examination. As improvements were made to the telescope, descriptive astronomy emerged, and the features of Mars and the Moon could be characterized for the first time. Mercury, though, continued to be elusive, and it was not until the beginning of the nineteenth century that serious attempts were made to map its surface.

Figure 2. Drawing of Mercury by Francesco Fontana, Novae coelestium, terrestriumque rerum observations, Naples, 1646.

One of the earliest known efforts to describe the features of Mercury was by Johann Hieronymous Schröter, working in Lilienthal, Germany, in 1800 (Figure 3). He recorded seeing a mountain extending 20 km in height, and deduced a rotational period of just over 24 hours (Denning, 1906). Others, such as Étienne Léopold Trouvelot and William Frederick Denning, made similar observations during the latter half of that century. Denning compared Mercury's surface with that of Mars, and derived a rotation period of 25 hours (Chapman, 1988a). Based on more than 200 observations made in Milan, Italy, between 1881 and 1889, Giovanni Virginio Schiaparelli produced a relatively advanced map of Mercury, which recorded observed features relative to a coordinate system for the first time (Antoniadi, 1934, per Davies, *et al.*, 1978). Based on comparison of the observed surface features throughout his campaign, he deduced that Mercury's rotation was synchronous with its 88-day orbital period and that the same side always faced the Sun (Holden, 1890). Although there was resistance at first to this conclusion, it was generally accepted by the end of the century and not successfully refuted until the 1960s.

As the capability of the telescope increased, higher-fidelity observations were possible, although their interpretations did not always match that quality. Using the 61-cm refracting telescope in Flagstaff, Arizona in 1896-7, Percival Lowell described a Mercury surface covered with long, linear, canal-like features. Needless to say, such features have not been confirmed to exist. A relatively sophisticated map of Mercury was, however, drawn by Eugené Michael Antoniadi in 1934 based on observations taken in Meudon, France (Figure 4). His results were consistent with the proposed 88-day rotation, and he justified it in terms of tidal forces.

A new approach to the study of Mercury was taken in June 1962 when Vladimir Kotelnikov and colleagues were the first to obtain radar echoes from Mercury (Thompson, 1963). Three years later, radar observations by Gordon Pettengill and Rolf Dyce using the 300-meter Arecibo Observatory radio telescope in Puerto Rico showed conclusively

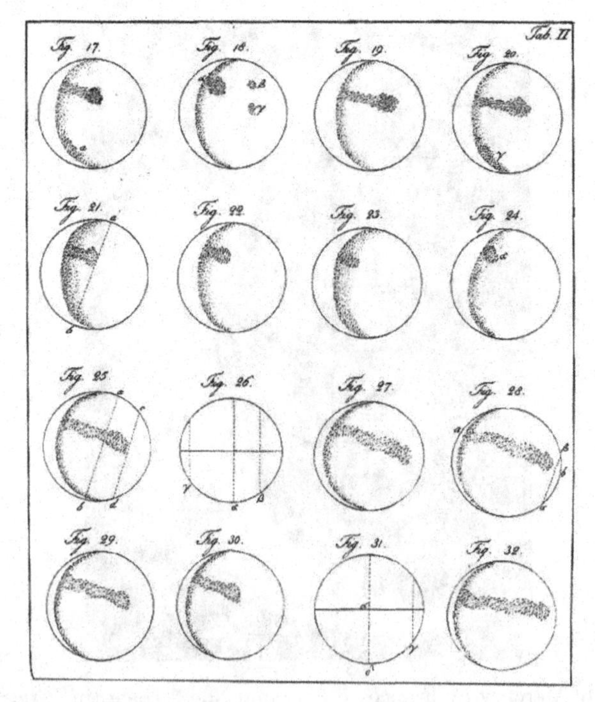

Figure 3. The first known map of Mercury by Johann Hieronymous Schröter (1745-1816).

that Mercury's rotational period was about 59 days (Pettengill and Dyce, 1965). The widely held theory that Mercury's rotation was Sun-synchronous would require its dark face to be extremely cold, but measurements of radio emission revealed temperatures much higher than expected. Astronomers were reluctant to drop the synchronous rotation theory and searched for evidence of an atmosphere that might transfer heat from the day to nighttime surfaces (Murray and Burgess, 1977).

Italian astronomer Giuseppe Colombo noted that the rotation value was about two-thirds of Mercury's orbital period, and proposed that the planet's orbital and rotational periods were locked into a 3:2 rather than a 1:1 resonance (Colombo, 1965). Despite this correction in the rotational period of Mercury, the early visual observations were not completely invalid. Because the planet rotates three times for every two orbits about the Sun, after three synodic periods, the same face of the planet presents itself at the same phase. Because the conditions for viewing Mercury are favorable every three synodic periods, most early observations were made at those times. As a result, the same face of Mercury was indeed being mapped.

With the advent of the Space Age came the first opportunity to study Mercury at close range. The desire to understand the innermost planet had been demonstrated for centuries, and by the 1960's the technology (and the funding) was finally available to visit the planet for the first time. Although multiple-planet orbits were first considered in the 1920's and 1930's, the first systematic development of the technique did not occur until the 1960's, at the Jet Propulsion Laboratory (JPL) in Pasadena, California. Soon thereafter, JPL trajectory designers discovered Earth-Venus-Mercury trajectory opportunities for launches in 1970 and 1973, and the spacecraft exploration of Mercury became a viable possibility for the first time. It was found that by flying through Venus' gravitational field, a spacecraft's trajectory could be altered causing it to fall in toward the Sun and, with careful timing, enable it to cross Mercury's path and encounter the planet. Without the use of Venus to "slingshot" the spacecraft toward Mercury, a much larger launch vehicle would be needed, and it would only be able to fly by the planet once on a trajectory toward the Sun.

The National Academy of Sciences Space Science Board conducted a planetary exploration study in 1968 in which they endorsed a mission to Mercury, via Venus, recommending a 1973 launch and suggesting scientific experiments that could be carried by the

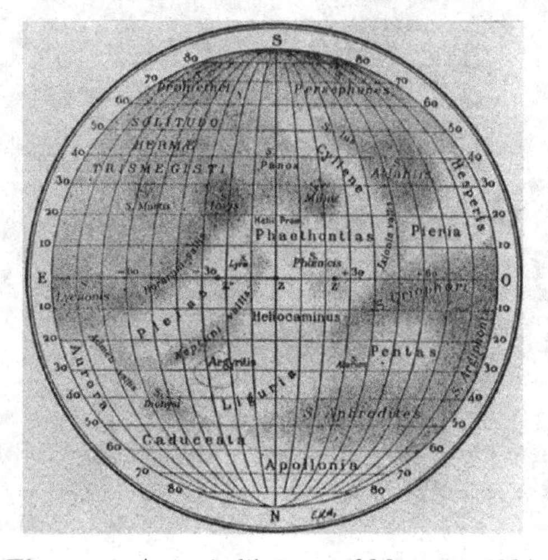

Figure 4. Antoniadi's map of Mercury, 1934.

spacecraft. NASA approved the mission in 1969, and by January 1970, a Venus/Mercury project office had been established at JPL (Dunne and Burgess, 1978).

At a conference on the new mission held at JPL in 1970, Giuseppe Colombo noted that once the proposed Mariner 10 spacecraft passed Mercury, its orbital period would be approximately twice that of the planet itself. If the point at which it passed Venus were well chosen, the craft would make repeated flybys of Mercury. JPL mission designers performed analyses that confirmed this, and determined that three flybys could be achieved before the mission exhausted its supply of propellant.

The Mariner 10 spacecraft was the seventh successful launch in a series that had previously explored Venus and Mars from 1962 to 1971 (Murray and Burgess, 1977). The spacecraft mass at launch was 533.6 kg, which included 29 kg of hydrazine propellant and a scientific payload of 78 kg. Unlike the previous Mariner spacecraft, Mariner 10 had to be able to survive much closer to the Sun, requiring modifications that included a sunshade, louvers and thermal blankets, and the ability to rotate the solar panels along their long axes in order to keep their temperature relatively stable as they were carried inward through the Solar System. In order to meet its objective to explore Mercury as thoroughly as possible, more science instruments were carried than on previous spacecraft in the program. The payload consisted of seven science experiments, including television photography, extreme ultraviolet spectroscopy, infrared radiometry, energetic particle, plasma, and magnetic-field detectors, and radio science (Figure 5).

On 2 November 1973, the Mariner 10 spacecraft was launched from Cape Kennedy in Florida. Within the first week of flight, data from the Earth and Moon were returned as calibration tests for the Mercury encounters. In January 1974, Mariner 10 was able to take ultraviolet scans of the comet Kahoutek that were not possible from Earth. Despite some issues with the onboard guidance and control system, the encounter with Venus went as planned, on 5 February 1974, with a closest approach of 5790 km (3600 mi). The gravity assist maneuver was successful and Mariner 10 was on its way to Mercury.

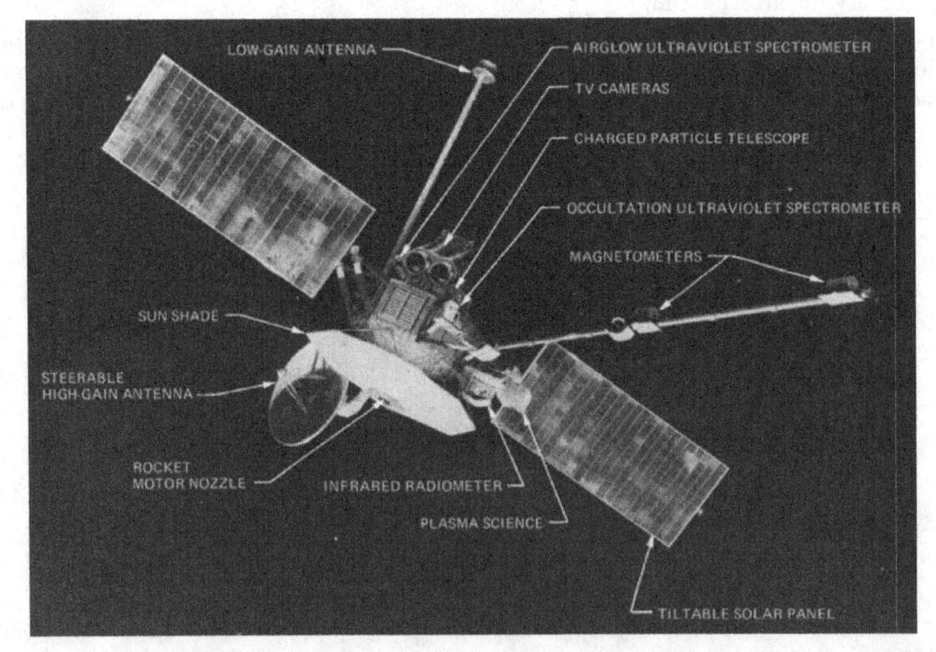

Figure 5. The Mariner 10 spacecraft (NASA/JPL).

Only a few weeks later, on 24 March 1974, the first blurry images of Mercury were sent back to Earth. As the spacecraft approached the planet, more and more detail appeared in the view of the television cameras revealing the planet to be a heavily cratered world similar to the Moon and parts of Mars. The first flyby was over the night hemisphere of Mercury, with a closest approach of 705 kilometers (438 miles). This meant that the imaging team was only able to acquire images of the Mercury crescent as the spacecraft approached the planet, then as it departed, and was not able to join the two sets of images together. In addition, the images were highly foreshortened because of the viewing geometry. Nevertheless, it was revealed that Mercury contained abundant impact features, including multi-ringed basins, secondary crater chains, and bright rays (Murray *et al.*, 1975). Cliffs or scarps up to 3 km (2 mi) high and 500 km (300 mi) in length were visible across much of the surface, their lobate form suggesting they were the result of compressional forces. These scarps were thought to be due to a readjustment of the surface in response to a slight shrinking of the planet's core.

A huge basin was discovered lying across the terminator such that less than half of its eastern portion was visible. The basin was estimated to be about 1300 km in diameter, larger than the Imbrium basin on the Moon. Because of its situation at one of Mercury's "hot poles" - locations that are closest to the Sun at perihelion - the basin was named Caloris, which is the Greek word for hot. In between many of the heavily cratered areas were smooth, lightly-cratered plains. Based on observations of the Moon, it was not

Figure 6. M10 early image of Mercury, acquired by Mariner 10 on March 28 at 952,000 km (590,240 mi).

known whether these plains were volcanic in origin, or whether they were the result of fluidized impact ejecta, a question that was to remain unresolved for over three decades.

Gravity measurements determined the mass of Mercury to two orders of magnitude greater than previously possible, and the surface temperature range was measured by Mariner 10's infrared radiometer to vary from 90 K (-297°F) on the nightside to 460 K (369°F) on the dayside (although the temperature can go even higher - 650 K (1170°F) - at perihelion, higher than any other planet). Perhaps the most surprising measurement was that Mercury possessed a magnetic field, estimated to be about $1/60^{th}$ as strong as Earth's, the source of which was a mystery (see review in Ness *et al.*, 1979).

Six months after this first flyby, on 21 September 1974, Mariner 10 returned to Mercury, flying by on the sunlit side at a distance of 47,000 kilometers (29,200 miles), allowing images to be taken to fill in the missing areas of the surface from the first encounter, and thereby increasing coverage to 75% of the illuminated hemisphere. Because of this favorable viewing geometry, this second encounter was primarily devoted to imaged science, and it did not disappoint. Images were obtained of Mercury's south pole, showing that the compressional scarps extended into this area. A number of images were obtained to determine the angular separation between Mercury and stars, which showed for the first time that the technique of optical navigation was viable for planetary missions, and that reliance on Earth-based radio measurements was no longer essential (Dunne & Burgess, 1978).

Mariner 10's third encounter with Mercury took place six months later, on 16 March 1975. This flyby was primarily focused on obtaining more information about Mercury's magnetic field, and the closest approach point was only 327 kilometers (203 miles) from the surface. Data from this flyby showed unequivocally that Mercury's magnetic field was not created by the solar wind, but is intrinsic to the planet itself, and that the magnetosphere is very like a scaled-down version of Earth's (see review in Ness *et al.*, 1979). Predictions were made as to when the spacecraft would pass through the bow shock, magnetopause, and maximum field, and the actual times of these events were almost exactly as expected. In addition, Mercury, like Earth, was found to have a magnetically neutral tail. Gravity measurements from this and the previous flybys were interpreted to show that Mercury has a surprisingly large core compared to its radius (Chapman, 1988b). The close approach to Mercury allowed some spectacularly high-resolution images to be obtained, enabling features as small as 137 m (450 ft) to be identified, such as small fresh craters and detail of ridges and fractures in the floor of the Caloris basin. Eight days after Mariner 10's third encounter with Mercury, the spacecraft's supply of nitrogen maneuvering gas was exhausted and commands were sent to turn off its transmitter. The Mariner 10 spacecraft had vastly increased scientific knowledge of Mercury.

The next step in Mercury exploration was widely recognized to be an orbiter, to investigate the planet's interior, improve imaging coverage, and determine the chemical composition of the surface (COMPLEX, 1978). However, it was thought that conventional propulsion systems were insufficient to enable the change in spacecraft velocity that would be required for orbit insertion about the planet. This view persisted until the mid-1980's, when multiple gravity-assist trajectories were discovered that would allow Mercury orbit insertion with chemical propulsion systems (Yen, 1985, 1989). However, it was almost 15 more years before such a mission became a reality.

In the meantime, the study of Mercury did not rest. In 1991, radar experiments designed to image the half of Mercury not photographed by Mariner 10 (Slade *et al.*, 1992) revealed highly reflective regions near the planet's poles. The similarity of the radio echoes to those of icy regions of Mars and icy outer-planet satellites strongly suggested that, despite the harsh environment, ice exists in the permanently shadowed craters near Mercury's poles.

The Mariner 10 mission answered many questions about Mercury, yet much remained to be learned. The MESSENGER mission was first proposed to NASA's newly-created Discovery program in 1996, and was eventually selected for flight in July 1999 (McNutt *et al.*, 2006). The spacecraft was built and the mission managed by the Johns Hopkins University Applied Physics Laboratory (APL), in Laurel, MD. The spacecraft takes advantage of lightweight materials, miniaturized electronics, and an ingenious trajectory design to achieve orbit insertion around Mercury under the constraints of a relatively low NASA budget (Santo *et al.*, 2001; Leary *et al.*, 2007). In order to withstand the searing heat at Mercury, the MESSENGER spacecraft employs a combination of thermal blankets, coolers, and a 2-m × 2.5-m ceramic-fabric sunshade that protects the wiring, electronics, and science instruments.

The MESSENGER payload consists of seven scientific instruments and a radio science experiment (Gold *et al.*, 2007). The instruments include the Mercury Dual Imaging System (MDIS), the Mercury Atmospheric and Surface Composition Spectrometer (MASCS), the Mercury Laser Altimeter (MLA), the Gamma-Ray and Neutron Spectrometer (GRNS), the X-Ray Spectrometer (XRS), the Magnetometer (MAG), and the Energetic Particle and Plasma Spectrometer (EPPS).

The MESSENGER spacecraft was launched from Cape Canaveral Air Force Station, Florida, on 3 August 2004. The long cruise phase includes six planetary flybys - one of Earth, two of Venus, and three of Mercury - as part of a 7.9-billion-km journey that includes more than 15 orbits around the Sun (McAdams *et al.*, 2007). The Earth and Venus flybys afforded opportunities for flight tests of the instruments in preparation for the spacecraft's first encounter with Mercury. At the time of writing, MESSENGER has recently completed its third and final Mercury flyby, and is en route for orbit insertion on 18 March 2011.

Although the primary purpose of MESSENGER's three Mercury flybys was to achieve the gravity assists needed to place the spacecraft in orbit about Mercury, they proved

Figure 7. This MESSENGER image of Mercury's previously unseen hemisphere was acquired about 80 minutes after the spacecraft's closest approach to Mercury on the first flyby, from a distance of about 27,000 kilometers (about 17,000 miles). The giant Caloris basin is the large bullseye feature in the top right of the image, with the bright center surrounded by a dark annulus (NASA/JHUAPL).

to be tremendously valuable in terms of scientific return. During the first flyby, on 14 January 2008, acquisition of science data began weeks before the closest approach and continued throughout the encounter and for weeks afterward. During the approach to Mercury, high-resolution measurements were made of the planet's exosphere tail, extending hundreds of thousands of kilometers anti-sunward (McClintock *et al.*, 2008). Mariner 10 had only been able to view one hemisphere of Mercury, but much of the opposite hemisphere was sunlit during the first encounter, enabling MESSENGER to image almost half of the planet that had never before been viewed by a spacecraft (Figure 7). The first ever measurements of ions at Mercury revealed a complex environment resulting from a mixture of solar wind plasma and species originating from the surface. Although Mercury's magnetosphere was found to be surprisingly calm, it was still found to possess an array of dynamic plasma physical processes, similar to Earth's (Slavin *et al.*, 2008). Magnetometer measurements were made throughout the flyby, and provided the best assessment yet of the field at the equator, suggesting that it is predominantly dipolar, as would be expected if it was produced by a dynamo in a molten outer core (Anderson *et al.*, 2008).

The closest approach on the first flyby was at a mere 201 km (125 mi). The close flyby allowed the first ever laser altimeter profile of Mercury's surface to be obtained. The profile indicated that Mercury's craters are shallower than similar-sized craters on the Moon, probably because of the higher surface gravity (Zuber *et al.*, 2008). The close approach also allowed high-resolution (200-m/pixel) images to be captured, and much of the surface was imaged in 11 colors, providing the most comprehensive color data of Mercury. The western portion of the Caloris basin was imaged for the first time, and it was found to be 250 km larger than previously thought, measuring 1550 km in diameter (Murchie *et al.*, 2008). Lobate scarps were found to be widespread across the surface, indicating that compression had been global, rather than confined to one hemisphere. High spectral-resolution data at infrared and near-infrared wavelengths were obtained, providing new clues as to the composition of surface minerals (Robinson *et al.*, 2008). One question that had been unresolved for over 30 years was whether Mercury's smooth plains were the result of volcanic processes or whether they were formed by impact ejecta. MESSENGER images showed a number of different surface features that were interpreted to be volcanic in nature, such as flooded and embayed impact craters and volcanic vents, thereby firmly establishing volcanism as a significant process in Mercury's evolution (Head *et al.*, 2008).

MESSENGER's first encounter with Mercury was a resounding success, but the spacecraft would soon be back for more. Nine months later, on 6 October 2008, the spacecraft flew by Mercury a second time, dipping to only 199 km (124 miles) above the surface. The closest approach point was on the opposite hemisphere than was seen during the first encounter, allowing more of the surface to be imaged for the first time. During the approach, the first simultaneous measurements of sodium and calcium tails in Mercury's exosphere were made, showing that their spatial distributions are complementary to each other, and that they vary according to solar wind conditions (McClintock *et al.*, 2009). Measurements of the magnetic field throughout the flyby showed that the types of interactions between the magnetic field and the solar wind at Mercury that produce auroras and magnetic storms are similar to those found on Earth (Slavin *et al.*, 2009). The MESSENGER spacecraft returned the only magnetic data to date from the planet's western hemisphere, key to understanding the geometry of Mercury's internal field. These data, combined with earlier results, showed that the planet's magnetic moment is closely aligned with its rotation axis to within 2° (Anderson *et al.*, 2008).

In total, MESSENGER had by this time imaged about 80% of the surface of the planet. A large new basin, Rembrandt, was discovered on the opposite hemisphere from that viewed by Mariner 10 (Watters *et al.*, 2009). At 715 km in diameter, this basin is half the size of the Caloris basin, yet has a floor which bears witness to a complex geological history of volcanic flows and tectonic deformation. Soon after closest approach, when MESSENGER was still relatively close to the surface, images were acquired in 11 colors, the highest resolution color data yet obtained, allowing clear compositional variations to be distinguished. It appears that Mercury has a complex and heterogeneous crust created over time from numerous volcanic flows (Figure 8), impact crater events, and tectonic deformation (Denevi *et al.*, 2009). Further geological information was acquired by the laser altimeter, which took profiles over lobate scarps and impact craters.

MESSENGER's third and final flyby of Mercury took place almost a year later, on 29 September 2009. On this occasion, the closest approach point and lighting were similar to that of flyby 2, but offset by ~20 degrees of longitude. Speeding past the surface at an altitude of 229 km (142 miles), this flyby provided the final boost needed to keep the spacecraft on track for Mercury orbit insertion 18 months later. Although the gravity assist was executed flawlessly, the science observation sequence was interrupted minutes before closest approach by the fault protection system, which halted it as a protective measure, which later proved unnecessary. Despite this, some substantial new scientific data were acquired during Mercury approach, including observations of a newly-discovered basin which appears to contain the youngest identified volcanic deposit on Mercury (Prockter *et al.*, 2010) as well as observations of the loading and unloading of Mercury's magnetotail (Slavin *et al.*, 2010) and detection of ionized calcium from Mercury (Vervack *et al.*, 2010). Between Mariner 10 and MESSENGER, 98% of the surface of Mercury has been imaged by a spacecraft.

The MESSENGER team is now preparing for Mercury orbit insertion on 18 March 2011. At the time of writing, the spacecraft is healthy, the instruments are performing well, and the trajectory is on course. The spacecraft will remain in orbit about Mercury for at least one Earth year, orbiting Mercury twice each day. These orbits are highly elliptical in order to protect the spacecraft from the extreme thermal environment at Mercury. At periherm the spacecraft will pass the planet at distances ranging from 200 to 500 km at 60-70° north latitude, and at apoherm up to 15,200 km, south of the planet. Because

Figure 8. Close-up view of the crater Rudaki, showing examples of craters in the plains that appear to have been significantly flooded with volcanic lava, leaving only their circular rims preserved.

of the orbital motion and spin of Mercury, the spacecraft will, at different times, reside in a "dawn-dusk" orbit, where it essentially flies over the terminator, which is ideal for monochrome imaging, or a "noon-midnight" orbit, which, when on the dayside, is ideal for multispectral imaging. The MESSENGER orbital observation campaign will seek to characterize the planet's interior, surface, exosphere, and magnetosphere, answering questions about the nature of the planet and its history.

The next step in the study of Mercury is already under development. The BepiColombo mission, named after the same man who explained Mercury's 3:2 spin orbit resonance and suggested that Mariner 10 could achieve multiple flybys, is an international collaboration between the European Space Agency and the Japan Aerospace Exploration Agency (Grard *et al.*, 2000; Anselmi and Scoon, 2001). The mission comprises a pair of spacecraft that will be placed in orbit about Mercury, one to study the magnetosphere and its interactions with the solar wind, and one to characterize the planet with a variety of different instruments (Hayakawa *et al.*, 2004; Schulz and Benkhoff, 2006). The two spacecraft are scheduled to launch in 2014 on a single rocket and will be placed in coplanar polar orbits in 2020.

The story of man's quest to understand Mercury is one of perseverance and ingenuity. Just as the telescope in the seventeenth century transformed the planet from a featureless light in the sky to a cratered world cloaked in mystery, robotic space probes three hundred years later have uncovered its true nature as a planet of intriguing extremes. The six flybys of Mercury have revealed a great deal, but much more remains to be learned. MESSENGER's yearlong observation campaign will return a wealth of information not previously available, but will undoubtedly leave questions to be answered by BepiColombo and perhaps future missions.

Acknowledgments: we owe grateful thanks to Dr. Robert Strom, whose 2009 talk at the Geological Society of America on the history of Mercury exploration inspired this conference paper.

References

Anselmi, A. & Scoon, G. E. N. 2001, BepiColombo, ESA's Mercury Cornerstone mission, *Planet. Space Sci.*, 49, 1409-1420

Anderson, B. J., Acuña, M. H., Korth, H., Purucker, M. E., Johnson, C. L., Slavin, J. A., Solomon, S. C., & McNutt, R. L. Jr. 2008, The Structure of Mercury's Magnetic Field from MESSENGER's First Flyby, *Science*, Vol. 321., pp. 82-85

Antionadi, E. M. 1934, *La Planete Mercure et la Rotation des Satellites.* Paris: Gauthier-Villars. English Translation by P. Moore and K. Reid, Shaldon, 1975

Beaver, D. deB. & Walther, B. 1970, Innovator in Astronomical Observation, *J. History Astronomy*, 1, 39-43

Caspar, M. 1993, *Kepler*, Dover Publications, Inc., New York

Chapman, C. R. 1998a, Mercury: Introduction to an end-member planet, in *Mercury*, edited by F. Vilas, C. R. Chapman and M. S. Mathews, pp. 1-23, University of Arizona Press, Tucson

Chapman, C. R. 1988b, *Mercury's heart of iron*, Astronomy, November, p. 22-35

Colombo, G. 1965, Rotational period of the planet Mercury, *Nature*, 208, 575

COMPLEX, Strategy for Exploration of the Inner Planets: 1977-1987, (Committee on Lunar and Planetary Exploration, National Academy of Sciences, Washington), 97 pp.

Davies, M. E. *et al.* 1978, *Atlas of Mercury, NASA Science and Technical Information Office*, NASA SP-423, Washington DC

Denevi, B. W., Robinson, M. S., Solomon, S. C., Murchie, S. L., Blewett, D. T., Domingue, D. L., McCoy, T. J., Ernst, C. M., Head, J. W., Watters, T. R., & Chabot, N. L. 2009,

The Evolution of Mercury's Crust: A Global Perspective from MESSENGER, *Science* 324, 613-615

Denning, W. F. 1906, The planets and planetary observation, *The Observatory*, Vol 29, 101, 458-462

Dunne, J. A. & Burgess, E. 1978, *The Voyage of Mariner 10*, NASA SP-424, Washington DC

Fontana, F. 1646, *Novae coelestium terrestriumq[ue] rerum observationes, et fortasse hactenus non vulgatae*, 208

Grard, R., Novara, M., & Scoon, G. 2000, BepiColombo—A multidisciplinary mission to a hot planet, *ESA Bull. 103*, 11-19

Hayakawa, H., Kasaba, Y., Yamakawa, H., Ogawa, H., & Mukai, T. 2004, The Bepi-Colombo/MMO model payload and operation plan, *Adv. Space Res.*, 33, 2142-2146

Head, J. W., Murchie, S. L., Prockter, L. M., Robinson, M. S., Solomon, S. C., Strom, R. G., Chapman, C. R., Watters, T. R., McClintock, W. E., Blewett, D. T., & Gillis-Davis, J. J. 2008, Volcanism on Mercury: Evidence from the First MESSENGER Flyby, *Science*, 321, 69-72

Head, T. L. 1921, *A History of Greek Mathematics: From Thales to Euclid*, Volume 1, Clarendon Press, Oxford

Holden, E. S. 1890, Announcement of the Discovery of the Rotation Period of Mercury [by Professor SCHIAPARELLI], *Publications of the Astronomical Society of the Pacific*, Vol. 2, No. 7, p.79

Jones, A. 2006, Ptolemy's Ancient Planetary Observations, *Annals of Science*, Vol. 63, No. 3, 255-290

Klostermaier, K. K. 1989, *A Survey of Hinduism*, State University of New York Press, Albany, p. 417

Leary, J. C. et. al. 2007, The MESSENGER spacecraft, *Space Sci. Rev.*, 131, 187-217

McAdams, J. V., Farquhar, R. W., Taylor, A. H., & Williams, B. G. 2007, MESSENGER Mission Design and Navigation, *Space Sci. Rev.*, 219-246

McClintock, W. E., Bradley, E. T., Vervack, R. J., Killen, R. M., Sprague, A. L., Izenberg, N. R., & Solomon, S. C. 2008, *Mercury's Exosphere: Observations During MESSENGER's First Mercury Flyby*, Science, 92-94

McClintock, W. E., Vervack, R. J. Jr, Bradley, E. T., Killen, R. M., Mouawad, N., Sprague, A. L., Burger, M. H., Solomon, S. C., & Izenberg, N. R. 2009, MESSENGER Observations of Mercury's Exosphere: Detection of Magnesium and Distribution of Constituents, *Science*, 610-613

McNutt Jr, R. L., Solomon, S. C., Gold, R. E., & Leary, J. C. and the MESSENGER Team, 2006. The MESSENGER mission to Mercury: Development history and early mission status, *Adv. Space Res.* 38, 564-571

Makemson, M. W. 1957, *The miscellaneous dates of the Dresden Codex, Publications of the Vassar College Observatory*, number 6, June

Morrison, D. 1976, IAU Nomenclature for Topographic Features on Mercury *Icarus*, 28, 605-606

Murchie, S. L., Watters, T. R., Robinson, M. S., Head, J. W., Strom, R. G., Chapman, C. R., Solomon, S. C., McClintock, W. E., Prockter, L. M., Domingue, D. L., & Blewett, D. T. 2008, Geology of the Caloris Basin, Mercury: A View from MESSENGER, *Science*, 321, 73-76

Murray, B. & Burgess, E. 1977, *Flight to Mercury*, Columbia University Press

Murray, B. C., Belton, M. J. S., Danielson, G. E., Davies, M. E., Gault, D. E., Hapke, B., O'Leary, B., Strom, R. G., Suomi, V., & Trask, N. 1974, Mercury's Surface: Preliminary Description and Interpretation from Mariner 10 Pictures *Science*, 185, 169-179

Murray, B. C., Strom, R. G., Trask, N. J., & Gault, D. E. 1975, Surface History of Mercury: Implications for Terrestrial Planets *J. Geophys. Res.*, 80, 2508-2514

Ness, N. F. 1979, The magnetosphere of Mercury, in *Solar System Plasma Physics, vol II*, edited by C. F. Kennel, L. J. Lanzarotti, E. N. Parker (North-Holland, New York), pp. 185-206

Pettengill, G. H. & Dyce, R. B. 1965, A Radar Determination of the Rotation of the Planet Mercury, Letters to the Editor *Nature*, 205

Prockter, L. M., Ernst, C. M., Denevi, B. W., Chapman, C. R., Head III, J. W., Fassett, C. I., Merline, W. J., Solomon, S. C., Watters, T. R., Strom, R. G., Cremonese, G., Marchi, S., &

Massironi, M. 2010, Evidence for young volcanism on Mercury from the third MESSENGER flyby *Science*, 329, 668-671

Robinson, M. S. & Lucey, P. G. 1997, Recalibrated Mariner 10 Color Mosaics: Implications for Mercurian Volcanism *Science*, vol. 275, p.197-200

Robinson, M. S., Murchie, S. L., Blewett, D. T., Domingue, D. L., Hawkins III, S. E., Head, J. W., Holsclaw, G. M., McClintock, W. E., McCoy, T. J., McNutt, R. L. Jr, Prockter, L. M., Solomon, S. C., & Watters, T. R. 2008, Reflectance and Color Variations on Mercury: Regolith Processes and Compositional Heterogeneity *Science*, 66-69

Santo, A. G. *et al.* 2001, The MESSENGER mission to Mercury: spacecraft and mission design *Planet. Space Sci.*, 49, 1481-1500

Schaefer, B. E. 2007, The Latitude and Epoch for the Origin of the Astronomical Lore in MUL.APIN, American Astronomical Society Meeting 210, #42.05 *Bulletin of the American Astronomical Society*, Vol. 38, p.157

Schulz, R. & Benkhoff, J. 2006, BepiColombo: Payload and mission updates, *Adv. Space Res.*, 38, 572-577

Slade, M. A., Butler, B. J., & Muhleman, D. O. 1992, Mercury Radar Imaging: Evidence for Polar Ice *Science*, 258, no. 5082

Slavin, J. A., Acuña, M. H., Anderson, B. J., Baker, D. N., Benna, M., Gloeckler, G., Gold, R. E., Ho, G. C., Killen, R. M., Korth, H., Krimigis, S. M., McNutt, R. L. Jr, Nittler, L. R., Raines, J. M., Schriver, D., Solomon, S. C., Starr, R. D., Trávníček, P., & Zurbuchen, T. H. 2008, Mercury's Magnetosphere After MESSENGER's First Flyby *Science*, 85-89

Slavin, J. A., Acuña, M. H., Anderson, B. J., Baker, D. N., Benna, M., Boardsen, S. A., Gloeckler, G., Gold, R. E., Ho, G. C., Korth, H., Krimigis, S. M., McNutt, R. L. Jr, Raines, J. M., Sarantos, M., Schriver, D., Solomon, S. C., Trávnícek, P., & Zurbuchen, T. H. 2009, MESSENGER Observations of Magnetic Reconnection in Mercury's Magnetosphere *Science*, 606-610

Slavin, J. A., Anderson, B. J., Baker, D. N., Benna, M., Boardsen, S. A., Gloeckler, G., Gold, R. E., Ho, G. C., Korth, H., Krimigis, S. M., McNutt, R. L. Jr, Nittler, L. R., Raines, J. M., Sarantos, M., Schriver, D., Solomon, S. C., Starr, R. D., Trávnícek, P., & Zurbuchen, T. H. 2010, MESSENGER Observations of Extreme Loading and Unloading of Mercury's Magnetic Tail, *Science*, 329, 665-668

Strom, R. 1987, *The Elusive Planet*, Smithsonian Institution Press, Washington, D.C., 197 p.

Thompson, J. H. 1963, Planetary Radar, *Quarterly Journal of the Royal Astronomical Society* Vol 4, 347

Van Helden A. 1976, The Importance of the Transit of Mercury of 1631 *Journal for the History of Astronomy* 7, 1-10

Vervack, R. J., McClintock, W. E., Killen, R. M., Sprague, A. L., Anderson, B. J., Burger, M. H., Bradley, E. T., Mouawad, N., Solomon, S. C., & Izenberg, N. R. 2010, Mercury's Complex Exosphere: Results from MESSENGER's Third Flyby *Science*, 329, 672-675

Watters, T. R., Head, J. W., Solomon, S. C., Robinson, M. S., Chapman, C. R., Denevi, B. W., Fassett, C. I., Murchie, S. L., & Strom, R. G. 2009, Evolution of the Rembrandt Impact on Mercury *Science* 618-621

Yen, C.-W. 1985, Ballistic Mercury orbiter mission via Venus and Mercury gravity assists, *AAS/AIAA Astrodynamics Specialist Conference* AIAA 83-346, San Diego, CA

Yen, C.-W. 1989, Ballistic Mercury orbiter mission via Venus and Mercury gravity assists *J. Astronaut. Sci. 37* 417-432

Zuber, M. T., Smith, D. E., Solomon, S. C., Philips, R. J., Peale, S. J., Head III, J. W., Hauck, S. A., McNutt, R. L. Jr, Oberst, J., Neumann, G. A., Lemoine, F. G., Sun, X., Barnouin-Jha, O., & Harmon, J. K. 2008, Laser Altimeter Observations from MESSENGER's First Mercury Flyby *Science* 77-79

Galileo's Medicean Moons: their impact on 400 years of discovery
Proceedings IAU Symposium No. 269, 2010
C. Barbieri, S. Chakrabarti, M. Coradini & M. Lazzarin, eds.

doi:10.1017/S1743921310007362

Jupiter and the other Giants: A Comparative Study

Thérèse Encrenaz
LESIA, Observatoire de Paris
F-92190 Meudon, France
email: therese.encrenaz@obspm.fr

Abstract. The four giant planets - Jupiter, Saturn, Uranus and Neptune - have common properties which make them very different from the terrestrial planets: located at large distances from the Sun, they have big sizes and masses but low densities; they all have a ring system and a large number of satellites. These common properties can be understood in the light of their formation scenario, based upon the accretion of protosolar gas on an initial icy core. Giant planets have been explored by space missions (Pioneer 10 and 11, Voyager 1 and 2, Galileo and Cassini) but also by Earth-orbiting satellites and ground-based telescopes. There are still open questions related to the origin and evolution of the giant planets, in particular their moderate migration, the origin of the cold planetesimals which formed Jupiter, the origin of the atmospheric dynamics in Jupiter and Saturn, and the differences in the internal structures of Uranus and Neptune.

Keywords. Planets and satellites: formation, Planets and satellites: general; Solar system: formation

1. Introduction

The four giant planets - also called jovian planets - are located in the outer solar system, at heliocentric distances ranging between 5 and 30 AU. As shown in Table 1 and 2, their orbital and physical properties (mass, radius, density), as well as their rings and satellite systems make them very different from the terrestrial planets (Mercury, Venus, the Earth and Mars) which all orbit within 2 AU from the Sun.

2. Early observations

Jupiter and Saturn, as naked-eye objects, have been known since Antiquity. However, their physical properties remained unknown until Galileo, in 1610, started to look at them with his new telescope. His major discovery was the identification of four satellites around Jupiter (the "Medicean moons", later renamed galilean satellites), which came in support of the heliocentric system proposed a few decades earlier by Copernic. In addition to other spectacular results - the relief on the Moon, the phases of Venus, the multitude of stars which populate the Milky Way -, Galileo also observed the changing aspect of Saturn's disk, depending on the orientation of its rings with respect to the Earth, but could not identify the explanation of this variation. The answer was given in 1659 by Christiaan Huygens, who also discovered Saturn's largest satellite Titan. With the apparition of large observatories at the end of the XVIIth century, astronomical observations of Jupiter and Saturn were made on a regular basis. Cassini monitored the Great Red Spot of Jupiter and its system of zones and bands; he identified a gap in Saturn's rings, later called Cassini's Division, and discovered several of Saturn's icy satellites.

Table 1. Orbital properties of solar-system planets.

Name	Semi-major axis (AU)	Eccentricity	Inclination over the Ecliptic (°)	Revolution period (years)
Terrestrial Planet				
Mercury	0.39	0.205	7.00	0.241
Venus	0.72	0.007	3.39	0.615
Earth	1.00	0.017	0.00	1.000
Mars	1.52	0.093	1.85	1.881
Giant Planet				
Jupiter	5.20	0.054	1.30	11.856
Saturn	9.54	0.047	2.48	29.424
Uranus	19.2	0.086	0.77	83.747
Neptune	30.1	0.008	1.77	163.723

Table 2. Physical properties of solar-system planets.

Name	Mass (M_E)	Equatorial radius (R_E)	Density (g/cm^3)	Rotation period	Obliquity (°)
Terrestrial Planet					
Mercury	0.055	0.382	5.43	58.646 d	0.0
Venus	0.815	0.949	5.20	243.018 d	177.33
Earth	1.000	1.000	5.52	23.934 h	23.45
Mars	0.107	0.532	3.93	24.623 h	25.19
Giant Planet					
Jupiter	317.9	11.21	1.33	9.925 h	3.08
Saturn	95.16	9.45	0.69	10.656 h	26.73
Uranus	14.53	4.00	1.32	17.24 h	97.92
Neptune	17.14	3.88	1.64	16.11 h	28.80

At the end of the XVIIIth century, a new step was achieved in planetary exploration with the discovery of a new planet beyond Saturn. In 1781, William Herschel, using a larger telescope of improved quality, detected an unknown object which was called Uranus, thus extending the frontiers of the solar system. At the same time, a new science, celestial mechanics, was developing rapidly, on the basis of Newton's law of universal gravitation. In 1821, precise calculations of the giant planets' orbits showed evidence for an anomaly in Uranus' orbit; the only possible explanation had to be the orbit's gravitational perturbation by a more distant planet. Two astronomers, John Couch Adam in England and Urbain Le Verrier in France, independently found the coordinates of the object. Le Verrier sent the coordinates of the new planet to Johannes Galle who immediately found it within 1° of its predicted position. The discovery of this new planet, called Neptune, marked the triumph of celestial mechanics.

3. The space exploration

Giant planets were first explored by two NASA-led missions in the 1970s, first by the Pioneer 10 and 11 spacecraft, then by the Voyager mission. Pioneer 10 encountered Jupiter in 1973; Pioneer 11 encoutered Jupiter in 1974 and Saturn in 1979. The Pioneer mission provided us with the first amazing images of Jupiter's meteorology and Saturn's rings, and with the first exploration of the planets' magnetospheres. Soon after Pioneer, the very ambitious and successful Voyager mission has been a major milestone in the space exploration of the giant planets. Two identical spacecraft, Voyager 1 and Voyager 2, successively encountered the four giant planets between 1979 and 1989. Voyager 1 flew by Jupiter in 1979 and Saturn in 1980, with a close approch of Titan; Voyager 2 flew by Jupiter in 1979, Saturn in 1981, Uranus in 1986 and Neptune in 1989. Among the Voyager highlights are the discoveries of Io's volcanism, Europa's probable subsurface water ocean, the complex structure of Saturn's rings, Miranda's traces of tectonic activity, Triton's cryovolcanism, and the complex structure of the giant planets' magnetospheres.

Ten years later, the Galileo mission was launched by NASA for an in-depth exploration of the jovian system. On December 7, 1995, a descent probe entered Jupiter's atmosphere and transmitted data down to a pressure level of 22 bars. An orbiter performed an in-depth exploration of the planet and the galilean satellites from 1995 to 2003. Ten years later, Saturn's system was approached by the Cassini mission, launched in 1997 by NASA in collaboration with ESA. On January 14, 2005, the Huygens probe, led by ESA, successfully landed on Titan's surface and provided the world with the first images of Saturn's biggest satellite. The Cassini orbiter, in operation since 2005, is planned to continue its exploration of Saturn's system until 2017.

In addition to in-situ planetary exploration, our knowledge of the giant planets has benefited from Earth-orbit observations : the International Ultraviolet Explorer (IUE) and the Hubble Space Telescope (HST) in the UV, the HST in the visible, the Infrared Space Observatory (ISO) and Spitzer in the IR, and more recently Herschel in the far-infrared and submillimeter range. Infrared and millimeter ground-based observations have also been precious for investigating the chemical composition of the giant planets.

Figure 1. Jupiter (left) and Saturn (right) as observed by the Cassini spacecraft in 2000 and 2004 respectively ((c) NASA).

4. The giant planets' formation scenario

Why do we have two distinct classes of planets in the solar system? The answer can be found in the formation scenario of the solar system, now widely accepted in the scientific community. This scenario is based on a few simple observations: all planets rotate in the same direction around the Sun on nearly circular, coplanar and concentric orbits. These facts led to the concept of the Primordial Nebula model, proposed in the XVIIIth century, first by Immanuel Kant and later by Pierre-Simon de Laplace : a fragment of interstellar cloud collapses into a disk, as a result of its own gravity and rotation. At the center, matter accretes to form the young Sun while, within the turbulent disk, planets form from the accretion of solid particles. This scenario, observed on a number of nearby young stars and protoplanetary disks, is believed to be common the Universe; it is even more supported with the discovery of several hundreds extrasolar planets over the past fifteen years. Particles gather into small embryos as an effect of turbulence and multiple collisions, and accrete into km-size planetesimals; later, the largest cores sweep the surrounding material and grow by gravity. Numerical simulations show that a small number of planet-sized objects can emerge within a few million years or less.
Among the protosolar disk, the size of the planet finally depends upon the amount of solid material available to build the core; this is where the difference between terrestrial and giant planets appears. The protosolar disk is mostly composed of light elements (hydrogen and helium). The abundances of other elements (C, N, O, ...) follow the cosmic values, i.e. the heaviest ones are the least abundant. Near the Sun, the temperature is high enough for the simple molecules (H_2O, CH_4, NH_3, CO_2, H_2S...), most abundant after H_2 and He, to be gaseous. The only solid material is thus made of silicates and metals. As their cosmic abundance is small, only small and dense objects can be formed: they are the *terrestrial planets*. In contrast, at larger heliocentric distances, where the temperature is below about 200 K, the simple molecules mentioned above are in the form of ices and big cores can be formed. Models predict that, when a critical mass of about 10-15 terrestrial masses is reached, their gravity is sufficient for the surrounding nebula (mostly made of gaseous hydrogen and helium) to collapse on them and to form a sub-disk along the cores' equatorial planes. As a result, big planets of low density are formed: the *giant planets*. Within the sub-disks, solid matter accretes into several bodies: the regular satellites of the giant planets. Ring systems are found in the immediate

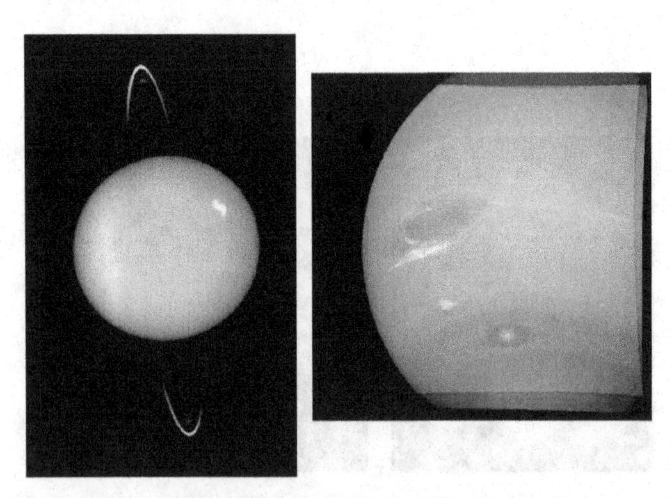

Figure 2. Left: Uranus, as observed by the HST in 2005. Right: Neptune, as observed by Voyager 2 in 1989 ((c)NASA)

proximity of the planets, within the Roche limit, where tidal forces prevent the accretion of solid bodies larger than planetesimals. The limit beween the terrestrial planets and the giant planets is called the "snowline" which corresponds to the distance of condensation of the ices. Water plays a major role for two reasons: first, being made of two abundant atoms, H and O, it is the most common ice; second H_2O is the first molecule to condense as the temperature decreases; other ices (NH_3, CO_2, H_2S, CH_4...) condense at larger heliocentric distances. This explains the composition of outer satellites, mostly made of water ice.

Table 2 shows that the giant planets fall into two categories. Jupiter and Saturn, with masses close to 300 and 100 terrestrial masses, are mostly made of protosolar gas; they are called the gaseous giants. In contrast, Uranus and Neptune, with masses of about 15 terrestrial masses, are mostly made of their icy core; they are called the icy giants. What is the origin of this difference? A plausible explanation might be that Uranus and Neptune, formed at larger heliocentric distances, needed more time to accrete their icy core, possibly more than 10 million years. The critical mass of 10 terrestrial masses might have been reached after the dissipation of the protosolar gas. Indeed, the observation of nearby protoplanetary disks shows that their lifetime is usually shorter than 10 My.
The study of protoplanetary disks and extrasolar planets also suggests that many giant exoplanets, found close to their parent star, have encountered inward migration, as a result of their interaction with the turbulent disk. Recent dynamical calculations, performed at the Nice Observatory by A. Morbidelli and his colleagues, suggest that in the solar system, giant planets exhibited a moderate migration, with Jupiter moving inward and the three other giants moving outward. A signature of this migration appears to be the Late Heavy Bombardment (LHB), observed on the surfaces of all bare solar-system objects as a strong maximum of impact craters dated at -3.8 Gy, i.e. about 800 My after the planets' formation. Numerical simulations show that, at that time, the Jupiter-System may have crossed the 2:1 resonance (Saturn's revolution period being twice Jupiter's one), which must have led to a huge perturbation in the orbits of small bodies, with high eccentricities and inclinations. Now, why was the giant planets' migration moderate in the solar system, while it appears to be so efficient around nearby stars? This question is still open.

5. Atmospheric composition and structure

The thermal structure of the giant planets shows different regions. In the convective trospophere, the temperature decreases as the altitude increases, following the adiabatic gradient, up to the tropopause where the temperature is minimum. The pressure level at the tropopause is about 100 mbar for all giant planets; the temperature at this level is 110 K for Jupiter, 90 K for Saturn, and about 50 K for Uranus and Neptune. In the stratosphere, above the tropopause, the temperature increases again with altitude, due to the absorption of solar radiation by methane and aerosols. In this region, the photodissociation of methane leads to the formation of several hydrocarbons (C_2H_2, C_2H_6...).
In the hydrogen-dominated atmospheres of the giant planets, most of the species are in reduced form. Tropospheric species include CH_4, NH_3, H_2O, PH_3, GeH_4 and AsH_3. In the case of Uranus and Neptune, only methane is observed, because the condensation levels of the other species are too deep to be observable. In 1997, the ISO satellite has detected oxygen species in the stratospheres of all giant planets (H_2O, CO_2). The origin of this external oxygen source might be either local (rings, satellites) or interplanetary (comets, micrometeorites). CO has also been detected; its origin (internal or external, or both) is still a matter of debate. In the case of Neptune, unexpected stratospheric

abundances of CO and HCN have been found; a possible origin might be a cometary impact. Such event (the collision of comet Shoemaker-Levy 9 with Jupiter in 1994) is believed to be at least partly responsible for the presence of the stratospheric water observed during the following years, as well as the CO and HCN high stratospheric contents.

The cloud structure of the giant planets can be studied by the images of the planetary disks. The Great Red Spot, identified on Jupiter since the XVIIth century, is believed to be a giant anticyclonic feature; its stability over such a long time is still not fully understood. Saturn is also dynamically very active, with huge cyclones at both poles and a peculiar hexagonal convective feature around the north pole. The mechanism of such activity is still an open problem. The alternative structure in zones and belts, easily observed on Jupiter and Saturn, is the signature of a Hadley-type convective circulation in a fast rotating gaseous sphere. The zones are cloudy regions of ascending motions and the belts are dry regions of subsidence. Additional information on the cloud structure of giant planets is given by models of thermochemical equilibrium. In the case of Jupiter and Saturn, a NH_3 cloud is expected (and indeed observed as a white-yellow layer) at a temperature level of 145 K (P = 0.5-1 bar). Clouds of NH_4SH (T= 210 K) and H_2O (T=270 K) are expected at deeper levels. In the case of Uranus and Neptune, the temperature is low enough for CH_4 to condense at a level of about 1 bar (T = 80 K). A H_2S cloud is expected at about 3 bars (T = 120 K), and clouds of NH_4SH and H_2O are expected at deeper levels. In the stratosphere, hydrocarbon aerosols are expected, coming from the methane photolysis.

In the case of Jupiter, the Galileo probe, in 1995, entered one of the dry "hot spots" which probe the deep troposphere. Elemental abundance ratios were mesured by mass spectrometry for several elements. For six of them (C, N, S, Ar, Xe, Kr) an enrichment by a factor 4 with respect to hydrogen was mesured, as compared with the protosolar values. This enrichment was found to be in full agreement with the predicted values derived from the nucleation model of the giant planets, assuming an initial icy core of 12 terrestrial masses. This result, however, raised an unexpected problem. The uniform enrichment by a factor 4 implies that all the measured elements were equally trapped in

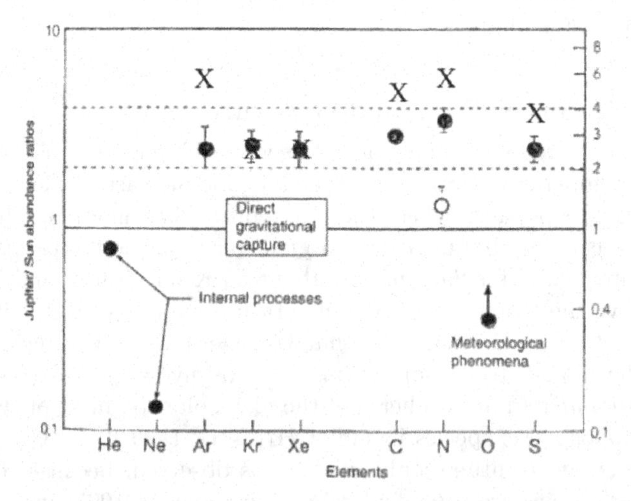

Figure 3. Enrichments of heavy elements in Jupiter as measured by the Galileo Probe Mass Spectrometer. The crosses indicate updated values, following a reestimation of the solar abundances. He and Ne are believed to be depleted by condensation within the metallic hydrogen of Jupiter's interior. O is depleted in the hot spot that the probe explored by convective motions. The figure is adapted from Owen *et al.* (1999) and Owen and Encrenaz (2006).)

ices. Still, this is not expected for Ar and N unless the temperature is very low (< 40 K), much lower than the expected temperature at Jupiter's orbit (about 110 K). What is the origin of the cold planetesimals which formed Jupiter? This is still an open question. What is the situation for the other giants? Unfortunately, we have no in-situ measurement; however, C/H is known from CH_4 measurements, and here again, the measured C/H ratio is in full agreement with the predictions of the nucleation model: an enrichment factor of 9 for Saturn and about 30 for Uranus and Neptune. This conclusion is also supported by the measurement of the D/H ratio in the giant planets. This ratio, which is enriched in ices at low temperatures, is indeed diagnostic of the temperature formation of these objects. D/H in the giant planets has been measured from the observation of HD infrared lines by ISO. D/H in Jupiter and Saturn was found to be consistent with its protosolar value ($2.1\ 10^{-5}$), which was expected as these planets are mostly made of protosolar gas. In contrast, D/H in Uranus and Neptune was found to be enriched by a factor of about 2.5, in agreement again with the core accretion model. What is the temperature formation of the planetesimals which formed Saturn, Uranus and Neptune? The answer will be given by future space missions, when descent probes are sent into their tropospheres.

6. Similarities and differences among the giant planets

With comparable chemical compositions, one would expect the spectra of the gaseous giants (Jupiter and Saturn) to be similar, as well as those of the icy giants (Uranus and Neptune). It is indeed the case of the near-infrared solar component (1-3 μm) where all spectra are dominated by methane, a very active spectroscopic agent. In the mid-infrared thermal range, spectra show significant differences which are the signature of intrinsic differences in their atmospheric dynamics. A first example is shown in Figure 4, in the 7-12 μm range. The spectra of Jupiter and Saturn show drastic differences : in particular, a strong NH_3 band dominates the Jovian spectrum while Saturn's exhibits CH_3D and PH_3 signatures. There are two main reasons for these differences. The first is the colder environment of Saturn, where ammonia is mostly condensed in clouds. The second reason is a higher vertical dynamical activity on Saturn. This is illustrated

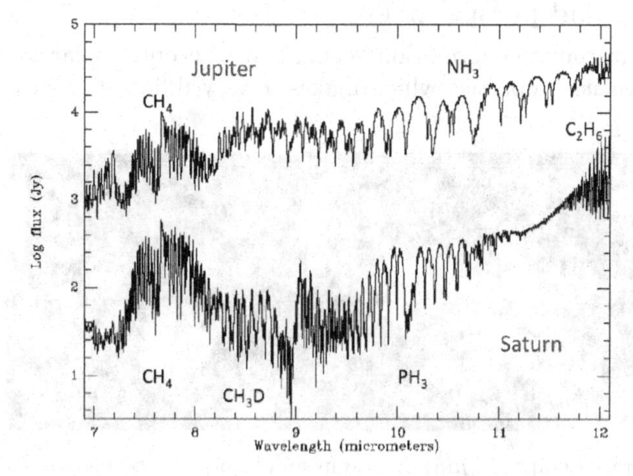

Figure 4. The spectra of Jupiter and Saturn, as observed in 1997 by the ISO Short-Wavelength Spectrometer (SWS). Adapted from Encrenaz (2003).

by the large abundance of phosphine (PH_3), a disequilibrium species which should not be observable according to thermochemical equilibrium models. Its presence in large abundances in Saturn's troposphere is the signature of strong upward motions which carry it from deep interiors (the same effect applies on Jupiter but to a weaker extent). A third difference between Jupiter and Saturn is the larger methane abundance in Saturn (for the reason mentioned above, see Section 5) which translates into a larger content of stratospheric hydrocarbons.

Another example can be found in the temperature maps of Uranus and Neptune, recorded at the VLT in 2006. Figure 5 mid-infrared maps of the two planets at 17 μm; this wavelength probes their tropospause. The map of Uranus is more or less symmetrical with respect to the equator, as expected for the present season since Uranus is close to equinox. Neptune, in contrast, exhibits a strong maximum at the south pole.

A possible explanation is that, since Neptune's southern summer solstice was in 2004, its south pole has been continuously illuminated for 40 years. This polar heating has an important consequence: gaseous methane, usually trapped in a cloud below the tropopause, can escape in the stratosphere and spread over the planet. This effect explains the anomalously high contents of methane and hydrocarbons in Neptune's stratosphere (Figure 6). Another difference between the two planets enhances this effect: just as Saturn with respect to Jupiter, Neptune shows a stronger vertical dynamical activity than Uranus, as shown by its active meteorology.

What could be the reason for these different dynamical behaviours? In the case of Uranus, the lack of atmospheric activity observed at the time of the Voyager encounter (1986) could have been due to the peculiar geometry of the planet, as its rotation axis was pointing toward the Sun. Indeed, we observed in 2005, just before equinox, some signs of meteorology (Figure 2). Another explanation, possibly connected with the previous one, is the lack of internal energy on Uranus, as compared to the other giants. In the case of Jupiter and Saturn, Voyager has measured an energy excess of 1.7 and 1.8 respectively, with respect to the absorbed solar energy; in the case of Neptune, the excess factor is 2.6. The most plausible origin of this internal energy is the planet's contraction and cooling following the planet's accretion phase. Why is this mechanism not observed on Uranus? Possibly, convection is inhibited in Uranus' interior for some unknown reason.

7. Conclusions and perspectives

In spite of their common formation scenario in the outer solar system, each giant planet shows intrinsic properties which makes it very different from the others. Many

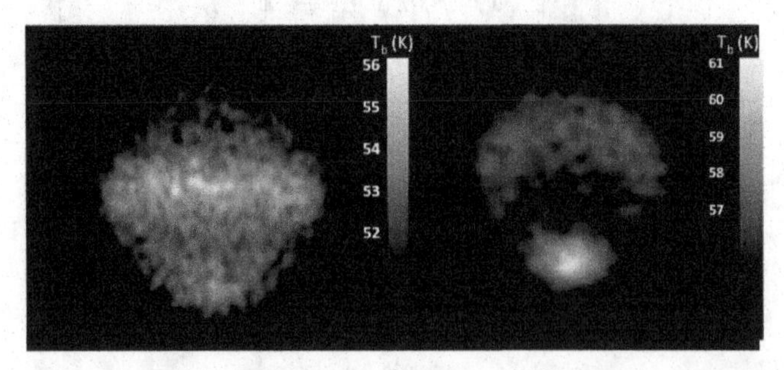

Figure 5. Mid-infrared maps (17 μm) of Uranus and Neptune recorded in September 2006 with the VISIR instrument at the VLT. The 17 μm wavelength probes the tropopause at a pressure level of about 100 mbar. Adapted from Orton *et al.*, 2007.

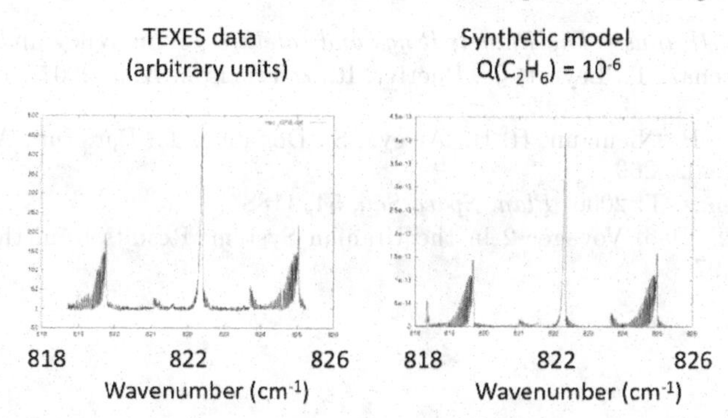

Figure 6. Infrared emission of ethane (C_2H_6) on Neptune, observed with the TEXES instrument at IRTF in 2003. Left: TEXES data; right: synthetic model. These data illustrate the capabilities of ground-based infrared spectroscopy for studying the atmospheric composition of planets. ((c) T. Greathouse, J. Lacy and the TEXES Team).

pending questions still arise about their origin and evolution : What is the origin of the cold planetesimals which formed Jupiter? What is the enrichment in heavy elements in Saturn, Uranus and Neptune relative to hydrogen, as compared with their protosolar value? Did the giant planets migrate in the outer solar system, and by how much? If so, why did the migration stop? Why is Neptune more massive than Uranus, although at a larger heliocentric distance? Why is Uranus' rotation axis tilted to the ecliptic plane? Why are the internal structures of Uranus and Neptune different?

To answer these questions, new investigations are needed. Two ambitious missions are presently under study at NASA and ESA: The Europa and Jupiter System Mission (EJSM), which might be launched about 2020, and later the Titan and Saturn System Mission (TSSM) for horizon 2025. However, there are presently no plans to explore Uranus and Neptune, nor to send in-situ probes to measure the elemental composition of the giant planets. Still, these measurements are key diagnostics to understand the formation processes of these objects. This research is of special importance now, as it might help us to better understand the origin and history of the hundreds of exoplanets recently discovered around nearby stars.

References

Bagenal, F., Dowling, T., & McKinnon, W. 2007, *Jupiter: The planet, satellites and magnetosphere*, Cambridge Planetary Science, Cambridge University Press

Bergstrahl, J. T., Miner, E. D., & Matthews, M. S. 1990, *Uranus*, University of Arizona Press

Bézard, B. 2001, Neptune *Encyclopedia of Astronomy and Astrophysics*, P. Murdin, Edt. IoP Publishing, p.798

Cruikshank, D. P. 1995, *Neptune and Triton*, University of Arizona Press

Drossart, P. 2001, Saturn *Encyclopedia of Astronomy and Astrophysics*, P. Murdin, Edt. IoP Publishing

Encrenaz, T. 2001, Jupiter *Encyclopedia of Astronomy and Astrophysics*, P. Murdin, Edt. IoP Publishing, p. 1330

Encrenaz, T. 2003, *Plan. Space Sci.*, 51, 89

Gehrels, T., & Matthews, M. S. 1984, *Saturn*, University of Arizona Press

Herbert, F. 2001, Uranus and Neptune: Atmospheres, Ionospheres ad Magnetospheres, *Encyclopedia of Astronomy and Astrophysics*, P. Murdin, Edt. IoP Publishing, p. 3408

Lumine, J. I. 1993, The atmospheres of Uranus and Neptune, *Ann. Rev. Astron. Astrophys.*, 31, 217

Miner, E. D. 1998, *Uranus - The Planet, Rings and satellites*, John Wiley and Sons, New York

Orton, G. S., Encrenaz, T., Leyrat, C., Puetter, R., & Friedson, A. J. 2007, *Astron. Astrophys*, 473, L5

Owen, T., Mahaffy, P., Niemann, H. B., Atreya, S., Donahue, T., Bar Nun, A., & de Pater, I. 1999, *Nature*, 402, 269

Owen, T. & Encrenaz, T. 2006, *Plan. Space Sci.*, 54, 1188

Smith, B. A. *et al.* 1986, Voyager 2 in the Uranian System: Results from the Imaging Team *Science*, 233, 97

Galileos Medicean Moons: their impact on 400 years of discovery
Proceedings IAU Symposium No. 269, 2010
C. Barbieri, S. Chakrabarti, M. Coradini & M. Lazzarin, eds.

doi:10.1017/S1743921310007374

Spectroscopic and spectrometric differentiation between abiotic and biogenic material on icy worlds

Kevin P. Hand[1], Christopher P. McKay[2] and Carl B. Pilcher[3]

[1]Jet Propulsion Laboratory, California Institute of Technology,
MS 183-601, 4800 Oak Grove Drive, Pasadena, CA 91109
email: khand@jpl.nasa.gov

[2]NASA Ames Research Center
Moffett Field, CA 94035
email: chris.mckay@nasa.gov

[3]NASA Astrobiology Institute
NASA Ames Research Center, Moffett Field, CA 94035
email: carl.b.pilcher@nasa.gov

Abstract. The ability to differentiate abiotic organic material from material of a biological origin is a critical task for astrobiology. Mass spectrometry and spectroscopy provide key tools for advancing this task and are two techniques that provide useful and highly complementary compositional information independent of a specific biochemical pathway. Here we address some of the utility and limitations of applying these techniques to both orbital and in situ exploration of icy moons of the outer solar system.

Keywords. Astrobiology, Europa, Moons, Biosignatures, Exobiology, Spectroscopy.

1. Introduction

Astrobiology is the study of the origins, evolution, distribution, and future of life in the Universe, including life on Earth. As with the study of any phenomenon in the Universe - from neutrinos to earthquakes - our knowledge and understanding is limited by our capability to observe and measure the phenomenon in question. The challenge that empiricism imposes has lead to some of the most advanced technical achievements of our civilization, the large hadron collider built by the European Organization for Nuclear Research (a.k.a. CERN) perhaps representing the most extraordinary such achievement to date. In the decades to come, the challenge of astrobiology will continue to push the technological frontier in the pursuit of new knowledge.

As we move forward with our exploration of the Solar System and beyond, the search for, and detection of life on distant worlds persists as one of our top priorities (NRC 2003). For the first time in the history of humanity, we have the technological capability to actively pursue this endeavor. We can now build the spacecraft needed to orbit, land, and investigate the many worlds in our Solar System that may have been, or may currently be, habitable. Nevertheless, our understanding of life as a phenomenon remains largely qualitative and poorly constrained. Indeed, as a phenomenon it lacks a singular universal definition (Cleland and Chyba 2002), though several operational definitions have been proposed (NRC 2003).

Furthermore, the science of biology is not a science based on first principles from which one can then derive an understanding of observations (*e.g.* as one can do with the Law of Gravity and orbital dynamics). In biology, specific predictions are difficult, though

Darwinian evolution provides a context for expectations. Given our current understanding of life and the origin of life, the closest we can come to a predictive equation is (Hand *et al.* 2009),

$$\text{Liquid water} + \text{Biologically essential elements} + \text{Energy} + \text{Catalytic surface} \rightarrow \text{Life}$$

The lack of a definition for 'life' presents considerable challenges for astrobiology. Joyce (1995) humorously notes that the popular definition for life may well be 'that which is squishy'. Of course this definition leaves much to be desired from a measurement standpoint. Though somewhat broad in scope, the 'chemical Darwinian' definition that "life is a self-sustained chemical system capable of undergoing Darwinian evolution" is often offered by biologists as a useful starting point (Joyce 1995; Chyba and McDonald 1995). Darwinian evolution includes three processes: amplification, that is, replication of a genetic description of a prototype; mutation to introduce variation during amplification; and selection of replicates produced during amplification to be used for further amplification (Joyce 1995). This is useful for assessing extant life over time, but when designing an instrument suite for an orbiting or landed spacecraft the prospect of observing Darwinian evolution poses obvious challenges. Furthermore, the detection of life includes extinct life; fossil evidence must be included in our operational definition.

An engineering perspective is useful here. We may think of a circuit containing a battery as an analogue for life. If the circuit is left open a charged battery will slowly lose charge; if a circuit is closed between the terminals, the battery will dissipate energy faster than in the open-circuit scenario. Life closes the chemical and radiation energy circuits available in the environment, advancing the second law of thermodynamics. Depending on the dynamics of the environment, the life structure that provided the 'closed-circuit' can persist in the environment long after the energy has been dissipated, thus providing potential fossils for detection. As presented by McKay (2004) and in detail by Summons *et al.* (2008) and other workers throughout the history of organic geochemistry, the structures of life arise from a relatively small set of universal building blocks. Thus when we search for life we might be well served by looking for patterns indicative of life's structural biases. McKay (2004) referred to this as the 'Lego principle', in reference to the toy building blocks that come in specified basic structural units.

Ice covered worlds like Europa and Enceladus are highly unlikely to have life on their surfaces, though some interesting concepts have been proposed. Instead, our first dedicated missions to these worlds will need to be able to assess their habitability and characterize in detail their surface compositions. Here we examine the utility of two broad methods, spectrometry and spectroscopy, for differentiation between abiotic and biogenic material on ice covered moons of the outer Solar System. By abiotic we mean compounds and complexity not generated by life. By biogenic, we mean of or derived from life. Europa is presented as the type example of a possibly habitable ice-covered ocean world.

2. On the Habitability of Europa

The evidence for a sub-surface ocean on Europa is strong and consistent with a broad range of compositional and geophysical models (Greeley *et al.* 2004; Schubert *et al.* 2004). Arguably the most compelling data is the combination of ground-based Doppler RADAR measurements of the Galileo spacecraft, and the magnetometer measurements made from Galileo (Anderson *et al.* 1998; Khurana *et al.* 1998; Zimmer *et al.* 2000). The combination of these observations necessitate a near-surface conducting layer of unit density material below the surface ice of Europa. This set of empirical constraints is hard to satisfy without the presence of a salty, global, sub-surface ocean approximately 100 km thick. Though subject to considerable debate, geological and compositional data also supports the ocean hypothesis (Pappalardo *et al.* 1998; McCord *et al.* 1998).

The putative liquid water ocean of Europa is a top priority for future astrobiological investigation (NRC 2003). Though photosynthetic organisms are unlikely due to the thickness of the ice shell (estimated to be a few to < 20 km in thickness, Hand and Chyba (2007) and references therein), numerous chemosynthetic pathways for life in Europa have been proposed based on models for the availability of the required compounds (Zolotov and Shock 2004). The temperature and pressures predicted for the europan ocean are comparable to those found in the depths of the Earth's oceans, and McCollom (1999) has argued that an active seafloor on Europa could support considerable biomass. If Europa were uninhabited, such hydrothermal activity would nevertheless likely drive the production of organics and other prebiotic compounds on Europa. Key regions of interest for habitability are the ice-water interface of the ice shell and the water-rock interface of the seafloor (Figueredo *et al.* 2003). Finally, the irradiated surface ice of Europa may also serve as an important source for compounds of both prebiotic and biological interest (Chyba 2000; Kargel *et al.* 2000; Hand 2007).

Exchange of ocean material with the surface has not been observed on Europa (see *e.g.* Phillips *et al.* (2000)). Nevertheless, delivery of oceanic material - and any life or prebiotic compounds contained in the water - to the surface of Europa is hypothesized to occur in several different ways. If the ice of Europa is thin (a few kilometers), fracturing through the entire shell can occur and direct communication between the ocean and surface is possible. In this case, emplacement of material onto the surface would occur through cryovolcanic eruptions and extensional fracturing that exposes fresh oceanic material to the surface. If the ice shell of Europa is thick ($\sim$15 km or more) a thin brittle ice shell may overlay a thick convecting ductile ice layer (Pappalardo and Head III 2001). In this case, convective cells within the ductile layer may serve to deliver oceanic material to the base of the brittle ice layer. Tectonic activity in the brittle layer would then be required to bring subsurface material to the surface. Carlson *et al.* (2009) provide a summary of various mechanisms for such exchange.

The surface of Europa receives an average of 125 mW m^{-2} of energetic electron and ion irradiation (Cooper *et al.* 2001). This radiation drives chemical modification and weathering of the surface, creating a considerable challenge for differentiating abiotic organic chemistry driven by radiolysis on the surface of Europa from degraded biological chemistry resulting from delivery of oceanic material to the surface. An important aspect of differentiating such exogenous surface chemistry from endogenous biosignatures will be mapping of surface ages and regions of recent activity on Europa. Once such regions are identified based on geological interpretation (*e.g.* crater counts and monitoring of fractures) then chemical mapping of organic compounds containing nitrogen, phosphorous and sulfur will be a an important step towards identifying possible surface biosignatures of subsurface life.

3. The detection of Life on Earth

The technological challenge of detecting life beyond Earth brings into question our capabilities for detecting life on Earth. The development of such technologies, and the extension of such innovations to other worlds, is one of the primary goals of NASA's Astrobiology Program.

The inherent bias, of course, is that many of the tools and techniques to study terrestrial biology are tethered to the specific attributes of life on Earth, which, to the best of our knowledge comprises one contiguous genetic tree and one universal biochemistry (that of DNA, RNA, and proteins). Perhaps the best example of this bias is the polymerase chain reaction (PCR) technique that has opened the floodgates to understanding the microbial biosphere. Though revolutionary for the study of RNA/DNA-based life, PCR essentially uses a RNA/DNA baseline to find more RNA/DNA life forms. If you do not know *a priori* that you might find RNA/DNA-based life, then PCR does you little good. Many other biochemical techniques suffer similar pitfalls.

Mass spectrometry is one of the few *in situ* techniques that provides useful information on chemical complexity without being tethered to the RNA-DNA-Protein paradigm. The limitation, as discussed in the following section, is that mass spectrometry by itself and without any information on the geological or chemical context of the sample, is not sufficient for answering the binary question 'Is life present?.' If coupled with measurements of isotopic fractionation (*e.g.* δ^{13}C) and measurements of a chiral excess of one enantiomer (*e.g.* in amino acids and/or sugars), then a very compelling case can be made for biology having played a role in the production of the sample material.

Detecting life from orbital or airborne platforms presents a host of challenges. Foremost among these challenges is that spectroscopic techniques are largely limited to using photosynthesis as a proxy for the bulk biosphere; very little attention has been given to non-photosynthetic spectroscopic biosignatures. When considering the question of life detection on Europa or Enceladus, one can start with the much more basic question: "Could we detect life on Earth if the ocean was frozen over?" The answer is a resounding 'Yes', but is contingent on photosynthetic cyanobacteria and the absorption features associated with chlorophyll-a, -b, and carotenoids. As discussed below, the use of orbital spectroscopy for the search for life on icy worlds - where photosynthesis is likely a marginal niche, at best - necessitates the identification and development of alternative spectroscopic features to serve as biosignatures for non-photosynthetic life forms. This endeavor advances our capability to search for life beyond Earth while simultaneously enhancing our capability to better monitor life here on Earth.

Finally, spectroscopic and mass spectrometric techniques are highly complementary. Vibrational (*i.e.* infrared) spectroscopy yields information about the functional groups, branches, and aromatic structure of a given compound. Spectra reveal the relationship of atoms to atoms and groups of atoms to other atoms within a compound. Mass spectrometry reveals the size of the compound and the size of the subunits generated by ionizing the compound (yielding the 'cracking pattern' of the compound). Together, this spectroscopic and mass spectrometric information permit a detailed understanding of the specific compounds under observation.

4. Spectrometric Signatures

For over 50 years mass spectrometry has served as a critical tool for assessing the complexity of organic matter (Peters *et al.* 2005). From oil exploration to the Viking Landers, the ability to measure the mass distribution of sampled compounds provides useful information about the sample composition. When searching for petroleum reserves, the mass distribution provides a metric of thermal maturity and the potential yield of various types of hydrocarbons from a given well. When searching for evidence of life on Mars, the Viking gas chromatograph and mass spectrometer set an upper limit on the concentration of organics in the lander regions - no organic compounds were detected at the parts per billion level (Biemann *et al.* 1977). Interestingly, however, we note that the Viking experiments were insufficient for detecting the expected background flux from meteorite infall (Benner *et al.* 2000).

As a tool for detecting life as we know it, mass spectrometry serves two key roles: 1) detecting organics , and 2) providing information on the complexity of organics. The first yields a binary answer (or rather an upper limit) that informs further investigation. The second yields information that may serve to distinguish abiotic organic chemistry from biologically generated organics. McKay (2004) has argued that the spectrometric distinction can be characterized as the 'Lego Principle'; biology preferentially uses specific organic subunits to build larger compounds while abiotic organic chemistry proceeds randomly. According to the 'Lego Principle', mass spectra associated with life are punctuated by peaks indicative of the subunit preferences while those associated with abiotic chemistry have a broad Gaussian spread of peaks with no obvious structure.

Though the 'Lego Principle' is broadly correct, caution must be taken when comparing abiotic organic chemistry to a rather featureless Gaussian distribution of organics. First, significant differences exist between the processes of abiotic synthesis and that of decomposition of organic material. Breakdown of organic material forms, to varying degrees, Gaussian distributions in mass spectra. As McKay (2004) notes, patterns distinctive of biology are typically lost when exposed to heat and other types of radiation. For example, the degree to which an oil well shows a Gaussian mass spectrum is used in part as a metric of the thermal maturity of the reservoir (Peters *et al.* 2005); more time and heat causes more of the original material to break down. Nevertheless, even the most thermally mature oil reservoir is not generally interpreted to be of abiotic origin (though some debates do persist).

Abiotic synthesis can often proceed by sequential addition of carbon atoms or CH_2, thus yielding a relatively smooth Gaussian mass spectrum, but other synthesis pathways exist that can yield considerably more complex spectra. Here we examine two experiments that reveal some of the complexity that can emerge from abiotic chemistry, and some of difficulty in distinguishing that chemistry from biological processes.

In their analysis of lipid formation via Fischer-Tropsch-Type (FTT) reactions under hydrothermal conditions, McCollom *et al.* (1999) demonstrated facile production of long chain alkanoic acids and alcohols - compounds found in many biological membranes. Starting with oxalic or formic acid, FTT production of lipids proceeded to yield compounds containing over 35 carbon atoms ($>C_{35}$). To explain the abundance of acids and alcohols in their results, McCollom *et al.* (1999) argue that FTT synthesis proceeds through addition of a COO group to the end of a hydrocarbon followed by reduction to an alcohol. Alkenes can then be formed via dehydration of the alcohol and formation of the double bond. In other words, abiotic synthesis does generate organic subunits and in some cases those subunits can be polymerized.

Importantly, though distinct patterns persist in the mass spectra from FTT synthesis, McCollom *et al.* (1999) reported that differentiation from biogenic lipids was possible because, unlike biology, FTT synthesis yielded: 1) few isoprenoids, a common baseline biomarker for life on Earth (though not necessarily for life on other worlds), 2) few branched chain lipids, 3) no even/odd carbon number abundance in the alkane and alkene series, 4) an abundance of saturated n-alkanols and n-alkanoic acids, and 5) a decreasing abundance of heavy carbon compounds beyond C_{15}.

FTT synthesis has long been appreciated as an important abiotic process for gas and liquid phase carbon reactions, but on worlds like Europa, the intense radiation environment and cold (100K) icy surface provides an unfamiliar setting for organic chemistry. Hand and Carlson have recreated the surface environment of Europa in the lab, complete with a 100 keV electron gun for irradiating ice mixtures (Hand 2007; Hand *et al.* 2007). Attached to the vacuum chamber is a quadrupole mass spectrometer used to scan organics up to an m/z of 100. Figure 1 shows the mass spectrum integrated over the warming

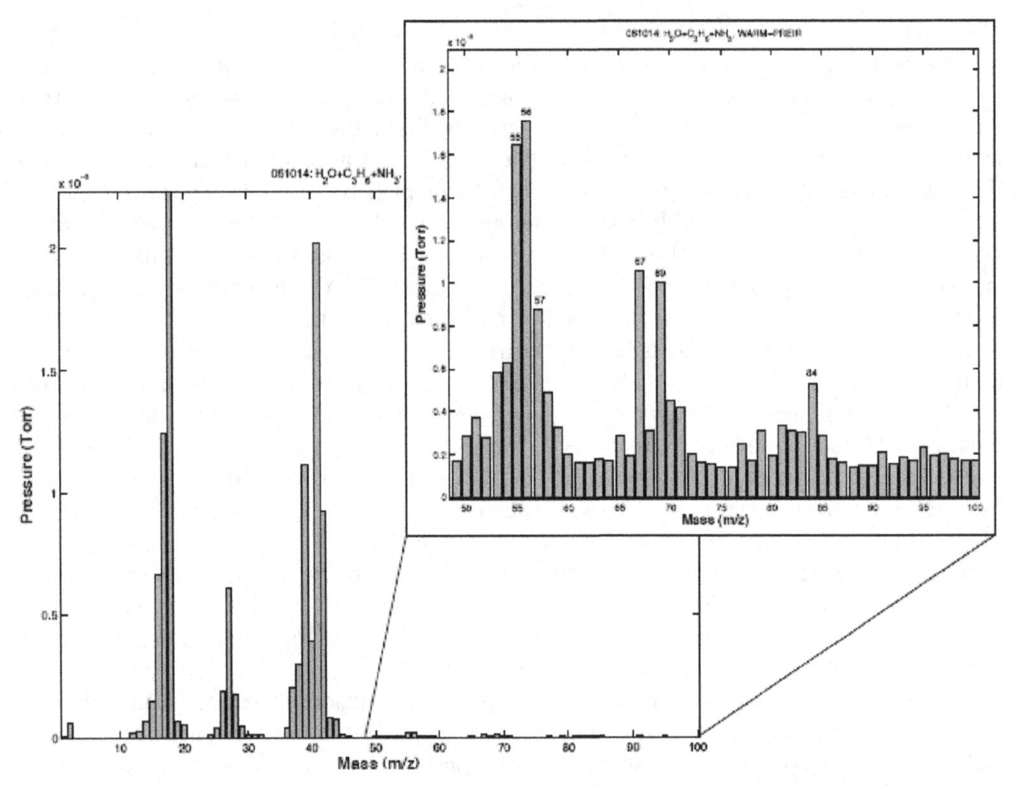

Figure 1. Mass spectrum results integrated over the warming period from 70-300 K after 10 keV electron irradiation of a H_2O+NH_3+C_3H_6 ice film (adapted from Hand and Carlson, *in prep.* The inset shows detail of the m/z region 50-100. Abiotic radiolysis shows preferential formation of certain hydrocarbons, not just a broad distribution of all variations. This complicates the use of the 'Lego Principle' (McKay 2004) when working to differentiate abiotic chemistry from biological chemistry on icy worlds.

from 70 K to 300 K of products evolved from an irradiated H_2O+NH_3+C_3H_6 ice film (modified from Hand and Carlson, *in prep*). Clearly mass spectra that extend to much larger m/z are critical for astrobiology, but the spectrometer used in this work provides a useful comparison to flown spectrometers (*e.g.* on *Cassini* and *Huygens*).

The cluster of peaks around m/z of 18 is from water and associated ions. At m/z = 27 lies a strong peak from HCN. Peaks at m/z = 41 and 39 were likely due to the OCN^- ion and to synthesized nitriles such as isocyanomethane (C-N≡C) and acetonitrile (C-C≡N). The peak at 39 is a related ion, HCNC.

At larger m/z, a cluster of peaks around m/z = 56 was seen. Much of this was likely due to the nitrile of propane (isocyanopropane, C-C-C≡N). Hydrocarbons without nitrogen, such as propenal may have also contributed to this peak. For the heaviest ions, those at m/z = 67, 69, and 84 (Figure 1 inset), several complex nitriles serve as candidates. For the peak at 84, formamide with a cyanomethyl- group is consistent with both the mass spectra and infrared data. Similarly, 2-cyanoacetamide contains similar bonds (though it hosts an NH_2 group, unlike cyanomethylformamide) and yields an m/z of 84. Loss of an NH, NH_2, or NH_3 would then yield the peaks at 67 and 69.

The critical observation for the present manuscript is that complexity exists even within the low mass range of this radiolysis experiment. The pattern of subunit synthesis is difficult to distinguish in this range, but based on matrix assisted laser desorption ionization analyses of the refractory residue left on the irradiated target, Hand and

Carlson (*in prep*) argue for polymers of isopropyl-propionate ($C_6H_{12}O_2$). In a similar experiment, Bernstein *et al.* (1995) reported synthesis of hexamethylenetetramine (HMT, or $C_6H_{12}N_4$) in their GCMS analysis of residues resulting from UV-photolysis of H_2O + CH_3OH + CO + NH_3 ice. Again, the observation was that abiotic synthesis can lead to preferential formation of select compounds.

As per McCollom *et al.* (1999), information within the structure of the mass spectra can often be used to distinguish abiotic from biogenic, at least for the case of known biochemistry. However, for carbon-based life on distant worlds, the lack of *a prior* knowledge of a preferential organic biochemistry could lead to false positives in environments where abiotic synthesis leads to complex patterns of organic polymers.

5. Spectroscopic Signatures

The elucidation of spectroscopic biosignatures has been the subject of considerable work, however, much of that work has focused on the gas phase, *i.e.* planetary atmospheres (des Marais *et al.* 2002; Pilcher 2003; Kiang *et al.* 2007b,a). Atmospheric biosignatures largely fall into two categories, 1) the detection of species in chemical disequilibrium, and 2) isotopic fractionation. Though of great utility for the study of worlds like Mars, Venus, and extrasolar planets, gas-phase biosignatures are generally not applicable to the vacuum environments present on the surfaces of many icy worlds. Indeed, at Europa the thin, radiolytically produced oxygen atmosphere (Hall *et al.* 1995) serves as an important reminder of the potential for false positives when employing the chemical disequilibrium argument.

For spectroscopic biosignatures of material in the solid phase, *e.g.* ices, work has been largely limited to pigments such as chlorophyl-a, chlorophyl-b, carotenoids, and other chromophore molecules (Painter *et al.* 2001). Blooms of the the algae *C. nivalis*, also known as 'snow algae', show a reflectance spectrum with absorption by carotenoids in the range $0.4 - 0.64$ μm and a local reflectance minimum at 0.68 μm due to absorption by chlorophyll-a and -b. Using ground calibration, the depth and breadth of the 0.68-μm feature was found to increase with increasing algal concentration, permitting mapping of algal concentrations in the snowpack from 20 km altitude (Painter *et al.* 2001).

The detection threshold for algal concentrations was determined to be ~2,300 cells ml^{-1}. Assuming that measurments of algal concentrations represented the mean for the top 10 cm, Painter *et al.* (2001) estimated the total biomass to be 16.55 kg for the ~0.75 km^2 region surveyed.

The detection threshold of Painter *et al.* (2001) is two orders of magnitude below the concentration of algae in the sea surface and over three orders of magnitude below that of microbes in hot springs and hydrothermal vents (Hand *et al.* 2009). Were life on Europa to contain biological pigments comparable to chlorophyll and carotenoids, detection from orbit at concentrations far below terrestrial oceanic concentration is feasible. Without such compounds, however, prospects for detection from orbit are very limited.

Spectral bands associated with the covalent bonds of biological structures are not as strong as the absorptions due to pigments and chromophores. Nevertheless, the systematic detection and mapping of these weaker spectral features will be our first tool for assessing biosignatures on the surfaces of icy worlds. In the sections that follow we divide spectroscopic features into 'Spectroscopic Signposts', 'Compelling Chemistry', and finally 'Complexity'. By 'spectroscopic signposts' we mean strong spectral features that are due to compounds that could be associated with biological processes. The strength of the feature makes it easy to see and the compound or bond revealed by the feature provides some indication that further examination of the region may be warranted. The features deemed as indicative of 'compelling chemistry' are those resulting from bonds between biologically essential elements such as carbon, nitrogen, phosphorous and sulfur. Finally, the features that fall into the 'complexity' class are those spectral features that result from added structural complexity in a molecule, *e.g.* shifts in the band position of a functional group resulting from polymerization and refinements of the vibrational modes available to the functional group in polymeric form.

5.1. *Spectroscopic Signposts*

Carbon dioxide has a very strong band at 4.27 μm and though CO_2 is not in and of itself a biosignature, this band can be used as an important sign post for carbon chemistry. When organic matter is exposed to the radiation environment of Europa's surface, some of that material is destroyed and some of the liberated carbon goes into CO_2. By mapping the geographic distribution of this spectral feature we will be able to generate a broad map of surface regions containing carbon chemistry, and possibly biochemistry.

The first spectroscopic clue toward detection of organic chemistry on Europa will be the $\sim$ 3.44 μm bands of the C-H stretch. This feature appears on the shoulder of the broad 3 μm water band and has been detected on Ganymede, Callisto, Enceladus, and comets (McCord *et al.* 1997; Brown *et al.* 2006; Chyba and Sagan 1987; A'Hearn *et al.* 2005). Both abiotic and biological hydrocarbons show this feature, but detection of this feature provides the basis for searching for weaker signatures indicative of carbon chemistry.

The second and third overtones of the C-H stretch ($2\nu_3$ and $3\nu_3$) for organics in ice occur at 1.67μm and 1.19μm respectively (Clark *et al.*, *in press*). Features of methane and ethane in ice occur between 2.27–2.46 μm. Methane has a stronger line at 7.67 μm and since methane is a by product of many chemosynthetic ecosystems on Earth, detection of methane features serves as a useful signpost. Finally, lipids and proteins display C-H absorption features near 2.3 μm (Dalton *et al.* 2003).

After water and carbon, nitrogen is often considered to be the next key elemental signpost for habitability and life detection (Mancinelli and Banin 2004; Capone *et al.* 2006). The weak N-H band at $\sim$3.1 μm is important to look for, but it overlaps with the broad water feature at 3 μm. The triple bond of carbon to nitrogen yields a strong band in the 4.56–4.60 μm region and this may be the best band for identifying the presence of nitrogen chemistry. The band occurs in abiotic radiolyitc chemistry experiments, as well as on planetary and astrophysical bodies, including Ganymede and Callisto (McCord *et al.* 1997).

5.2. *Compelling Chemistry*

After mapping strong and relatively easy features to identify, the next step is to search for features that can be associated with biochemistry. The carbonyl bond (C=O) has a strong band that appears at $\sim$5.9 μm and is associated with carbonic acid, esters, and aldehydes, all of which can be produced abiotically under europan conditions (see e.g. Hand *et al.* (2007)). If observed at slightly longer wavelengths (5.83–5.95 μm) then this band can be associated with the carbonyl stretch in nucleic acids (Naumann *et al.* 1996; Maquelin *et al.* 2002). The shift in position results from the ordered structure of biopolymers and is discussed in more detail below.

Detection of carbon bonded to carbon is clearly of astrobiological interest. The C-C σ-bond shows several features in the region 6.62–7.37 μm. The strongest features are at 6.82 μm and 7.25 μm and have very little overlap with water features.

The thiol functional group (-S-H attached to carbon) is important for metabolic and structural components (De Duve 2005). Mapping of thiols and any association with carbon and nitrogen features could serve to differentiate endogenous biologically utilized sulfur from either exogenous radiolytically processed sulfur or endogenous sulfate salts.

5.3. *Complexity*

Complexity in spectroscopic analysis here refers to complex bonds and shifts in band position due to polymerization and extended molecular backbones. The π-bond (C = C) of a carbon-carbon linkage results in a strong band at 6.08 μm. Alkenes in ice show this feature, as well as several weaker lines in the region of 6.9 μm. The π-bond is easily destroyed by irradiation, causing many of these bands to disappear or be greatly reduced in strength (Hand 2007). Detection of the 6.08 μm feature could be interpreted as complex endogenous organic chemistry on a geologically young surface.

The C-N bond, and bands associated with the amide structure (O=C-N-H), provide strong potential for spectroscopic biosignatures. Dalton *et al.* (2003) froze samples of the microbes *Sulfolobus shibatae*, *Deinococcus radiodurans*, and *Escherichia coli* to characterize spectral signatures under Europan conditions of vacuum and low temperature (100 K), funding that the 2.05 and 2.17 μm C-N bands of proteins were subtle but distinguishable in the laboratory.

Experiments by Hand and Carlson (Hand 2007; Hand *et al.* 2009) in Europa simulation chambers have shown that electron irradiated spores of *Bacillis pumilus* can be spectroscopically distinguished from the products of abiotic radiolysis. It is important to note that the total dose in these experiments was sufficient to disrupt every bond in the sample. The distinguishing features arise from shifts in the amide bands associated with amino acids. Amino acids in proteins are arranged in ordered sheets and helical structures resulting in a shift of the band position. Amino acids and other amide structures produced radiolytically do not have this added layer of structural complexity.

The amide I band of α-helical structures in proteins occurs at 6.04 μm and results from the carbonyl bond in the right-handed spiral of the helical structure. In abiotic radiolysis experiments this band was seen at 5.99 μm. The amide I band seen in β-pleated sheets of proteins occurs at 6.11 μm (Naumann *et al.* 1996), and is also due to the carbonyl stretch interacting with the amide N-H, but in β-pleated sheets the strands of linked amino acids form a sheet by hydrogen bonding with neighboring strands. This difference in structural geometry shifts the amide I in β-sheets to the longer wavelength.

The amide II bond was also seen by Hand and Carlson to persit with subtle but distinguishable differences between the abiotic and biological radiolysis experiments. The amide II band comes from bending of N-H coupled to C-N stretch (Hayashi and Mukamel 2007). In the irradiated biological samples the band was seen at 6.5 μm, whereas in the abiotic samples it was seen at at 6.27 μm. In the abiotic material the feature results from NH_2 scissoring and uncoordinated amide bonds.

Detection and mapping of the above described amide and phosphodiester features could yield clues toward the detection of complex organic chemistry associated with structures of known importance to life here on Earth. Working through the complete set of spectral signposts, features of compelling chemistry, and those indicative of complexity could point the way toward possible biochemistry on icy worlds. Mapping the spatial distribution of these spectroscopic features across geological fractures, ridges, and chaos terrain could permit differentiation between endogenous organic chemistry in the subsurface and organic chemistry driven by surface radiolysis and exogenous delivery.

6. Conclusions

The ability to differentiate abiotic organic material from material of a biological origin is a critical task for astrobiology. Mass spectrometry and spectroscopy provide key tools for advancing this task and are two techniques that provide useful and highly complementary compositional information independent of a specific biochemical pathway (*e.g.* DNA-RNA-Proteins).

Patterns in the mass spectrometric distribution of organics can, to first order, be applied to differentiate abiotic from biogenic, as per the 'Lego Principle' (McKay 2004). However, much caution is warranted as both Fischer-Tropsch Type synthesis and radiolytically driven chemistry can lead to complex mass spectra. For terrestrial biochemistry, our knowledge of the lipids and proteins utilized by biology make it possible to use additional metrics within a mass spectrum to distinguish abiotic from biogenic material (McCollom *et al.* 1999). The biochemistry of an alien ecosystem may reveal important clues in the pattern and complexity of a mass spectrum, but the lack of an *a priori* comparison will limit the utility of the 'Lego Principle.'

The analyses made possible by infrared spectroscopy reveal composition (*e.g.* atoms of C, H, N, O, P, S), connectivity (*e.g.* single, double, or triple bonds) and concentration. Though compositional information and geographic mapping of concentrations will provide an important context for astrobiology, the complexity revealed by functional group connectivity is arguably the most useful information. The bonds associated with the polymers of life, such as amides of proteins and phosphodiesters of nucleic acids, will serve as powerful spectroscopic biosignatures, especially if coupled to mass spectra consistent with such polymerization.

Finally, we note that the detection of life from orbit – by a spacecraft designed for studying the Jovian system – is not unprecedented. During the Galileo spacecraft's gravity assist fly-by of the Earth, Sagan *et al.* (1993) turned the instrument payload toward Earth and detected several signs of life, including an atmosphere out of chemical equilibrium and regions of the surface rich with photosynthetic pigments.

Though it was the Earth being observed, the Galileo experiment provided a useful proof-of-concept: we are technologically capable of detecting life on an inhabited planet. However, as Sagan *et al.* (1993) wisely advised, as we move forward with our search for life beyond Earth, we must use caution and invoke life only as 'the hypothesis of last resort'.

Acknowledgement

The research by K. P. Hand was carried out at the Jet Propulsion Laboratory, California Institute of Technology, under a contract with the National Aeronautics and Space Administration. Government sponsorship acknowledged.

References

A'Hearn, M. S., Belton, M. S. J., Delamere, W. A., Kissel, J., Klaasen, K. P., McFadden, L. A., Meech, K. J., Melosh, H. J., Schultz, P. H., Sunshine, J. M. *et al.* 2005, Deep Impact: Excavating Comet Tempel 1. *Science*, 310(5746):258–264

Anderson, J. D., Schubert, G., Jacobson, R. A., Lau, E. L., Moore, W. B., & Sjorgen, W. L. 1998, Europa's Differentiated Internal Structure: Inferences from Four Galileo Encounters. *Science*, 281(5385):2019–2022

Benner, S. A., Devine, K. G., Matveeva, L. N., & Powell, D. H. 2000, The missing organic molecules on mars. *Proceedings of the National Academy of Sciences*, 97(6):2425

Bernstein, M. P., Sandford, S. A., Allamandola, L. J., Chang, S., & Scharberg, M. A. 1995, Organic Compounds Produced by Photolysis of Realistic Interstellar and Cometary Ice Analogs Containing Methanol. *The Astrophysical Journal*, 454:327

Biemann, K., Oro, J., Toulmin III, P., Orgel, L. E., Nier, A. O., Anderson, D. M., Flory, D., Diaz, A. V., Rushneck, D. R., & Simmonds, P. G. 1977, The search for organic substances and inorganic volatile compounds in the surface of mars. *Journal of Geophysical Research*, 82

Brown, R. H., Clark, R. N., Buratti, B. J., Cruikshank, D. P., Barnes, J. W., Mastrapa, R. M. E., Bauer, J., Newman, S., Momary, T., Baines, K. H., *et al.* 2006, Composition and Physical Properties of Enceladus' Surface. *Science*, 311(5766):1425–1428

Capone, D. G., Popa, R., Flood, B., & Nealson, K. H. 2006, Follow the nitrogen. *Science (Washington, D. C.)*, 312(5774):708–709

Carlson, R. W., Calvin, W. M., Dalton III, J. B., Hansen, G. B., Hudson, R. L., Johnson, R. E., McCord, T. B., & Moore, M. H. 2009, *Europa*, chapter Europa's Surface Composition, pages 283–327. University of Arizona Press

Chyba, C. & Sagan, C. 1987, Infrared emission by organic grains in the coma of comet Halley. *Nature*, 330(6146):350–353

Chyba, C. F. 2000, Energy for microbial life on Europa. *Nature*, 403(6768):381–2

Chyba, C. F. & McDonald, G. D. 1995, The Origin of Life in the Solar System: Current Issues. *Annual Review of Earth and Planetary Sciences*, 23(1):215–249

Cleland, C. E. & Chyba, C. F. 2002, Defining 'life'. *Origins of Life and Evolution of Biospheres*, 32(4):387–393

Cooper, J. F., Johnson, R. E., Mauk, B. H., Garrett, H. B., & Gehrels, N. 2001, Energetic Ion and Electron Irradiation of the Icy Galilean Satellites. *Icarus*, 149:133–159

Dalton, J. B., Mogul, R., Kagawa, H. K., Chan, S. L., & Jamieson, C. S. 2003, Near-infrared detection of potential evidence for microscopic organisms on Europa. *Astrobiology*, 3(3): 505–29

De Duve, C. 2005, *Singularities: Landmarks on the Pathways of Life.* Cambridge University Press

des Marais, D. J., Harwit, M. O., Jucks, K. W., Kasting, J. F., Lin, D. N. C., Lunine, J. I., Schneider, J., Seager, S., Traub, W. A., & Woolf, N. J. 2002, Remote sensing of planetary properties and biosignatures on extrasolar terrestrial planets. *Astrobiology*, 2(2):153–181

Figueredo, P. H., Greeley, R., Neuer, S., Irwin, L., & Schulze-Makuch, D. 2003, Locating Potential Biosignatures on Europa from Surface Geology Observations. *Astrobiology*, 3(4): 851–861

Greeley, R., Chyba, C. F., Head III, J. W., McCord, T. B., McKinnon, W. B., Pappalardo, R. T., & Figueredo, P. H. 2004, Geology of Europa. In F. Bagenal, T. E. Dowling, and W. B. McKinnon, editors, *Jupiter. The planet, satellites and magnetosphere*, chapter 15, pages 329–362. Cambridge University Press, first edition

Hall, D. T., Strobel, D. F., Feldman, P. D., McGrath, M. A., & Weaver, H. A. 1995 Detection of an oxygen atmosphere on Jupiter's moon Europa. *Nature*, 373(6516):677–679

Hand, K. P. 2007, *On the Physics and Chemistry of the Ice Shell and Subsurface Ocean of Europa.* PhD thesis, PhD Thesis, Department of Geological and Environmental Sciences, Stanford University, Stanford, CA

Hand, K. P. & Chyba, C. F. 2007, Empirical constraints on the salinity of the europan ocean and implications for a thin ice shell. *Icarus*, 189(2):424–438

Hand, K. P., Carlson, R. W., & Chyba, C. F. 2007, Energy, chemical disequillibrium, and geological constraints Europa. *Astrobiology*, 7(6):1006–1022

Hand, K. P., Chyba, C. F., Priscu, J. C., Carlson, R. W., & Nealson, K. H. 2009 *Europa*, chapter Astrobiology and the Potential for Life on Europa, pages 589–629. University of Arizona Press

Hayashi, T. & Mukamel, S. 2007, Vibrational-Exciton Couplings for the Amide I, II, III, and A Modes of Peptides. *J. Phys. Chem. B*, 10.1021/jp070369b(111):11032–11046

Joyce, G. F. 1995, The RNA world: Life before DNA and protein. In B. Zuckerman and M. Hart, editors, *Extraterrestrials: Where are they?*, pages 139–151. Cambridge University Press

Kargel, J. S., Kaye, J. Z., Head, J. W., Marion, G. M., Sassen, R., Crowley, J. K., Ballesteros, O. P., Grant, S. A., & Hogenboom, D. L. 2000, Europa's Crust and Ocean: Origin, Composition, and the Prospects for Life. *Icarus*, 148(1):226–265

Khurana, K. K., Kivelson, M. G., Stevenson, D. J., Schubert, G., Russell, C. T., Walker, R. J., & Polanskey, C. 1998, Induced magnetic fields as evidence for subsurface oceans in Europa and Callisto. *Nature*, 395(6704):749–751

Kiang, N. Y., Segura, A., Tinetti, G., Blankenship, R. E., Cohen, M., Siefert, J., Crisp, D., & Meadows, V. S. 2007, pectral signatures of photosynthesis. ii. coevolution with other stars and the atmosphere on extrasolar worlds. *Astrobiology*, 7(1):252–274

Kiang, N. Y., Siefert, J., & Govindjee, B. R. E. 2007, Spectral signatures of photosynthesis. i. *Astrobiology*, 7(1):222–251

Mancinelli, R. L. & Banin, A. 2004, Where is the nitrogen on Mars? *International Journal of Astrobiology*, 2(03):217–225

Maquelin, K., Kirschner, C., Choo-Smith, L. P., van den Braak, N., Endtz, H., Naumann, D., & Puppels, J. P. 2002, Identification of medically relevant microorganisms by vibrational spectroscopy. *Journal of Microbiological Methods*, 51(3):255–271

McCollom, T. M. 1999, Methanogenesis as a potential source of chemical energy for primary biomass production by autotrophic organisms in hydrothermal systems on Europa. *Journal of Geophysical Research*, 104(E12):30729–30742

McCollom, T. M., Ritter, G., & Simoneit, B. R. T. 1999, Lipid synthesis under hydrothermal conditions by fischer-tropsch-type reactions. *Origins of Life and Evolution of Biospheres*, 29(2):153–166

McCord, T. B., Carlson, R. W., Smythe, W. D., Hansen, G. B., Clark, R. N., Hibbitts, C. A., Fanale, F. P., Granahan, J. C., Segura, M., Matson, D. L., *et al.* 1997, Organics and other molecules in the surfaces of Callisto and Ganymede. *Science*, 278(5336):271–275

McCord, T. B., Hansen, G. B., Martin, P. D., & Hibbitts, C. 1998, Non-water-ice constituents in the surface material of the icy Galilean satellites from the Galileo near-infrared mapping spectrometer investigation. *Journal of Geophysical Research*, 103(E4):8603–8626

McKay, C. P. 2004, What is life—and how do we search for it in other worlds? *PLoS Biology*, 2(9)

Naumann, D., Schultz, C. P., & Helm, D. 1996, What can infrared spectroscopy tell us about the structure and composition of intact bacterial cells? In H. H. Mantsch and D. Chapman, editors, *Infrared Spectroscopy of Biomolecules*, pages 279–310. Wiley-Liss: New York

NRC. *New Frontiers in the Solar System an Integrated Exploration Strategy.* National Academies Press, 2003.

Painter, T. H., Duval, B., Thomas, W. H., Mendez, M., Heintzelman, S., & Dozier, J. 2001, Detection and Quantification of Snow Algae with an Airborne Imaging Spectrometer. *Applied and Environmental Microbiology*, 67(11):5267–5272

Pappalardo, R. T. & Head III, J. W. 2001, The Thick-Shell Model of Europa's Geology: Implications for Crustal Processes. *32nd Annual Lunar and Planetary Science Conference, March 12-16, 2001, Houston, Texas, abstract no. 1866*

Pappalardo, R. T., Head, J. W., Greeley, R., Sullivan, R. J., Pilcher, C., Schubert, G., Moore, W. B., Carr, M. H., Moore, J. M., Belton, M. J. *et al.* 1998 Geological evidence for solid-state convection in Europa's ice shell. *Nature*, 391(6665):365–368

Peters, K. E., Walters, C. C., & Moldowan, J. M. 2005, *The Biomarker Guide Vol. II: Biomarkers and isotopes in petroleum systems and Earth history.* Cambridge University Press

Phillips, C. B., McEwen, A. S., Hoppa, G. V., Fagents, S. A., Greeley, R., Klemaszewski, J. E., Pappalardo, R. T., Klaasen, K. P., & Breneman, H. H. 2000, The search for current geologic activity on Europa. *Journal of Geophysical Research*, 105(E9):22579–22598

Pilcher, C. B. 2003, Biosignatures of early earths. *Astrobiology*, 3(3):471–486

Sagan, C., Thompson, W. R., Carlson, R., Gurnett, D., & Hord, C. 1993, A search for life on Earth from the Galileo spacecraft. *Nature*, 365(6448):715–721

Schubert, G., Anderson, J. D., Spohn, T., & McKinnon, W. B. 2004, *Interior composition, structure and dynamics of the Galilean satellites*, pages 281–306. Cambirdge University Press

Summons, R. E., Albrecht, P., McDonald, G., & Moldowan, J. M. 2008, Molecular biosignatures. *Space Science Reviews*, 135(1):133–159

Zimmer, C., Khurana, K. K., & Kivelson, M. G. 2000, Subsurface oceans on Europa and Callisto: constraints from Galileo magnetometer observations. *Icarus*, 147(2):329–347

Zolotov, M. Y. & Shock, E. L. 2004 A model for low-temperature biogeochemistry of sulfur, carbon, and iron on Europa. *J. Geophys. Res*, 109:2003JE002194

Galileo's Medicean Moons: their impact on 400 years of discovery
Proceedings IAU Symposium No. 269, 2010
C. Barbieri, S. Chakrabarti, M. Coradini & M. Lazzarin, eds.

doi:10.1017/S1743921310007386

Other Worlds, Other Civilizations?

Guy Consolmagno SJ
Specola Vaticana
Vatican City
email: gjc@specola.va

Abstract. Galileo's work had a profound influence on our understanding of the question of "other worlds" and the possibility of other intelligent life in the universe. When he saw the Moon with its mountains, and Jupiter with its moons, he implicitly recognized that these were physical places and thus could themselves be possible abodes for life. But some ancient and medieval scholars had already suggested as much, though without the empirical backing that Galileo's observations provided. Thus perhaps an even more important influence on the development of these ideas is that Galileo made them popular with the educated public, rather than merely the speculations of specialists. By inciting the popular imagination to take seriously the possibility of other worlds, he engaged subsequent generations of philosophers and storytellers to explore the possibilities and implications of life on those worlds.

Keywords. History of science, extraterrestrial life.

---

1. The Discarded Image

As so many of the presentations at this conference have confirmed, Galileo's discoveries with the telescope, epitomized by his discovery of satellites orbiting Jupiter, revolutionized astronomy. They also revolutionized our view of the universe, what has been come to be called our scientific "cosmology." And in the process, they gave a new emphasis to the whole question of other worlds and other civilizations.

To appreciate the impact that Galileo's observations had on our cosmic sense of ourselves and our place in the universe, it is important to begin with an accurate understanding of the old version that his observations overthrew. It is wrong to think that humanity's understanding of the universe was a simple Earth-centered view, that there were no other competing models in the ancient world, or that there was no speculation about other worlds and other civilizations before Galileo. And, indeed, as we know, Galileo did not propose a new model of the universe; that honor, of course, goes to Copernicus. Nor did he improve it, as Kepler did. Nor did he put it on a deeper physical basis, the way Newton did. Nor did his observations even demonstrate that it was true, in a mathematical sense, since certainly the Tycho Brahe model fit Galileo's data as well as the Copernican model did. Nonetheless, Galileo's observations and publications were crucial, indeed a pivotal event, in all of these developments. To see why, it is necessary to see the nature and the role of cosmologies in human culture.

C. S. Lewis (Lewis, 1964) has described the medieval view of the universe as "The Discarded Image". It was an image that underpinned all of art and literature, as the physical universe was thought to mirror the metaphysical. In that context, therefore, it is important to recognize that in the ancient world there were many different models for the universe, reflecting many different systems of metaphysics. While Eudoxus most famously proposed the concept that the planets were carried about the Earth in a series of concentric crystalline spheres, there were other rival ideas as well. Among them, the Atomists proposed that space was infinite, containing an infinite number of atoms, and thus there should be an infinite number of worlds. Aristotle obviously went with Eudoxus' vision and fashioned it into his understanding of the physics of motion. One obvious result of the Aristotelean system is that if the Earth is in a unique location as the natural resting point of the earthly elements, then almost by definition one would not expect there to be other Earths; the idea of other civilizations is ruled out.

Still, other thinkers even after Aristotle had other ideas. Thus, the Roman poet Lucretius, who was admired by Cicero and referenced by Virgil, could continue the ideas of the atomists and the multitude of other worlds. This quotation is from his one surviving poem, *De Rerum Natura*:

"Moreover, when there is an abundance of matter available, when there is space vacant, and no object or reason delays the process, then certainly shapes of reality must be combined and created. For there is such a huge supply of atoms that all eternity would not be enough time to count them; there is the force which drives the atoms into various places just as they have been driven together in this world. So we must realize that there are other worlds in other parts of the universe, with races of different men and different animals."

2. The Birth of Science Fiction

Stories of fantasy, including what we would now call science fiction, closely mirror the scientific advances and popular culture of the times when it is written. Among the few ancient stories of space travel, Cicero wrote a short piece called *The Dream of Scipio.* In it, he imagines Scipio actually physically traveling through the celestial spheres envisioned by Eudoxus. The purpose of this story was to reflect back on the nature of Earth from this celestial vantage point. Several hundred years later, around AD 165, Lucian of Samosata wrote a more adventurous story, *True History*, where he imagined Greek warriors traveling to the Moon where they do battle the King of the Moon and the King of the Sun, and go on to colonize Jupiter. This book was published only about ten years after Ptolemy's *Almagest* and it is reasonable to suggest that perhaps Ptolemy inspired this early bit of science fiction. What is fascinating to us, however, is that Jupiter, the Moon, and the Sun are all thought of not as mere dots of light demarking the location of crystalline spheres, but actual places that are big enough to visit and fight over. (For a further discussion of ancient science fiction, see Gunn, 1975).

The development of these ideas over the next thousand years is beyond the scope of this paper (see Hannam, 2009); it is sufficient here to note that even in the difficult times after the fall of Rome and the temporary loss of much of Greek science to the West (though it survived in the Byzantine and Arabic worlds) there were thinkers, like John Scotus Eriugena, who continued to ponder what we would call cosmology. Eriugena even considered a system of planets orbiting the Sun, which then orbited the Earth, similar to that proposed seven hundred years later by Tycho Brahe.

It is certainly true that by the medieval period the cosmology of Aristotle and Ptolemy was the common assumption in both philosophy and literature. Dante's *Inferno* is only the most famous of a large body of literature that assumed a single "world" made up of concentric spheres. Note, however, that in the medieval vision, the result of the marriage of Greek science with Christian sensibilities, the Earth is not at the center of this universe, as you might imagine from an image like this. Rather, it is at the bottom of the universe, only one step above the Damned and the Inferno. The eternal motions of the planets are occasioned by their proximity to the Prime Mover in the Firmament; Earth is so humble that it doesn't even get to move. Indeed, Lewis (1964) cites medieval thinkers who argue that the physical cosmology is in fact an exact *inverse* of the metaphysical reality...where God in the Firmament is the true center, and Earth is relegated to the furthest reaches.

But even with this cosmology predominating in medieval times, there were other ideas argued as well in the medieval university.

When 13th century philosophers at the University of Paris rediscovered Aristotle's ideas, like a universe without beginning or end, they came in conflict with the Christian teaching of an omnipotent creator God. In 1277 the Bishop of Paris, Stephen Tempier, responded by listing 219 philosophical propositions that he banned from his diocese. Among them was the assertion that "God could not have made other worlds." The Bishop reasoned that God is omnipotent, and so he must have the power to make other worlds. Given their vision of the universe, "other worlds" in this sense is more closely associated with what we would call "alternate universes."

The historian David Lindberg (1992) has noted: "...the articles that stressed God's unlimited creative power gave license to all manner of speculations about possible worlds and imaginary states of affairs that it was evidently within God's power to create. This led to an avalanche of speculative or hypothetical natural philosophy in the fourteenth century, in the course of which various principles of Aristotelian natural philosophy were clarified, criticized, or rejected."

Just as one example of this "avalanche", consider this passage from Bishop Nicholas of Cusa's *De Docta Ignorantia* of 1440 (Book II, Chapter 11):

Therefore, the earth, which cannot be the center, cannot be devoid of all motion...just as the earth is not the center of the world, so the sphere of fixed stars is not its circumference...Therefore, the earth is not the center either of the eighth sphere or of any other sphere... ... In like manner, we surmise that none of the other regions of the stars are empty of inhabitants...the one universal world is... in so many particular [parts] that they are without number except to Him who created all things...

John Buriden and Nicholas Oresme had written on the possibility that the Earth could be moving and spinning nearly 300 years before Galileo, and Cusa's work predated Galileo (and Giodarno Bruno, who is often credited with these ideas) by some 200 years. Even Copernicus's work was nearly 100 years old by the time of Galileo.

Ten years before Galileo got started, William Gilbert actively pushed Copernicanism in his book on the Earth's magnetic field, *De Magnete.* However, in his introduction, he specifically criticized those who would think to pander to the public by popularizing science, and so even in the religious turbulence of Elizabethan England, this barely caused a stir. By contrast, when Galileo dusted off these ideas, he wrote about them in Italian, and campaigned them like a 17th century Carl Sagan in the fashionable salons of Italy.

The result? Unlike the reaction to any of these other thinkers, people were shocked by Galileo's concepts. This reaction has fed the Renaissance (and modern) prejudice that the medieval times were a period of religious fundamentalism, some sort of dark age. In fact, it was a combination of Renaissance politics, bad theology, and bad Renaissance science that ganged up on Galileo. But, more crucially, it happened to Galileo and not any of his predecessors because he was the first to push these ideas into the popular culture.

This entry into the popular culture went beyond merely the discussion of Galileo's works among the educated gentry. It also is reflected, and carried further, by the creation of popular stories that took as their basic premise the new concept of the universe that the culture first learned about from Galileo.

In 1616, Kepler wrote a of a dream voyage to the Moon, which was finally published in 1634 as *Somnium.* It was based on contemporary astronomy, and the Moon depicted in the story was that seen in the telescope: his Moon had mountains, and the day there lasted 14 Earth days. Another pioneer of this sort of space fiction was John Wilkins, an Englishman who was one of the founders of the Royal Society and later became a bishop and noted theologian; his book, *The Discovery of a World in the Moon*, was published in 1638 and it also described traveling to our neighbor satellite. Likewise another bishop, Francis Godwin, the same year wrote about a trip to visit *The Man in the Moon.*

This sort of storytelling continued throughout the 17th century, including the 1657 adventure by Cyrano de Bergerac, *Voyages to the Moon and Sun*, and the 1686 publication by Bernard de Bovier de Fontenelle, *Conversations on the Plurality of Worlds.* The latter book is notable in that he discussed not only the possibility of visiting other planets but in fact the idea that the stars themselves might each have their own collection of planets: *"One vortex may have many planets that turn round its sun, another may have but a few; in one there may be inferior or lesser planets, which turn about those that are greater; in another, perhaps, there may be no inferior planets; here all the planets are got round about their sun, in the form of a little squadron, beyond which may be a large void space, which reaches even to the neighboring vortexes; in others, the planets may make their revolutions towards the extremity of their vortex, and leave the middle void..."*

This concept was illustrated in the 18th century in a marvelous painting on the ceiling of the "New Mathematical Hall" (dating from 1760) of the Jesuit college in Prague, the Klementinum (now part of the National Library of the Czech Republic.) The artist, believed to be Josef Kramolín, depicts a number of cherubs using a variety of astronomical and other scientific instruments to study the universe. One cherub, holding a banner reading "sapientissimi opus" ("the work of the wisest") is flying in a sky filled with stars and each star is surrounded by a series of elliptical planetary orbits, and radiating rays that may represent the tails of comets (Oulíková, 2006).

Not everyone in that era accepted the wisdom of dreaming of such voyages, however. The English poet John Milton, in *Paradise Lost*, has the archangel Raphael say to Adam (in Book 8, lines 172-178):

Heaven is for thee too high
To know what passes there; be lowly wise:
Think only what concerns thee in thy being;
Dream not of other Worlds,
what creatures there Live,
in what state, condition, or degree,
Contented that thus far hath been revealed
Not of earth only but of highest heaven.

It is not clear that Milton himself believes this; he was of course a great admirer of Galileo, citing him (referencing the "optic glass of the Tuscan artist") in this same epic poem. He had traveled to Florence when he was young and met with an aged Galileo after his trial. Maybe Milton is just quoting what he heard his detractors say to him about his love of that "crazy science fiction stuff".

Elsewhere (Consolmagno, 1996) I have discussed the mutual development of astronomy and science fiction. As each new development has occurred in the world of astronomy, an imaginative storyteller has been able to translate that idea into a story and thus bring that development to the attention of the general public. Some more famous examples include Jules Verne adapting the Moon maps of Beer and von Mădler into his *De la Terre á la Lune* (*From the Earth to the Moon*) and *Autour de la Lune* (*Around the Moon*);

Percival Lowell's vision of Mars becoming alive in H. G. Wells' *War of the Worlds*; and even the 1920s debates over the nature of our lens-shaped galaxy becoming the basis for E. E. Smith's *"Skylark of Space"* and "Lensman" books, the first classic space operas. And even today, most people probably first heard about black holes from watching *Star Trek*.

All of this delightful activity can ultimately be traced directly to Galileo. First, his observations made the Copernican system seem reasonable, if not precisely proven; and the Copernican system, unlike the crystalline spheres of Eudoxus, makes other planets around other worlds possible. Secondly, his popularization of those observations made the Copernican system well known; an idea must be known before it can be the basis of a story. And finally, Galileo's imagination allowed him to see the planets as indeed other worlds, not just in a philosophical sense but also a physical, tangible way: by seeing mountains on the Moon, and seeing moons around other planets, they became places where human beings could, conceivably, have adventures.

3. The Cetaceans of Europa?

After the Second World War, not only did the science fiction of alien civilizations mature into more respectable work; so did the science. Project Ozma, the attempt in 1960 to look for radio signals from other intelligent civilizations, was the beginning of a scientific treatment for the search for extraterrestrial intelligence. Soon thereafter, Frank Drake derived an equation that attempted to calculate the probability of us humans encountering other civilizations around other worlds by multiplying the odds of a favorable outcome to a number of functions necessary for such a detection. Among the key probabilities in this equation are the number of "earthlike" planets per star, and the fraction of those fit for life. The observational evidence to date still indicates that the number of intelligent civilizations in the universe is one, at most. As Enrico Fermi famously asked, if they exist, where are they?

Again, however, I suggest that Galileo's discovery may hold a clue to a new way of thinking about this problem.

In the early 1970s I was a student at MIT under professor John S. Lewis, who had just published (Lewis, 1971) a remarkable paper on the nature of the icy satellites orbiting Jupiter and the other giant planets, moons first discovered by Galileo 400 years ago. Lewis showed, with some simple but profound calculations, that a mixture of rock and ice with the density and size of an icy moon should have enough radioactive elements to cause the ice to melt. The result would be a differentiated moon with a rocky core, icy crust, and the possibility of a liquid water mantle.

His models were simple steady-state calculations; my task, for my master's thesis, was to create models to show how these moons would evolve with time, including such effects as the heat of fusion, convection in the liquid layers, and phase transitions between the different solid phases of water.

I was very young at the time... and indeed an avid science fiction fan. That no doubt is what led to this, the final sentence in my thesis: "Given the temperatures of the interiors, and especially of the silicate layers through which liquid will be percolating, the possibility exists of simple organic chemistry taking place, involving either methane from the ice or carbon in the silicate phase. However, we stop short of postulating life forms in these mantles; we leave such to those more experienced than ourselves in such speculations" (Consolmagno, 1975).

The Voyager (and later, Galileo) spacecraft confirmed that this was very likely a reasonable picture for Europa. They also confirmed, of course, that my models were fatally flawed. By considering radioactive heating alone, I had underestimated the heat input in Europa due to tidal heating by an order of magnitude; on the other hand, by neglecting solid state convection in the icy crust, I had also underestimated the heat flow out of Europa by an order of magnitude. My models gave the right answers, thanks to two huge but offsetting mistakes.

The "others more experienced" in such speculations were of course, Carl Sagan and his students. When I first mentioned this idea of sub-crustal life inside Europa at the Jupiter meeting in Arizona in 1975, Sagan himself challenged me by pointing out that such oceans would be too removed from sunlight, too dark to have an energy source to support life. But just a few years later, life forms such as tube worms were discovered in deep, dark parts of Earth's ocean which derive much of their energy from hydrothermal vents. Inspired by this discovery, Sagan's student Steve Squyres was among the authors proposing the habitability of Europa (Reynolds *et al.*, 1983).

If you can have life in a moon of a gas giant, then you can have life anywhere gas giants are found. This greatly increases the "number of habitable planets" in the Drake equation. Given the large number of gas giants already discovered around other stars, it seems now that life may be very possible indeed. But consider what sort of life! If you were an intelligent dolphin in Europa (perhaps breathing ammonia instead of oxygen), your universe would have a rocky center and be surrounded by concentric spheres of crystalline ice. Shades of Eudoxus! You would never know that there was a whole universe outside the upper sphere. No wonder you'd never think to communicate, much less travel to other planets around other stars. While the Drake number increases by including such possible habitats, the Fermi paradox is perhaps explained. Unless, of course, in one of those civilizations, an imaginative fantasy storyteller imagines a trip through the upper sphere. Imagine the shock if our dolphins should actually break through the crust to discover a universe above full of stars!

Acknowledgements

I wish to thank Bill Higgins for first introducing me to much of the research on the history of science fiction and for valuable discussions.

References

Consolmagno, G. J. 1975, *Thermal history models of ice satellites*, S. M. Thesis, Massachusetts Institute of Technology, 198pp

Consolmagno, G. J. 1996, Astronomy, science fiction and the popular culture *Leonardo*, 29, 127–132

Gunn, J. 1975, *Alternate worlds, The Illustrated History of Science Fiction*, Englewood Cliffs NJ: Prentice-Hall, 256 pp

Hannam, J. 2009, *God's Philosophers: How the Medieval World Laid the Foundations of Modern Science*, London: Icon Books, 320 pp

Lewis, C. S. 1964, *The Discarded Image: An Introduction to Medieval and Renaissance Literature*, Cambridge UK: Cambridge Univ. Press, 242 pp

Lewis, J. S. 1971, Satellites of the outer planets: their physical and chemical nature, *Icarus* 15, 174–185

Lindberg, D. C. 1992, *The Beginnings of Western Science: The European and Scientific Tradition in Philosophical, Religious, and Institutional Context, 600 B. C. to A. D. 1450*, Chicago: Univ. of Chicago Press., 456 pp. Quotation is from p. 239 of this edition

Oulíková, P. 2006, *The Klementinum A Guide*, (trans: K. Millerová and S. M. Miller)Prague: The National Library, 76 pp.

Reynolds, R. T., Squyres, S. W., Colburn, D. S., & McKay, C. P. 1983, On the habitability of Europa *Icarus*, 56, 246–254

Concluding Remarks

Let me first express my warmest acknowledgements to Cesare Barbieri for having taken the initiative of convening this symposium. These two days offered a unique opportunity to celebrate the scientific achievements and the legacy of Galileo Galilei. It allowed not only celebrating the scientist but also the philosopher and the human being. It was a fantastic journey in the past, present and future exploration of our universe and a fantastic retrospection into the Renaissance world which no better city than Padova would be able to offer. During these two days we could listen to a well balanced and well prepared set of excellent papers and presentations. All participants should be congratulated for their very active interactions during the discussions in the aula and also during coffee and lunch breaks.

The first session held in the historical Aula Magna of the University was the occasion of resuscitating Galileo 400 years after his master astronomical work. We were brought back in the universe of an inspired scientist. One who did not really want to be called an astronomer but rather preferred to be considered as an experimentalist. One who was keen to use physics as a tool to understand the world. We had a very sharp hint at the extraordinary talents of this Man who was able to observe Jupiter's moon with an unguided telescope, not only fighting against the diurnal motion which constantly displaced the image of the giant planet in the field of view, but also against the wind, against the mist condensing on the lens of his "cannocchiale" which he had to cleanse regularly during these chilly January nights, even against the beats of his heart, while at the same time using his hands and his eyes to draw on a piece of paper the details of the objects he was just observing. An impossible task indeed that he accomplished with incredible talent and dexterity!

The world of Galileo would not be unfamiliar to that of today's scientists: made of jealousy, of fights with the political authorities (the Church), of detractors. That world was also characterized by the internal dilemma between science and faith, and by the necessary negotiations for money and contracts, when you have to bow and to please often unpleasant people...

The rest of the symposium was devoted to the legacy of Galileo's work 400 years after his historical observations. The core of the presentations of course concerned the Medicean moons as observed from space. The richness of the progress of our observation capabilities could be assessed in perspective through the description of the remarkable production of missions such as Galileo and Cassini. A breathtaking legacy indeed! The meeting also dealt with the other objects that Galileo observed in the Solar System. Then it went on describing future missions to Solar System objects.

Unfortunately for me, the Sun was absent of these future plans. No doubt that if he were living today, Galileo would be exhilarated to see the daily images provided by the SOHO, STEREO and Hinode satellites. He would certainly appreciate the fantastic performance of Ulysses in its journey above the ecliptic plane, as well as the plans to send a Solar Probe diving into the Sun, a project also very dear to the heart of Beppi Colombo, another giant scientist from Padova.

The remarkable observation of the rotation of the Sun through the displacement of sunspots on the solar surface was timely in 1610. A few years later, 3 years after Galileo's death, the so-called Maunder Minimum started and no sunspots could be observed until 1715. A similar situation exists at this very moment, and would Galileo point his telescope today to the Sun, he would just see a pure unspotted disk. In a sense, he was lucky! Luck is often a key ingredient of discovery!

The scientific revolution started by Galileo was followed a little more than 300 years after by a second revolution when Edwin Hubble at Mount Wilson discovered the expansion of the Universe, a revolution amplified by the advent of space astronomy following the historical launch of Sputnik-1 on 4 October 1957. After that successful prowess, we could then on continue observing the same sky as Galileo but now from above the Earth atmosphere which blurred the view of his "cannocchiale" and limited his observations to the visible and near infrared part of the electromagnetic spectrum.

The most emblematic successor to the "cannocchiale" in the space era is undoubtedly the Hubble Space Telescope (HST), nearly more than 100 million times more powerful than its famous predecessor, due to its large mirror of 2.4 m and its using photoelectric detectors. Jupiter and its moons have then been observed in great detail and regularly. The existence of the moons could also be observed indirectly by HST through their signatures appearing as small dots in the Auroral oval around Jupiter's poles, the effect of the complex interaction between the solar wind, Jupiter strong magnetic field and the channeling of ionized gas flowing out of Io's volcanoes along the field's lines of force.

In the next 400 years, what will the next astronomy revolution be? It may start with the contribution of HST's successor, the James Web Space Telescope (JWST) and be thereafter fed with those of the more gigantic "cannocchiale" that are still on the drawing board today. The first task of these new instruments will be to respond to three presently very challenging problems, still unresolved with present day available instrumentation:

- What is causing the Universe to accelerate since about 7.5 billion years after the Big Bang? In other words, does Dark energy exist and can it explain that acceleration? Are we interpreting present observations rightly?
- What is "holding" the Universe together and what is exactly the nature of that still mysterious Dark matter discovered 80 years ago but which still escapes any real identification.
- Is there life on even a single one of the planets we are now discovering regularly orbiting other stars?

The present successors of Galileo are now designing or imagining the "cannocchiale" of the future that might provide responses to these major questions. Most of their concepts are still in the realm of Utopia: missions impossible? It is certainly the task of the engineers in industry and space agencies to make the impossible come through, exactly like Galileo did when he transformed an optical gadget into a scientific instrument which revolutionized our perception of nature and of science.

This is probably the best occasion for me to celebrate one of these engineers who has done so much for the success of the space missions to the outer planets of the Solar System. I would like that we applaud and acknowledge the unique contribution of John Casani who, at JPL, since the 1970's had led and managed the most prestigious missions of NASA: the Voyagers, Galileo and Cassini spacecraft among others. John has been sitting here quietly during these two days, enjoying I am sure the enthusiasm of his fellow scientists who thanks to him have got the most stunning insights into the little moons that Galileo observed several hundred years before with his "cannocchiale".

What lessons can we draw from Galileo's legacy? What have we learnt from his life and his achievements? There are many but I will concentrate on just a few which I consider as the most important ones.

- Discovery is the fruit of new technologies and new instruments. That has been well analyzed in the famous book of Martin Harwit *Cosmic Discovery: The Search, Scope, and Heritage of Astronomy* (Basic Books, NY, 1981; MIT Press, 1984), some 25 years ago. The problem space astronomers face today is to maintain and ensure the continuity of the expertise in instrument development and experimentation which made space astronomy the success it is today. That expertise is thinning and threatened of gradually disappearing as it is not regularly replaced.

• Discoveries are the ferment of scientific progress and the mark of the genius of the human brain. Like Galileo: Discover! Discover! Discover!

• The essence of Discovery, with only a few rare exceptions is the result of a scientific intuition and the use of new techniques and systems which integrate them and take advantage of their increased performances.

All geniuses are misunderstood or criticized and subject to jealousy, victims of sarcasms from their competitors or their peers. What affected Galileo is still at play today marking the life of some of the most productive and inventive of our colleagues. As Galileo did 400 years ago, we must resist this aggressiveness and fight as strongly as possible these syndromes. Scientific discovery is also an act of courage!

Although the number of potential "Galileo's" is increasing in proportion of the global population increase, -the World total population was reaching about 500 millions in 1610, it is presently more than 6.5 billions i.e. more than one order of magnitude higher- the appearance of a new "Galileo" may become a more exceptional event than it was 400 years ago. This is because the "impossible" lies further and further away. A new Renaissance is needed that should be born from the forthcoming dark ages, with very dangerous crisis looming at the horizon of our civilization. The gigantism of missions, which make them look like the cathedrals of the Middle age, with their gestation stretching over longer and longer periods of time, encompassing several generations of scientists, in a context of rapidly evolving technologies, logically will make them more costly and less and less frequent. The gradual disappearance of the synergies with the military will not ease the situation. In that context, international cooperation will become more than ever indispensable. This is another unavoidable effect of globalization. Fortunately, Science by its very nature would strongly benefit from the confrontation of ideas on the broadest basis. Hence, there is some room for optimism.

I would like again to thank Cesare Barbieri for having organized the symposium and for having offered me the opportunity to conclude this very moving and extraordinary meeting.

Roger M. Bonnet

POSTERS

Galileo's Medicean Moons: their impact on 400 years of discovery
Proceedings IAU Symposium No. 269, 2010
C. Barbieri, S. Chakrabarti, M. Coradini & M. Lazzarin, eds.

doi:10.1017/S1743921310007404

Observing mutual events of the trans-Neptunian object Haumea and Namaka from Brazil

Alexandre Bortoletto[1] **and Roberto K. Saito**[2]

[1]Ministério de Ciência e Tecnologia, Laboratório Nacional de Astrofísica,
R. Estados Unidos, 154, Itajubá, MG, CEP: 37504-364, Brazil
e-mail: abortoletto@lna.br

[2]Pontificia Universidad Catolica de Chile, Departamento Astronomia y Astrofisica, Office 206,
Campus San Joaquin, Vicuna Mackenna 4860, Casilla 306, Santiago 22, Chile
e-mail: rsaito@astro.puc.cl

Abstract. By pure coincidence, for the next few years the orbit of the satellite Namaka around the dwarf planet Haumea (formerly 2003 EL61) is nearly edge-on to our line-of-sight. This type of configuration does not last for long, because as Haumea travels around the sun in its 283 year orbit, we continuously see the Haumean system from different angles. It is only edge-on at the angle we see right now, and at the angle it will again be in 141 years – half of a Haumean year from now. In addition to being an interesting coincidence, the fact that the orbit of Namaka is nearly edge-on provides the opportunity to obtain an enormous amount of information about the Haumean system. We present measurements of the timing of these events observed from Laboratrio Nacional de Astrofsica (LNA), partner in an international campaign to observe these events from the most suitable mid-sized telescopes.

Keywords. Kuiper Belt, planets and satellites: fundamental parameters, methods: data analysis

1. Introduction

In 1610, the most important observational breakthrough was made by Galileo Galilei. In that year he turned a telescope to the skies for the first time in the human history. His new tool led to important discoveries of planetary science such as the phases of Venus, the moons of Jupiter, mountains on the Moon, and the rings of Saturn. The legacy of these findings are in the current research about extra-solar systems, outer solar-system, your formation and, possibly, life in such systems.

Kuiper Belt Objects (KBO) are icy bodies in orbits beyond Neptune. That name is in honor of Gerard Kuiper, who predicted the existence of these distant population of small bodies similar to the asteroid belt. Since the first KBO was discovered in 1992, the number of known objects increased to more than a thousand. The study of the orbital dynamics of KBO bodies can improve our understanding of the formation of the inner and outer solar system, as well as its evolution.

Haumea (formerly 2003 EL61), the third brightest known Kuiper belt object, is perhaps the most interesting object in the outer solar system. Its extremely rapid rotation and subsequent elongation immediately leads to the hypothesis that Haumea suffered a giant impact which removed much of the water ice and left the body with rapid spin. The subsequent discovery of two satellites (Hi'iaka and Namaka) in orbit around the body and the confirmation that the largest one has the spectral signature expected from a collisional fragment, confirms that the giant impact which gave Haumea its fast spin also

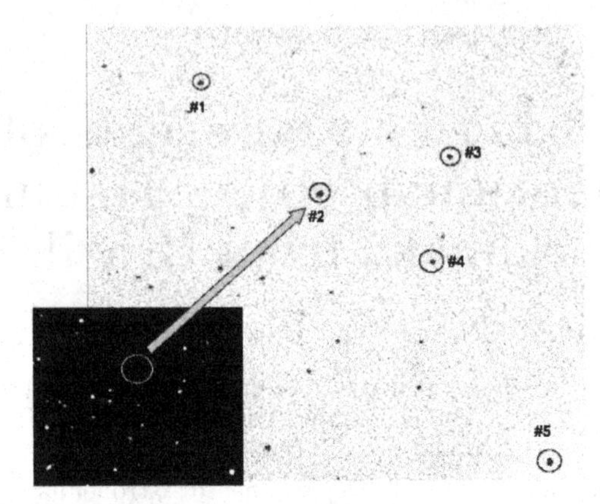

Figure 1. Finding charts for Haumea. In detail the chart showing the predicted position of Haumea at the precise time of the event. The large image was obtained at OPD during the event and shows Haumea and the comparison stars. The comparison star #1 was used to perform the differential photometry for the all the other objects.

shattered the icy mantle and ejected multiple fragments into orbit and beyond (Ragozzine & Brown (2009), Febrycky *et al.* (2008), Schaller & Brown (2008)).

For the next few years the orbit of the satellite Namaka around the dwarf planet Haumea is nearly edge-on to our line-of-sight. This type of configuration does not last for long, because as Haumea travels around the sun in its 283 year orbit, we continuously see the Haumean system from different angles. It is only edge-on at the angle we see right now, and at the angle it will again be in 141 years – half of a Haumean year from now. In addition to being an interesting coincidence, the fact that the orbit of Namaka is nearly edge-on provides the opportunity to obtain an enormous amount of information about the Haumean system. We present measurements of the timing of these events obtained from fast photometry conducted at Observatório Pico dos Dias (OPD), managed by Laboratório Nacional de Astrofísica (LNA), partner in an international campaign to observe these events from the most suitable mid-sized telescopes.

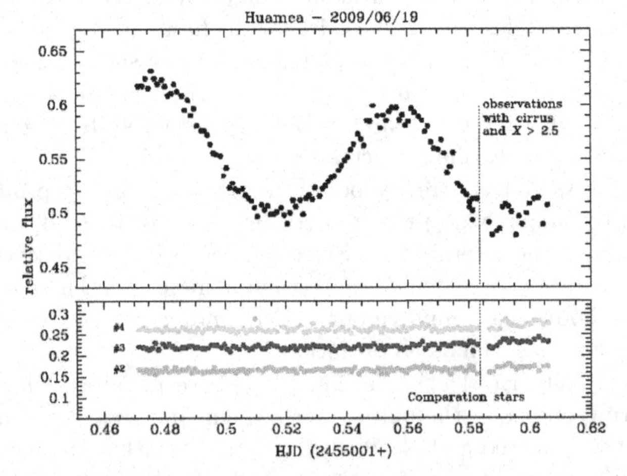

Figure 2. The upper panel shows the light-curve for Haumea obtained at OPD/LNA. A dotted line indicated the instant when clouds (cirrus) started to obscure the observations. These last images were also taken at high air-mass, $X > 2.5$, and not been used in the analysis. The lower panel shows the light-curves for the comparison stars, indicated in Figure 1.

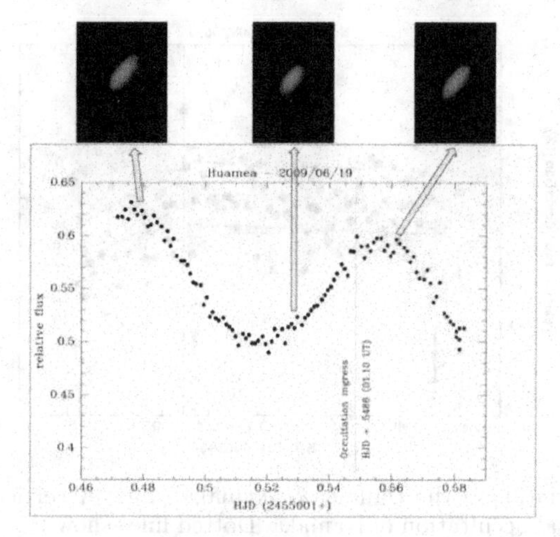

Figure 3. Light-curve analysis for Haumea. The modulation observed in the light curve is related with the geometry of the planet. The three upper panels illustrate the configuration for the system (Huamea + Namaka) in different instants. A dotted line is plotted to indicate the begin of the occultation of Namaka by Huamea.

2. Observation and Reduction

The data was secured in June 16, 2009 using the 1.6m Perkin-Elmer telescope at Pico dos Dias Observatory (OPD), the main optical observational facility in Brazil. OPD is operated and maintained by Laboratório Nacional de Astrofísica (LNA), an unit of Brazilian Ministry of Science (MCT). Figure 1 shows the chart with the predicted position of Haumea at the precise time of the event (marked as #2). The larger image was obtained at OPD during the event and shows Haumea and the comparison stars. The comparison star #1 was used to perform the differential photometry for the all the other objects.

In the Figure 2, the upper panel shows the light-curve for Haumea obtained at OPD/LNA. The lower panel shows the light-curves for the comparsion stars, indicated in the Fig 1. We can see that the modulation at the light curve of Haumea is intrinsical from object when compared with the light curve of the other objects. A dotted line indicated the instant when clouds (cirrus) started to obscure the observations. These last images were also taken at high air-mass, $X > 2.5$, and not been used in the analisys.

3. Data Analysis and Conclusions

The modulation observed in the light curve is related with the geometry of the planet. The three upper panels illustrate the configuration for the system (Haumea + Namaka) in different instants. A dotted line is plotted to indicate the begining of the occultation of Namaka by Haumea.

The relative deviation of the Haumea & Namaka's flux during the event as a fuction of time in HJD is showed in the Figure 4. The vertical dashed line indicates the beginning of the total occultation of Namaka. We split the relative deviations curve in 'before' and 'after' total occultation and we made linear fit to the both parts (showed like red dashed lines in the Figure 4). These adjustments indicate a decrement in light of the system, i.e. occultation of Namaka by Haumea. The typical value for the error bars is showed at the left side.

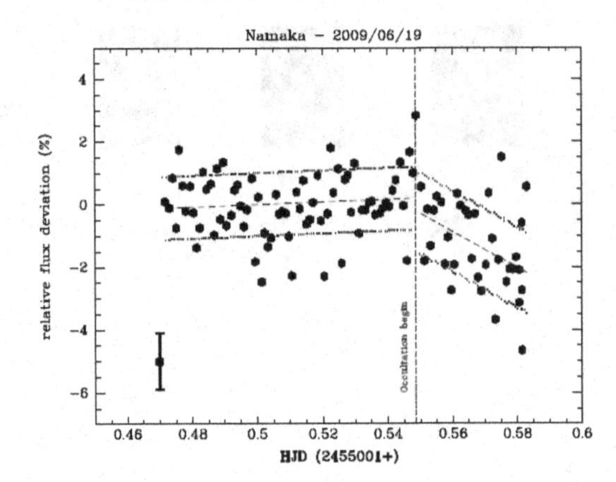

Figure 4. Relative deviation of the Haumea & Namaka's flux. A vertical dashed line indicates the beginning of the total occultation of Namaka. Dotted lines show the linear fit for the points. The adjustments indicate a decrement in light of the system, i.e. occultation of Namaka by Huamea. The typical value for the error bars is showed at the left side.

Acknowledgements

AB thanks the MCT/LNA for the PCI Scholarships 383342/2008-2 and the Scientific Organising Committee of IAUS269 for the partial grant. RS acknowledges financial support from CONICYT through GEMINI Project Nr. 32080016, BASAL PFB-06, and FONDAP Center for Astrophysics Nr. 15010003. AB thanks Mike Brown and Michele Bannister to participate in the campaign to observe mutual events of Huamea and Namaka.

References

Ragozzine, D. & Brown, M. E. 2009, *AJ*, 137, 4766

Fabrycky, D. C., Holman, M. J., Ragozzine, D., Brown, M. E., Lister, T. A., Terndrup, D. M., Djordjevic, J., Young, E. F., Young, L. A., & Howell, R. R. 2008, *Bulletin of the American Astronomical Society*, 40, 462

Schaller, E. L. & Brown, M. E. 2008, *ApJ*, 684L, 107

Galileo's Medicean Moons: their impact on 400 years of discovery
Proceedings IAU Symposium No. 269, 2010
C. Barbieri, S. Chakrabarti, M. Coradini & M. Lazzarin, eds.

doi:10.1017/S1743921310007416

Scientific literacy, pleasure and aesthetics: the collaborative virtual worlds that Galileo could not foresee

Daiana R. Bortoletto[1], **Mariângela de Oliveira-Abans**[1], **Propércio G. Guida Júnior**[1], **Iranderly F. de Fernandes**[1,3], **and Rafael Santos**[2]

[1]Ministério de Ciência e Tecnologia/Laboratório Nacional de Astrofísica,
R. Estados Unidos, 154, Itajubá, MG, CEP: 37504-364, Brazil
e-mail: dbortoletto@lna.br, mabans@lna.br, pguida@lna.br, ifernandes@lna.br

[2]Ministério de Ciência e Tecnologia/Instituto Nacional de Pesquisas Espaciais,
Av. dos Astronautas, 1.758, Jd. Granja, CEP: 12227-010, São José dos Campos, SP, Brazil
e-mail: rafael.santos@lac.inpe.br

[3]Universidade Estadual de Feira de Santana/Departamento de Física, Av. Transnordestina,
S/N, Novo Horizonte, CEP: 44036-900, Feira de Santana, BA, Brazil
e-mail: ifernandes@lna.br

Abstract. We present the various activities and infrastructure dedicated to educational scientific and technological outreach of the MCT/Laboratório Nacional de Astrofísica, Brazil and how useful they are in diminishing the scientific illiteracy of the layman, the young, the senior citizens and the handicapped. We also explore the future endeavors and commitments that scientists and educators are to face in order to bring enlightenment.

Keywords. astronomy, outreach, image bank, virtual observatory, remote observation.

1. Introduction: the desperate needs

These are times of strong contrasts: on one side, science and technology advance rapidly and influence civilization on a world-wide scale, on the other side, countries with poor socio-economical realities tend to fall more and more behind. It is important to mitigate these differences and provide the means for the common people to benefit from the technological advances and the chances offered by globalization. The MCT/Laboratório Nacional de Astrofísica - LNA, is a niche of highly specialized technology in the development of instruments for professional astronomical observatories and as such it is a knowledge and technology generator. It is commited to educational outreach because it is only through conveying information to the layman and media agents that individuals may be aware of the progress around them and so become fully conscious citizens, critic and capable of making those choices that are best for them.

Ground Astronomy has experienced extraordinary development and Brazil is perfectly inserted in that context through the LNA/Observatório do Pico dos Dias (OPD, web site: http://www.lna.br/opd/opd.html), the participation in the Gemini Observatory and the SOAR Telescope, and the use of the Canada-France-Hawai'i Telescope.

Astronomy is a powerful tool for promoting scientific literacy and pleasant education. This promising scenario, nonetheless, contrasts with the reality in Brazilian, where very rudimentary astronomical concepts are presented to school students by teachers educated in different areas. To make things worse, a large fraction of the professional scientific community is still far from the general public and frequently can not communicate very well with the media. This is one of the characteristics of social exclusion.

2. Galileo today

Four hundred years ago, Galileo employed the most advanced technology available at his time to begin the exploration of the skies. Through his telescope, he saw other worlds, and this opened the curtains of a larger and - who knows? - maybe a populous Universe. Today, scientists and technologists still pursue the scientific explanations and state-of-the-art technological solutions to challenges that Nature poses. Should Galileo live in these days, he would have been educated in schools with resources that were unimaginable in the past centuries. He would have such real and virtual infrastructure that he would certainly be seeking moons around extra-solar planets in very sophisticated ways. There are several scientific products and services that are offered to the community, media and especially to schools.

A first example is the Observatório no Telhado ("Observatory on the Roof"). Its 30-cm telescope will be part the Telescópios na Escola ("Telescopes at School") Brazilian Network and will also be used by schools of the Gemini Observatory and SOAR Telescope member countries.

A second one is the Virtual Image Bank (BIMA), which provides untrained users with a web interface that allows the display and use of astronomical images for, e.g., data reduction training and/or resources for the participants of the Olimpíada Brasileira de Astronomia, Astronáutica e Energia - OBA (Brazilian Astronomy, Astronautics and Energy Olympiads), or moments of aesthetic contemplation.

We have also developed the PocketVO with INPE (MCT/Instituto Nacional de Pesquisas Espaciais), a simple web-based tool for viewing images and data using the Virtual Observatory that can be used for both astronomical research and education.

3. Observatory on the Roof

A building of highly specialized workshops and laboratories dedicated to the design and construction of astronomical instruments was inaugurated at LNA's headquarters in 2006. A conference room for 80 people, the entrance hall and part of the roof area are used to display technical and astronomical exhibits and experiments, public talks, star gazing (see Figure 1), movies and so on.

The Observatório no Telhado (OnT - Observatory on the Roof) consists of a large open area surrounded by benches and garden, a telescope and dome tower and a control room, allowing for use either during the day or night. The 12-inch Meade telescope can be equipped with an SBIF CCD direct camera or a SGS SBIG self-guiding spectrograph. The OnT has already been used not only for star parties and the III Star Count, but also for talks and slides and video presentations for schools and general public. It should be operational by mid-2010.

What are the main objectives?

- complement the science contents of basic and high school of the surrounding regions;
- promote day and night sky gazing and observations of astronomical serendipitous and regular phenomena, such as comets and eclipses;
- allow for *in situ* and remote observations by schools in Brazil and the partner countries of the Gemini Observatory and SOAR Telescope;
- help teachers with correct astronomical concepts and recent discoveries;
- collaborate with teachers in arousing the interest in sciences from basic school level on, through the application of simple and atractive activities;
- allow handicapped people to have access to a telescope and astronomers;
- clarify the problem of light pollution to children and the layman;

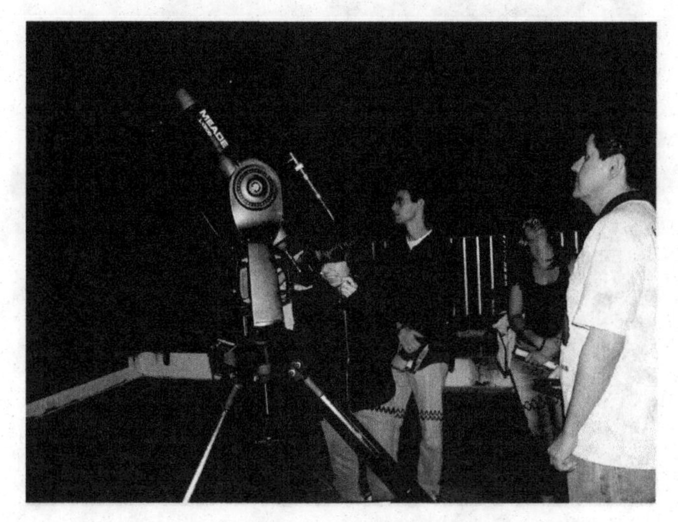

Figure 1. VI National Science and Techonolgy Week, OnT/LNA, 2009.

• take advantage of the interdisciplinarity of Astronomy to promote the integrated work of teachers of different areas of knowledge, different schools and different cities;
• to allow the use by amateurs, training technicians, physics undergraduates, offering them the chance of participating in scientific and educational reasearch.

4. Telescopes at School

TnE stands for Telescópios na Escola, and a full description of the project can be found at http://telescopiosnaescola.pro.br/. Through the remote use of small telescopes around the country, students and teachers become Universe explorers, choosing the objects to observe, planning the night runs, working with the data and interacting with TnE's support members. TnE and OnT share the same objectives and methodology.

Thousands of Brazilian students have already taken advantage of the TnE telescopes and instructional material available online, and have seen new small observatories grow throughout the country. These have become new true science centers.

One of the partner institutions (Instituto de Astronomia, Geofísica e Ciências Atmosféricas - IAGUSP) has facsimiles of the complete work of Galileo, in particular, his observations notes, which are invaluable.

5. Virtual Image Bank

In the context of the International Year of Astronomy - IYA2009 and its offspring, the Rede Brasileira de Astronomia (RBA, Brazilian Astronomical Network), LNA is a Local Node and so is committed to offering several services and products to the media and the community, mainly those related to formal and informal teaching.

A wide plan of improvement of LNA's outreach infrasctructure is presently being undertaken. The Outreach and Teaching of Astronomy Portal (Portal de Divulgação e Ensino – http://www.lna.br/~divulg/) received a boost through the implementation of the Virtual Image Bank (BIMA), which contains institutional, historical and astronomical low- and high-resolution images. One of us (D. Bortoletto) received a CNPq scholarship for the development of the BIMA site (see Figure 2).

Figure 2. A BIMA window projected onto the wall. The user browses the images with the help of an infrared pen and a Nintendo Wii control.

Internet is the most accessible way for the public and the media and so is a good tool for disseminating scientific information. BIMA should turn into a trustable window of Brazilian astronomical results and a suitable resource for educational projects.

The images have been organized in directories (telescopes, observatories, instruments, laboratories and workshops, research results, outreach events and various classes of astronomical objects) and the Bank is interesting for the public of all ages.

Based on the bank, learning materials (including playgames) are to be dedicated, as a first step, to teachers and students of fundamental and high school grades. At present, it is possible to solve puzzles online (see Figure 3), simply choosing one image from a suitably selected list and specifying the number of pieces, which allows its use from small children on. More products for various media are to be offered in the future.

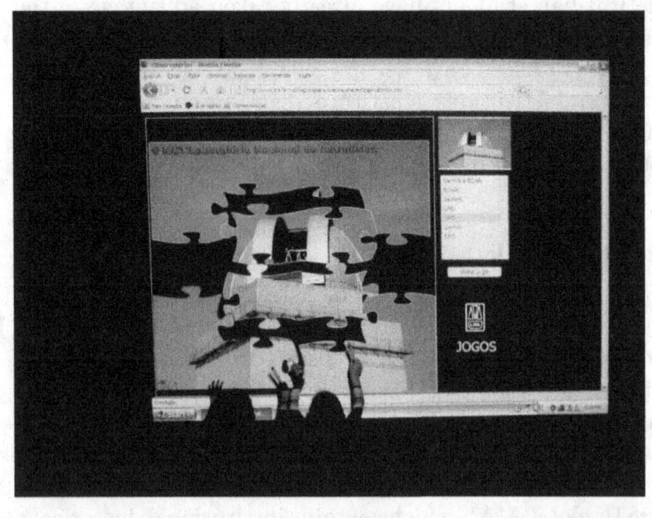

Figure 3. Children playing with virtual puzzle on the wall via infrared pen and a Nintendo Wii control.

6. PocketVO

We have developed the PocketVO (http://www.lac.inpe.br/~rafael.santos/pocketvo.jsp) with MCT/Instituto Nacional de Pesquisas Espaciais - INPE. It is a simple, easy-to-use, interactive tool to get images from different telescopes in different wavelenghts. Created specifically for untrained users, including teachers, students, amateur astronomers and researchers new to VO-enabled astronomy, it transparently allows access to information from different services through a unified interface. It is written in Java and works in any computer that has a Java Runtime Environment, version 5 or later, installed.

PocketVO's graphical user interface provides windows to see specific regions in the sky (see Figure 4), which can show one, two or three different wavelengths at the same time, so that users can easily see an object, e.g., in visible and X-Ray images. Regions can be entered directly in right ascension and declination coordinates or by object's name. Simple navigation controls allow the pan and zoom of the image. The various windows' configurations can be saved in files and shared with other users. Those files can be loaded either from local files or from URLs, making the sharing of data very simple.

Some possible extensions of these materials are: a digital whiteboard (already in use) and the development of a digital table with exchangeable modules according to the astronomical topics of interest. In the future, suitable interfaces will connect BIMA and PocketVO to the digital table and the whiteboard, a technology that will be available to schools, planetaria and science clubs due to its low cost.

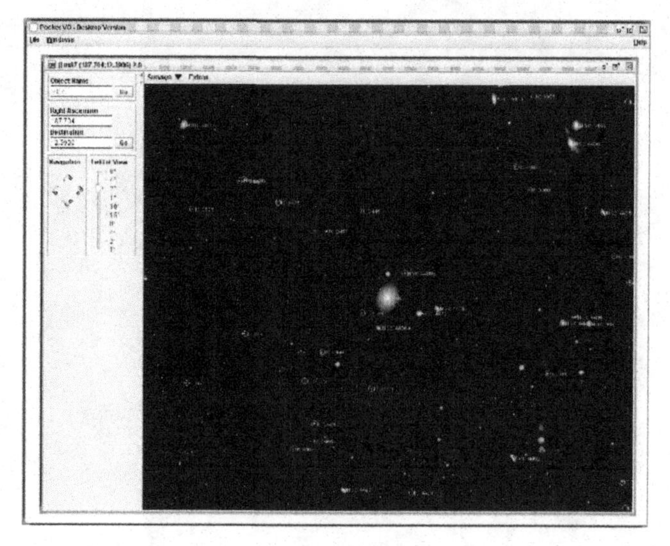

Figure 4. Pocket-VO's graphical user interface.

7. Concluding remarks

In the early 1600's, Galileo used a refracting telescope to observe celestial objects and may be considered the father of modern science. His astronomical research gave rise to a plethora of scientific endeavours, some of which are named after him.

The results of LNA's initiatives will be felt in various time scales. Beyond the present satisfaction of public outreach events and multimedia materials, the knowledge and experience aquired in the joint work with teachers and students will be gradually employed. The number of students interested in exact sciences will increase, and some of them will return to the LNA as undergraduates and graduates engaged in astronomical research and instrument development. As a byproduct, the results of the high school national evaluation, ENEM (http://www.enem.inep.gov.br/index.php), will be better and the participation in astronomical olympiads will be larger.

As time goes by, people will be aware of the science and technology their country produces and as such they will be citizens to the full extent of the word - and in pretty much less than 400 years.

Acknowledgements

We thank the MCT/LNA for the PCI Scholarships 38.0787/2009-1 and 381.671/2005-4, the John Hopkins University/NVO and the Scientific Organising Committee of IAUS269 for the partial grant.

Galileo's Medicean Moons: their impact on 400 years of discovery
Proceedings IAU Symposium No. 269, 2010
C. Barbieri, S. Chakrabarti, M. Coradini & M. Lazzarin, eds.

doi:10.1017/S1743921310007428

The planets and our culture a history and a legacy

Theodore C. Clarke[1] and Scott J. Bolton[2]

[1]New Renaissance Productions,
Altadena, CA 91001
email: tcclarke@earthlink.net

[2]Artistic Sciences, Inc.,
San Antonio, Texas 78257
email: sbolton@swri.edu

Abstract. This manuscript relates the great literature, great art and the vast starry vault of heaven. It relates the myths of gods and heroes for whom the planets and the Medicean moons of Jupiter are named. The myths are illustrated by great art works of the Renaissance, Baroque and Rococo periods which reveal poignant moments in the myths. The manuscript identifies constellations spun off of these myths. In addition to the images of great art are associated images of the moons and planets brought to us by spacecraft in our new age of exploration, the New Renaissance, in which we find ourselves deeply immersed.

Keywords. mythology; paintings; constellations; planets; moons; space exploration

1. Introduction

Since the beginning of human history man has sought in the vastness of infinite space a reflection and an understanding of the vastness of his own inner being. He has always identified with the celestial spheres. It was surely so for our ancestors in the Euphrates Valley over 5,000 years ago. There, at that time, thought that would shape our civilization was present about the land. To the east, near the meeting of the Tigris and Euphrates rivers was the site of biblical Eden and the biblical origins of man. To the north was Akkad, site of the origins of the star worship and mythology that would reach its zenith with the Greeks. To the west was Babylonia, site of the origins of astronomy. To the south was Ur, birthplace of Abraham, spiritual father of the Judeo, Christian, and Islamic religions. The same collective mind that was planting the seeds of these three great monotheistic religions was contemplating the nature of the universe and life, and was seeking answers in the heavens above. That same mind populated the firmament with gods, in a zodiac still used today in modern charts of the heavens, and invented great stories and myths of the deeds of those gods. The planets and many of their moons are named for the Greek and Roman gods.

2. Origins

Among the classical Greeks the most accepted account of the beginning of things was that given by the poet Hesiod, a near contemporary of Homer. Hesiod tells us in the "Theogony" that Chaos preceded all things (see Figure 1). Next came into being Earth personified as Gaea, and Love. From Mother Earth proceeded the starry vault of Heaven, personified as Uranus, where the gods would take up their abode. Then, in the heart of creation Love stirred making of things male and female, and bringing them together by natural affinity. We know today that this notion of opposites brought together by natural affinity is the basis for the subatomic structure of all matter. Uranus took Gaea to wife, and from their union were born the Titans, the Hecatonchires or hundred handed monsters, and the Cyclopes, all manifestations of violent upheavals on Earth. When Uranus tried to destroy the hundred handed monsters and the Cyclopes, whom he dreaded, Gaea, their mother, plotted with the Titans to strike down the tyranny of Uranus. Saturn, god of Time, stepped forward to take up his mother's cause.

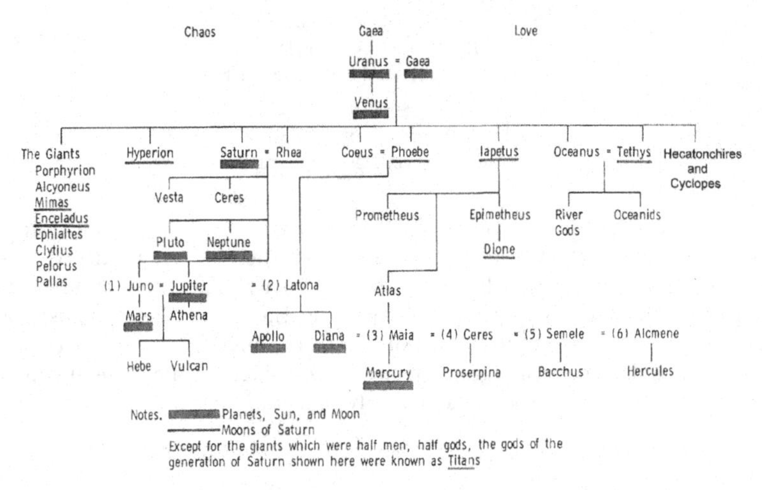

Figure 1. Geneology of the Gods.

Armed with an iron sickle provided him by his mother Gaea, Saturn lay in wait for his sire to come in for the night to overlay Earth. At the right time Saturn fell upon Uranus and grievously wounded him. Saturn thus seized power in Olympus in a bloody cosmic coup. From the bloody member cast into the sea arose Venus (Aphrodite - foam born) and from the blood that fell on Earth arose the Giants. Saturn and the Titans reigned during the golden age of Greek mythology.

Figure 2. Earth, Galileo 1990.

The Saturn moon Titan was discovered by Christian Huygens in 1655. Huygens was convinced no more celestial bodies would be found orbiting the sun, because Titan (which he named Luna Saturni) was the 12th such body. 12 was such a perfect number, he

Figure 3. The Mutilation of Uranus, Georgio Vasari and Christofano Gherardi, 16th cent. Italian Renaissance.

Figure 4. Saturn, Voyager 1 1980.

reasoned, it could only be the result of divine intent. Were there not 12 tribes of Israel, 12 Disciples of Christ, and 12 signs of the Zodiac? When Giovanni Cassini discovered two more moons of Saturn 16 years later, however, he hastened to point out to his patron Louis XIV that perhaps 14 was more divine than 12. John Herschel in 1858 named the 8 known Saturn moons for the Titans and for the Giants born of the blood that fell to Earth when Saturn castrated Uranus in the cosmic coup.

Saturn, having learned from an oracle that he would be overthrown by one of his own offspring, as he had overthrown his own sire, devoured his children as they were born to prevent this from occurring. Time, the ancients noted, destroyed that which it had created. Rhea, Saturn's sister-goddess-wife and mother of the children, was understandably distressed.

Figure 5. The Birth of Venus, Sandro Botticelli, 16th cent. Italian Renaissance.

3. The Reign of Jupiter

When her 6th child Jupiter was born Rhea passed off a rock wrapped in swaddling clothes in place of Jupiter to the god Saturn and spirited the infant Jupiter off to the island of Crete, where he was cared for by the nymphs Ida and Adrastea, and fed from the milk of the goat Amalthea.

Jupiter quickly grew to manhood (or godhood) and with the help of Gaea fed Saturn an emetic, which caused him to regurgitate his brothers and sisters. They then joined forces in an epic battle with Saturn and the Titans for supremacy of Heaven and Earth. The Jovian gods were triumphant and the defeated Titans were cast into Tartarus, the profound abyss beneath Earth.

Figure 6. Saturn Devouring His Children, Georgio Vasari, 16th Cent. Italian Renaissance.

Figure 7. The Nurture of Jupiter, Nicolas Poussin, 17th cent. French Baroque.

After the defeat of the Titans, Jupiter (Zeus) was chosen by all the gods of Olympus to be the supreme ruler of the Heavens, the Earth, and the gods. The Statue of Zeus by Pheidias is one of the Seven Wonders of the Ancient World. The image here is reconstructed from rare coins struck by the Roman Emperor Hadrian in the 2nd cent. A.D. The Stoic Dio Chrysostom, in a speech in 90 A.D., called Pheidias' Zeus "The most beautiful and the dearest to God of all works on Earth, the mere sight of which stills all sorrows."

Figure 8. Statue of Zeus, Pheidias, c. 430 B.C.

In time, the Giants, born by Earth from the blood of the mutilated Uranus, rose to challenge the Jovian gods for dominion over Olympus. With the help of Hercules the Jovian gods defeated the Giants. According to legend, in celebration of the victory of the Jovian gods over the Giants, Hercules instituted the first Olympic Games in 776 B.C.

Figure 9. Jupiter, Voyager 1, 1979.

4. Myths of Jupiter and those Lovers for Whom the Medicean Moons Are Named

The link between myths of the ancient past, art of the Renaissance, Baroque and Rococo periods, and space exploration of the present are here represented by the myths of Jupiter and those lovers for whom the Medicean moons are named. Three of the Jovian playmates, Io, Europa, and Callisto were descended from the river god Inachus, son of the Titans Oceanus and Tethys, as seen in the genealogy chart below.

Genealogies of Io, Europa, and Callisto
from
THE RACE OF INACHUS AND ITS BRANCHES

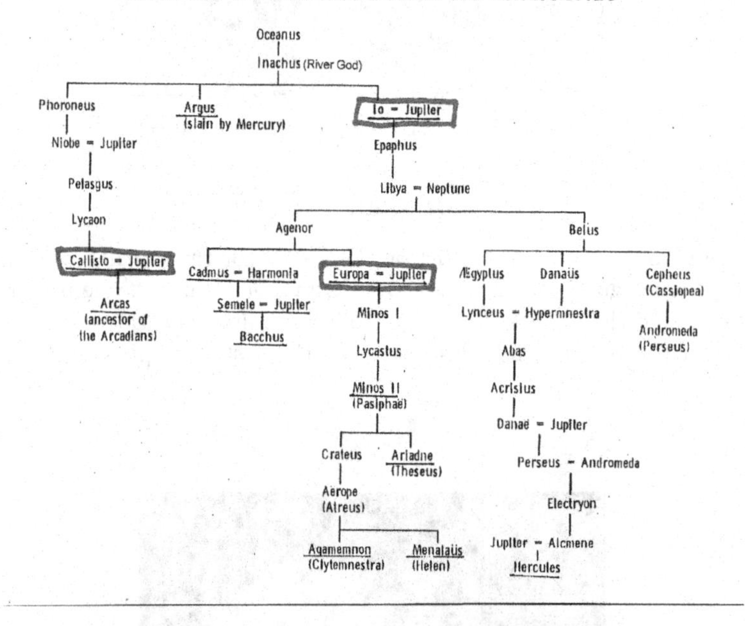

Figure 10. Genealogies of Io, Europa, and Callisto.

5. The Myth of Jupiter and Io

Io was a river nymph, daughter of the river god Inachus. One day Jupiter spotted the sweet and innocent Io on the banks of the Inachus River, near where it flows into the Argolic Gulf. As was his want, Jupiter fell in love with the beautiful Io. He came down from Olympus to be with her and spread a cloud over the land to conceal his dalliance from his jealous wife Juno.

In the painting Jupiter and Io Jupiter plants an immortal kiss on the ecstatic cheek of the river nymph Io. Jupiter had visited Io in her dreams and whispered to her as she slept:

Figure 11. Jupiter and Io, Antonio Allegri da Correggio, 16th cent. Italian Renaissance.

"Now to the meadow land of Lerna,
where thy father's pastures lie,
And the sleek cattle browse, do thou steal forth
Alone, and happly there thy yielding grace
May sooth the passion in the Sovereign's eye"
Aeschylus, *Prometheus Bound*

Figure 12. Juno Discovering Jupiter with Io, Pieter Lastman, 17th cent. Dutch artist.

Juno noted the cloud covering all of the Peloponnesus, and noted too that her husband Jupiter was not on his throne on Mt. Olympus. She became suspicious and decided to investigate. In the painting Juno Discovering Jupiter with Io we see Juno mounted on her peacock drawn chariot penetrating the clouds just as Jupiter, sensing her coming, changes Io into a heifer.

This tale is the basis for the name Juno given to the NASA space mission, which will

Figure 13. The Medicean Moon Io, Voyager 1, 1979.

be launched to Jupiter in 2011, will penetrate Jupiter's clouds with its instruments, and will learn Jupiter's secrets.

The names of the Ionian Sea and the Bosporus (fording of the ox) also derive from this myth. Note the prominent volcanic feature Loki in the Voyager image of Io, which looks amazingly like the hoof print of a heifer.

6. The Myth of Jupiter and Europa

From the myth of Jupiter and Europa comes the names of a major constellation and of a continent. Jupiter caught sight of Europa, the legendary princess of Phoenicia, as she picked flowers by the sea, and was instantly overwhelmed with love for her.

Jupiter took the shape of a bull, meandered over to where Europa played, and knelt by the princess. Europa was enthralled and climbed upon the bull's back and spread garlands of flowers about its neck. At that instant Jupiter the bull leapt into the sea with Europa clinging frantically to his horns, and carried her across the sea to the island of Crete. There he cast off his shape of a bull and tossed it into the heavens as the constellation Taurus.

Figure 14. Europa Picking Flowers, Fresco found in the ruins of Pompeii, 40 A.D., copy after Apelles, c. 430 B.C..

Legend has it that the continent of Europe was named for this event. Charlemagne, who conquered most of what we know today as Europe, was familiar with this legend, and some say it was he who officially gave the continent its name. According to the

Figure 15. The Abduction of Europa, Titian, 16th cent. Italian Renaissance.

myth Europa bore Jupiter 3 sons, among them Minos, legendary ancestor of the Minoan civilization, first of the great Greek civilizations.

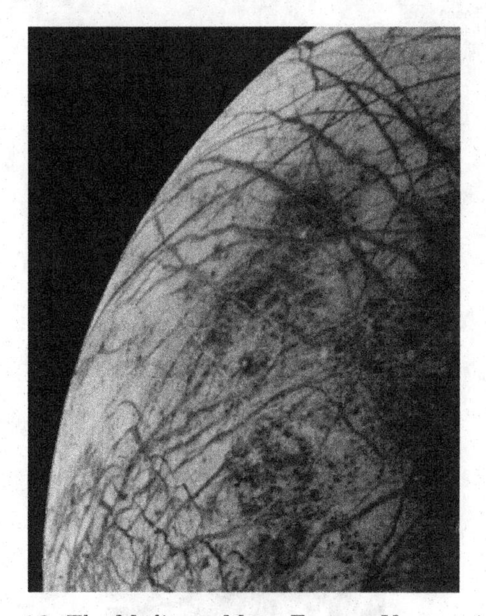

Figure 16. The Medicean Moon Europa, Voyager 1, 1979.

7. The Myth of Jupiter and Ganymede

According to the myth, one day as Jupiter was surveying his domain he spotted the boy Ganymede tending a flock of sheep on the slopes of Mt. Ida near Troy. Jupiter suddenly felt a profound passion for the boy, changed his shape into that of an eagle, swooped down and snatched the boy up and carried him off to Mt. Olympus to serve as cup bearer of the gods. However, cup bearer, or water bearer, was a position already held by the young goddess Hebe, daughter of Juno and Jupiter. In the painting *Ganymede and Hebe Vying to Serve Jupiter* we see Ganymede, with boyish facial hair, and Hebe, with adolescent breasts, vying to serve Jupiter while Juno stands nearby. Also shown, at the feet of Jupiter, is the goat Amalthea, whose milk nourished him as an infant. In the background is a scene depicting the abduction of Europa. Jupiter, in the shape of

THE ROYAL FAMILY OF TROY

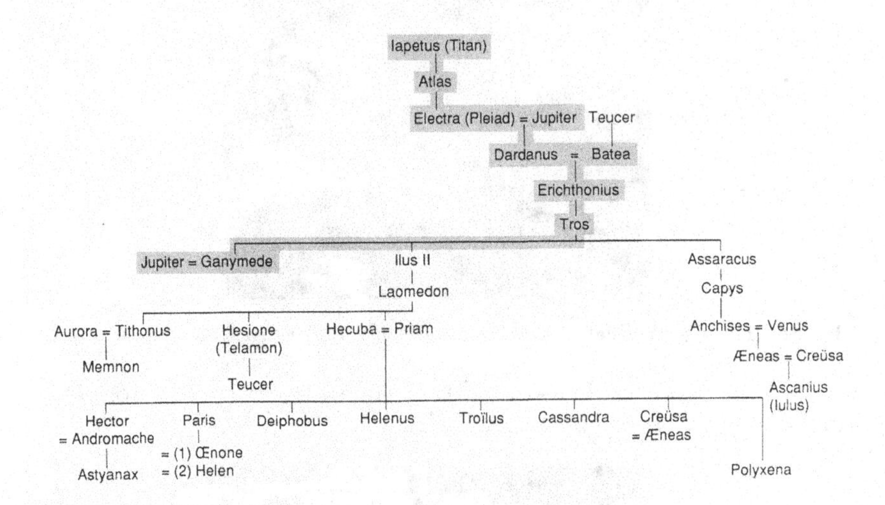

Figure 17. Genealogy of Ganymede.

Figure 18. The Abduction of Ganymede, Correggio, 16th cent. Italian Renaissance.

a bull, is plunging through the sea, with the pink clad Europa in much distress. Two constellations celebrate the myth of Jupiter and Ganymede: Aquarius the Water Bearer and Aquila the Eagle.

Figure 19. Ganymede and Hebe Vying to Serve Jupiter, Cristofano Gherardi, c. 1565.

Figure 20. The Medicean Moon Ganymede, Voyager 1, 1979.

8. The Myth of Jupiter and Callisto

Callisto was the favored companion of the virgin goddess Diana. Diana had warned her companions that if they wished to remain in her company, they must remain chaste. Callisto had accordingly vowed chastity. One day Jupiter saw the beautiful and chaste Callisto lying in the woods, waiting for Diana. He was aware of Diana's cautionary words to her companions, and was aware too of the special relationship Callisto had with Diana.

Figure 21. Jupiter in the Guise of Diana and the Nymph Callisto, Francois Boucher, 18th Cent. Rococo.

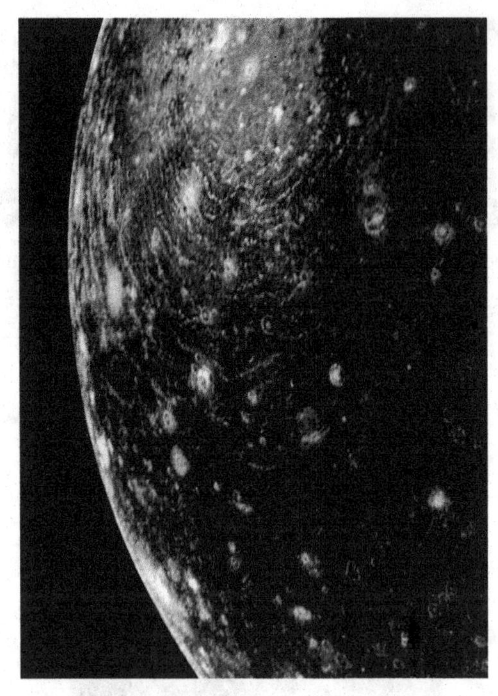

Figure 22. The Medicean Moon Callisto, Voyager 1, 1979.

So, in an inspired stratagem, Jupiter changed his shape into that of Diana, and went into the woods to where Callisto lay.

In Boucher's painting above, we see Jupiter as Diana, embracing the innocent Callisto. He whispers words of love to her:

"You are Callisto!
I behold in your eyes
The dawn of creation;
I behold in your smile
The innocence of surrender;
I behold the joy that you are,
And love is born in my breast
As stars are born in the arms of the galaxies,
And galaxies in the bosom of the Universe.
Fair Callisto,
I love you."

Callisto succumbs to Jupiter's words of love, and in their passionate embrace she conceives a child. Soon enough Callisto's condition becomes obvious and Diana forces her out of her company for having violated her strict rules of behavior. In time Callisto gives birth to a child, whom she names Arcas. Juno is outraged at this evidence of Jupiter's infidelity and changes Callisto into a bear. Callisto the bear is frightened and runs away into the woods, not to see her son for many years.

One day, when Arcas is a handsome and sturdy young man, he goes into these very woods to hunt, not realizing that his mother the bear resides there. Callisto sees her son Arcas, whom she hasn't seen for many years, and, forgetting she is a bear, rushes to embrace him. Arcas sees only a bear rushing down on him. He pulls out his bow and arrow and lets fly an arrow to the mark. At the last moment, just before the arrow strikes, Jupiter snatches up Callisto and her son Arcas and casts them into the heavens as the constellations Ursa Major and Ursa Minor, the big and little bears, the most recognizable parts of these constellations being the big and little dippers.

9. Conclusion

We have related myths of the great classics of Greek literature, illustrated them with masterworks of Renaissance, Baroque and Rococo art, identified the constellations, bodies of water and the continent that spun off these myths, and leapt into the present with images taken by our mighty spacecraft of the planets and Medicean moons named for the mythological characters of the myths related.

10. Epilogue

Epilogue is prologue. The history and legacy of one era becomes the starting point for the next. In this paper we have linked the past to the present and ride the crest of the New Renaissance into the future. From the dawn of time we have asked: Who am I? Where do I come from? We asked in the marshy Edens of Sumer, again in the glory that was Greece, again in the grandeur of the Renaissance, and again today. Today we are immersed in a New Renaissance and once again communications, the arts, mathematics and science are exploding in unison. The binary code, the internet, and space exploration are the masterworks of this new age.

Figures:

Fig. 1: Genealogy of the Gods, adapted from Table A, The Great Gods of Olympus, in Gayley's Mythology

Fig. 2: Planet Earth, image by Galileo taken during the first Earth encounter, 1990

Fig. 3: The Mutilation of Uranus, Georgio Vasari and Christofano Gherardi, 1560, Palazzo Vecchio, Florence, Italy

Fig. 4: Planet Saturn, image by Voyager 1 taken during the Saturn encounter, 1980

Fig. 5: The Birth of Venus, Sandro Botticelli, 1600, Uffizi Gallery, Florence, Italy

Fig. 6: Saturn Devouring His Children, Georgio Vasari, c. 1560, Palazzo Vecchio, Florence, Italy

Fig. 7: The Nurture of Jupiter, Nicolas Poussin, 1640, Dahlen State Museum, Berlin, Germany

Fig. 8: The Statue of Zeus, picture in Mythology, by Lady Hestia Evans, edited by Dugald Asteer, Candlewick Press, Doverville, MA, 2007

Fig. 9: Planet Jupiter, image by Voyager 1 taken during the Jupiter encounter, 1979

Fig. 10: Genealogies of Io, Europa, and Callisto, adapted from Table D, The Race of Inachus and its Branches, in Gayley's Mythology

Fig. 11: Jupiter and Io, Antonio Allegri da Correggio, c. 1530, Kunsthistorisches Museum, Vienna, Austria

Fig. 12: Juno Discovering Jupiter and Io, Pieter Lastman, 1618, The National Gallery, London, England

Fig. 13: The Medicean Moon Io, image by Voyager 1 taken during the Jupiter encounter, 1979

Fig. 14: Europa Picking Flowers, fresco found in the ruins of Pompeii, c 40 A.D., copy after Apelles, c. 430 B.C.

Fig. 15: The Abduction of Europa, Titian, 1562, Isabella Stewart Gardner Museum, Boston, MA
Fig. 16: The Medicean Moon Europa, image by Voyager 1 taken during the Jupiter encounter, 1979
Fig. 17: Genealogy of Ganymede, adapted from Table O(5), The Royal Family of Troy, in Gayley's Mythology
Fig. 18: The Abduction of Ganymede, Correggio, 1532, Kunsthistorisches Museum, Vienna, Austria
Fig. 19: Ganymede and Hebe Vying to Serve Jupiter, Cristofano Gherardi, c. 1565, Palazzo Vecchio, Florence, Italy
Fig. 20: The Medicean Moon Ganymede, image by Voyager 1 taken during the Jupiter encounter, 1979
Fig. 21: Jupiter in the Guise of Diana and the Nymph Callisto, Francois Boucher, 1759, Nelson Gallery - Atkins Museum, Kansas, MO
Fig. 22: The Medicean Moon Callisto, image by Voyager 1 taken during the Jupiter encounter, 1979

References

Aeschylus 1938, Prometheus Bound *The Complete Greek Drama*, edited by Whitney J. Oates and Eugene O'Neill, Jr., Random House, New York

Gayley, C. M. 1939, *The Classic Myths in English Literature and in Art*, Ginn and Company, Boston

1893, *The Iliad of Homer*, tr. by Andrew Lang, Walter Leaf, and Ernest Myers, MacMillan and Co., New York

1955, *The Metamorphoses of Ovid*, tr. by Mary M. Innes, Penguin Books, Baltimore

1950, *The Odyssey of Homer*, tr. by S. H. Butcher and A. Lang, Random House, The Modern Library, New York

1977, The Theogony of Hesiod, *Hesiod, The Homeric Hymns and Homerica*, tr. by Hugh G. Evelyn-White, Harvard University Press, Cambridge, MA

Clarke, T. C., 1979, A Love Poem from Jupiter to Callisto *Science of Mind Magazine*, August 1979

Galileo's Medicean Moons: their impact on 400 years of discovery
Proceedings IAU Symposium No. 269, 2010
C. Barbieri, S. Chakrabarti, M. Coradini & M. Lazzarin, eds.
doi:10.1017/S174392131000743X

Observing Mercury: from Galileo to the stereo camera on the BepiColombo mission

Gabriele Cremonese[1], Vania Da Deppo[2], Giampiero Naletto[2,3,4], Elena Martellato[4], Stefano Debei[4], Cesare Barbieri[4,5], Carlo Bettanini[4], Maria T. Capria[6], Matteo Massironi[7], Mirko Zaccariotto[4]

[1]INAF – Osservatorio Astronomico,
Vic. Osservatorio 5, I-35122 Padova, Italy
email: gabriele.cremonese@oapd.inaf.it

[2]CNR-IFN UOS Luxor, Padova, Italy
email: dadeppo@dei.unipd.it

[3] Dep. of Information Engeneering, University of Padova,
email: naletto@dei.unipd.it

[4]CISAS, University of Padova,
Via Venezia 15, 35131, Padova, Italy
email: elena.martellato@oapd.inaf.it
email: stefano.debei@unipd.it

[5]Astronomy Department, University of Padova,
Vic. Osservatorio 3, 35122, Padova, Italy
email: cesare.barbieri@unipd.it

[6]INAF – IASF, Via del Fosso del Cavaliere 100 00133 Roma, Italy
email: mariateresa.capria@iasf-roma.inaf.it

[7]Dep.Geosciences, University of Padova, Via Giotto, 35137 Padova, Italy
email: matteo.massironi@unipd.it

Abstract. After having observed the planets from his house in Padova using his telescope, in January 1611 Galileo wrote to Giuliano de Medici that Venus is moving around the Sun as Mercury. Forty years ago, Giuseppe Colombo, professor of Celestial Mechanics in Padova, made a decisive step to clarify the rotational period of Mercury. Today, scientists and engineers of the Astronomical Observatory of Padova and of the University of Padova, reunited in the Center for Space Studies and Activities (CISAS) named after Giuseppe Colombo, are busy to realize a stereo camera (STC) that will be on board the European (ESA) and Japanese (JAXA) space mission BepiColombo, devoted to the observation and exploration of the innermost planet. This paper will describe the stereo camera, which is one of the channels of the SIMBIOSYS instrument, aiming to produce the global mapping of the surface with 3D images.

Keywords. space vehicles: instruments, planets: Mercury

---

1. Introduction

The city and the University of Padova have a long standing love affair with the inner planets, Mercury in our case, since those very momentous days of fall - winter 1609 crowned by the Sidereus Nuncius, when Galileo Galilei made some of his most important discoveries using his 'cannocchiale'. Shortly after, in January 1611, he wrote to Giuliano de Medici stating that Venus and Mercury move around the Sun as the other planets do. Around the middle of the XX Century, Giuseppe Colombo, professor of Celestial Mechanics in the same University, made a decisive step to clarify the relationships between orbit and spin of the planet. To acknowledge such important role, the European Space Agency has named BepiColombo the forthcoming mission to Mercury devoted to the observation and exploration of the innermost planet. The names of Galileo and Colombo are thus associated to two of the most important space missions in the solar system. Today, scientists and engineers of the Astronomical Observatory and of the University of Padova, reunited in the Center for Space Studies and Activities (CISAS) named after Giuseppe Colombo, are working on a stereo camera (STC) that will be on board BepiColombo. This paper will describe the stereo camera (STC), which is one of the channels of the SIMBIOSYS instrument, aiming to produce the global mapping of the surface in 3D.

2. SIMBIOSYS

The Spectrometer and Imagers for MPO BepiColombo Integrated Observatory SYStem (SIMBIOSYS) (Figure 1) will be on board the Mercury Planetary Orbiter of the ESA-JAXA mission BepiColombo (Flamini *et al.* 2010). SIMBIOSYS includes a high resolution imaging channel (HRIC), providing images at a spatial resolution of 5 m/pixel at the periherm, the VIS-NIR spectrometer (VIHI) that will provide global mapping of Mercury's surface in the spectral range 400-2200 nm, with a spectral sampling of 6.25 nm, and a spatial resolution of 400 m/pixel at the periherm, and the stereo imaging channel (STC) (Cremonese *et al.* 2009) that will be described in the following sections.

3. General Scientific Objectives of STC

The analysis of the geological characteristics of Mercury's surface requires the mapping, at a resolution lower than 110 m per pixel, and the Digital Terrain Model (DTM) of the entire surface. The main scientific objectives can be summarized as follows:

• large scale Mercury's surface composition. The STC filters are located in the spectral range of 400-920 nm where the electronic processes in the transition elements give the major contribution to the absorption spectra of minerals. The color mapping of the STC will therefore allow to discriminate among rock-forming minerals primarily on the basis of the albedo and color contrasts;

• Mercury's lithostratigraphic units identification, as smooth plains, intercrater plains, heavily cratered terrains and hilly and lineated terrains. STC will constrain the relationship between different geological units in extended regions; therefore, a satisfactory knowledge of the global stratigraphy will be achieved;

• cratering record and surface age, which is an outstanding tool for geological dating and provides important information on the origin of impacting objects;

• volcanism. STC accuracy on 3D reconstruction will allow the detection of several volcanic structures and deposits (Massironi *et al.* 2008) that can be described in three main objectives:

  ◦ identification of different lava flows and volcanic deposits with definition of their emplacement mechanisms;
  ◦ identification of volcanic edifices, domes and dykes;
  ◦ identification of possible fissured vents.

• tectonic. The 3D restitution obtained by STC images should lead to a global mapping only slightly affected by directional interpretation errors since the derived DTM will reflect more directly the morphological pattern of the surface. Therefore, the despinning effect, cooling or dynamical loading that changed the planet shape during the early history can be better distinguished and quantitatively constrained.

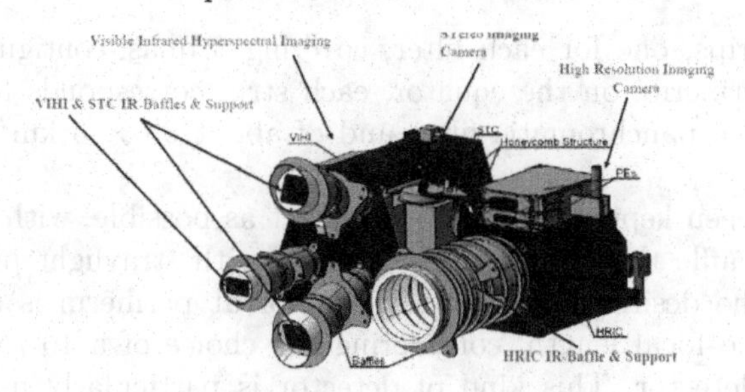

Figure 1. The SIMBIOSYS suite

Table 1. Summary of the STC nominal performance.

Parameter	**Value**
Scale factor	50 m/pix at periherm
Swath	38 km at periherm
Stereoscopic properties	±20 stereo angle with respect to nadir both images on the same detector
Vertical accuracy	80 m
EE	⩾70% inside 1 pix
MTF	⩾60% at Nyquist frequency
Wavelength coverage	410-930 nm (5 filters)
Filters	Panchromatic (700 ± 100 nm)
	420 ± 10 nm
	550 ± 10 nm
	750 ± 10 nm
	920 ± 10 nm

Table 2. Summary of the STC optical parameters

Parameter	**Value**
Optical concept	modified Schmidt telescope with rhomboid prisms and field corrector
Stereo solution(concept)	2 identical optical sub-channels; detector and most of the optical elements common to the two sub-channels
Focal length (on-axis)	90 mm
Pupil size (diameter)	15 mm
Focal ratio	*f*/6
Mean image scale	23 arcsec/px (111 μm)
FoV (cross track)	5.3°
FoV (along track)	2.4° panchromatic 0.4° color filters
Detector	Si PIN (format: 2048 × 2048; 10 μm square pix) 14 bits dynamic range

4. STC Optical Design

The stereo camera for the BepiColombo mission is based on a completely new optical design and acquisition technique, as the push-frame, instead of the push-broom techniques used by HRSC on board the Mars Express mission and the terrestrial satellites SPOT. We think that the push-frame technique allow us to achieve optimal results reducing the number of images to mosaic and having larger single images to find the matching points for the stereo reconstruction. The global FoV of each channel is 5.3° × 4.5°, subdivided in 3 strips, one for each filter, covering 3 quasi-contiguous strips on Mercury surface; at periherm, on the equator, each strip corresponds to an area of about 38×19 km^2 for the panchromatic filter and of about 38×3 km^2 for the color ones (Figure 2).

The design has been kept as short and compact as possible, with the need of having the possibility to baffle the two channels to cope with straylight problems due to the common optics. The desired 50 m/px scale factor at periherm is achieved with a 90 mm system effective focal length, considering the choice of a 10 μm pixel size hybrid SiPIN CMOS as detector. This kind of detector is particularly useful both in terms of radiation hardness, given the hostile Mercury environment, and for the capability of snapshot image acquisition, allowing very short exposure times of about 1 ms for STC. The optical sub-system consists of a modified Schmidt telescope plus a couple of rhomboid prisms positioned in front of the objective. In that solution, the classical Schmidt correcting plate positioned in the center of the curvature of the spherical mirror has been substituted with a correcting doublet positioned at about half distance: in this way the volume is reduced of about a factor of two with respect to the classical solution (Figure 3). All the optical elements, except the rhomboid prisms and the aperture stops (AS), are common to both channels. Common optics, together with the fact that the rhomboid prisms are not sensitive to the tilt, make the system free of co-registration error between the channels. Moreover, to optimize the separation between the optical paths of the two channels, an ad hoc baffling system is proposed and designed.

Table 1 and 2 summarize the optical characteristics of STC and its nominal performance respectively.

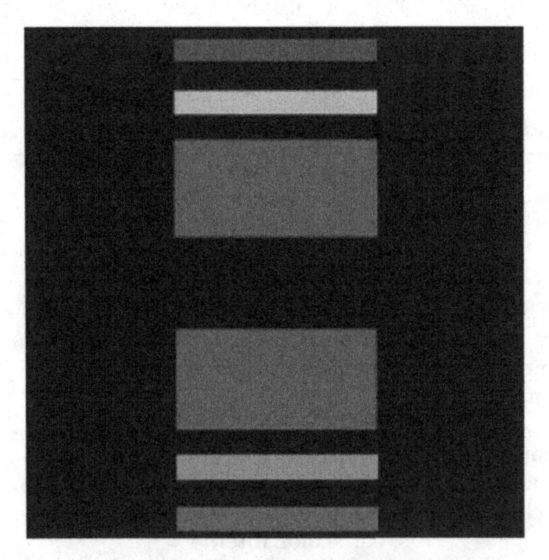

Figure 2. Position and size of the filter strips (depicted in light gray) on the full 2kx2k detector area (in black).

5. STC Mechanical Design

STC is mainly composed by 3 sub-units:

• focal Plane Assembly with dedicated Proximity Electronics to pilot detector in terms of integration time, windowing and binning;

• optical Module in which the optical elements, described in previous paragraph, are kept in stable position by means of a dedicated optical bench able to guarantee the optical

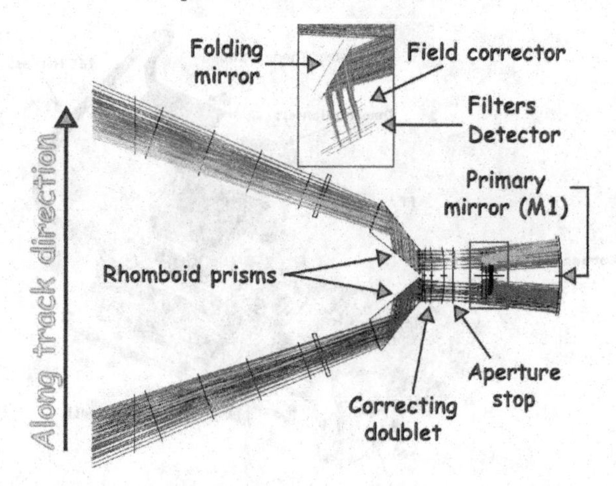

Figure 3. Schematic of the STC optical design

stability after launch, whose random vibration level are very demanding, and within the operative temperature range that is $-20 \div +30$ °C;

- external and internal baffling system.

Considering the extreme thermal environment of Mercury whose solar constant is up to 144000 W/m^2, and surface temperature of Planet varies from about 100 K (dark side) to 690 K (illuminated side), the thermal architecture is conceived in order to kept thermally decoupled the 3 sub-units from the Planet environment and the spacecraft (S/C). This is the reason why the external baffling is connected to a dedicated interface provided by S/C, while the STC Optical Bench and FPA with its PE, are connected to a high thermal resistance structure (shared with VIHI) made of sandwich panels with aluminum honeycomb and CFRP skins. To minimize the heat flux coming inside the telescope from the scene a heat rejection filter could be positioned on the honeycomb structure, in between the telescope entrances and the external baffles (Figure 4). The thermal architecture is completed by two Heat Pipes, provided by the S/C, reaching a maximum temperature of about + 16 °C in hot case, to dissipate the environmental and internally generated heat powers, and by thermal straps accurately designed to cold down respectively the PE and the detector. In this way the maximum operative temperature of PE is limited at about 40 °C, while the Detector maximum temperature at about 0 °C (Begin of Life) thanks to a dedicated Thermo Electric Cooler (TEC) Module. In particular the TEC module and the required electrical power are sized to provide margins on the minimum temperature on detector side. In fact detector sensitivity may change from the BoL up to the EoL mainly due to the interaction with the high energy particles: the possibility to cold down the detector up to -5 °C will allow the recovery of its performances.

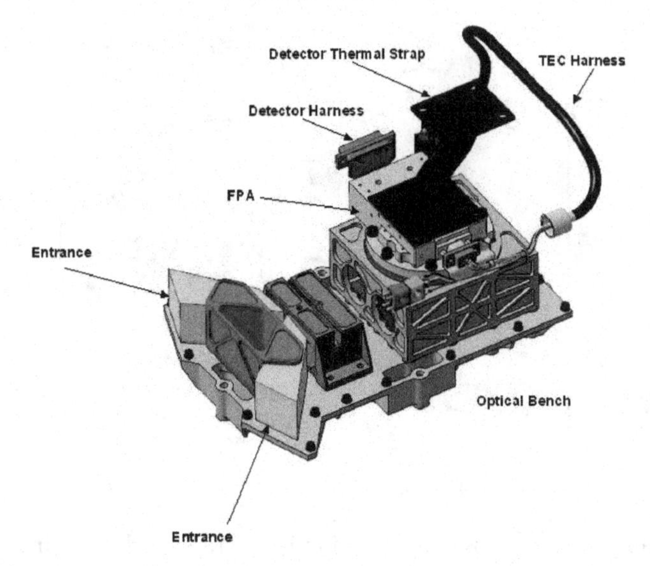

Figure 4. The optical bench of STC

Acknowledgement

References

Cremonese, G., Fantinel, D., Giro, G., Capria, M. T., Da Deppo, V., Naletto, G., Forlani, G., Massironi, M., Giacomini, L., Sgavetti, M., Simioni, E., Bettanini, C., Debei, S., Zaccariotto, M., Borin, P., Marinangeli, L., & Flamini, E. 2009, The stereo camera on the BepiColombo ESA/JAXA mission: a novel approach, Advances in Geosciences, 15, 305.

Massironi, M., Giacomini, L., Cremonese, G., Capria, M. T .,Da Deppo, V. , Forlani, G., Naletto, G., Pasquar, G., & Flamini, E. 2008, Simulations using terrestrial geological analogues of the Hermean surface to examine 3D rendering potentialities of the STereoscopic imaging Channel of the SIMBIO-SYS package (Bepi-Colombo mission), Plan.Space Scie., 56, 1079

Flamini, E., Capaccioni, F., Colangeli. L., Cremonese, G., Doressoundiram, A., Josset, J. L., Langevin, Y., Debei, S., Capria, M. T., DeSanctis, M. C., Marinangeli, L., Massironi, M., Mazzotta Epifani, E., Naletto, G., Palumbo, P., Eng, P., Roig, J. F., Caporali, A., DaDeppo, V., Erard, S., Federico, C., Forni, O., Sgavetti, M., Filacchione, G., Giacomini, L., Marra, G.,Martellato, E., Zusi, M., Cosi, M., Bettanini, C., Calamai, L., Zaccariotto, M., Tommasi, L., Dami, M., Ficai Veltroni, I., Poulet, F., Hello, Y., and the SIMBIO-SYS Team, 2010 SIMBIO-SYS: the Spectrometers and Imagers Integrated Observatory SYStem for BepiColombo Orbiter, Plan.Space Sci., 58, 125.

Galileo's Medicean Moons: their impact on 400 years of discovery
Proceedings IAU Symposium No. 269, 2010
C. Barbieri, S. Chakrabarti, M. Coradini & M. Lazzarin, eds.

doi:10.1017/S1743921310007441

Observations of mutual phenomena of Galilean's satellites at Catania

Daniele Fulvio and Carlo Blanco

Physics and Astronomy Department, Catania University, via S. Sofia 64, 95123 Catania, Italy
INAF - Catania Astrophysical Observatory, via S. Sofia 78, 95123 Catania, Italy
email: dfu@oact.inaf.it
c.blanco@oact.inaf.it

Abstract. The mutual phenomena between Jupiter and Saturn's satellites occur every half orbital period of these planets, when the Earth and the Sun cross their equatorial plane. At Physics and Astronomy Department of Catania University the events between Jupiter's satellites have been observed during the 1973, 1979, 1985/86, 1991, 1997 and 2009 campaigns and the ones between Saturn's satellites during the 1980/81 and 1995 campaigns. An overview of the main results obtained since 1973 is presented.

Keywords. techniques: photometric, telescopes, planets and satellites: Jupiter, planets and satellites: galilean satellites

1. Introduction

Prior to the Voyager missions the possibility to define the physical parameters and the surface morphology of the moons of the greatest Solar System planets was entrusted to the few photographs or to the drawings by astronomers. The lightcurves of the mutual phenomena give the objective data to better define the surface characteristics. As an example, the comparison between the albedo of the eclipsed areas and the drop in magnitude in different colors, the duration and the time of the minimum of light inferred, during the 1973 campaign, the existence of bright polar caps on Io, Europa and Ganymede, suggested by the drawings of many authors.

Since the Voyager missions have provided accurate physical parameters for all the satellites and described surface features and albedo variations, the observations of mutual events were best used for astrometric purposes. As an example, during the 1979 campaign, the relative separations in right ascension, $\Delta\alpha\cos\delta$, and declination, $\Delta\delta$, of a satellite pair at midevent were estimated.

The improvement of the integration time up to measures of the order of 0.1 sec, can supply relative positions with relevant residuals near 100 Km at Jupiter's mean distance, with the possibility, due to the great number of observed events, to evidence tidal effects on Io's orbit.

After the 1980 campaign, the observing conditions were completely changed by the incoming of the CCD detectors. Their spatial resolution of the order of arcsec per pixel, by the analysis of the lightcurve behavior during the event, also with small telescopes, makes it possible to obtain morphological details of the order of few kilometers. The use of several infrared wavelengths during Io's occultations by the other satellites, allows the detection of emitting hot spots associated to an intense volcanic activity on the first Galilean satellite of Jupiter.

2. Previous observational campaigns

At Physics and Astronomy Department of Catania University the events between Jupiter's satellites have been observed during the 1973, 1979, 1985/86, 1991, 1997 and 2009 campaigns. All the observations were carried out at M. G. Fracastoro Station of INAF - Catania Astrophysical Observatory.

During the 1973 campaign, the 61/41-cm Schmidt telescope equipped with a synchronous three-channel photometer able to obtain contemporaneous observations in the *UBV* bands (Blanco & Catalano 1974) was used. Two mutual occultation and one mutual eclipse were observed (see Table 1). The observed depth of all events is shallower than the predicted one by about $0^m.2$, apart from the JII occultation on October 22nd 1973, when the observed depth is much deeper than the predicted one. However, the observations of this event were made in the *B* band while the predictions refer to the *V* magnitude. The observed times of occurrence for all events are in close agreement with the predictions by Aksnes 1974.

Since the 1979 campaign, the 91-cm Cassegrain telescope was used and since the 1985 campaign it was equipped with photon counting photoelectric photometer.

Table 1. Relevant data on the events observed during the 1973 campaign (Blanco & Catalano 1974). Predicted values of the light minimum (UT) and the light loss were taken by Aksnes (1974).

Date	Event	Predicted minimum	Observed minimum	Predicted Light loss	Observed Light loss
OCT 22	3O2P	$17^h 43^m$	$17^h 43^m 45^s \pm 0^s.3$	$0^m.01$ (ΔV)	$0^m.08 \pm 0^m.04$ (ΔB)
NOV 15	4E3T	$18^h 10^m$	$18^h 09^m 45^s$	$0^m.80$ (ΔV)	$0^m.615 \pm 0^m.03$ (ΔV)
					$0^m.625 \pm 0^m.03$ (ΔB)
					$0^m.660 \pm 0^m.03$ (ΔU)
NOV 22	2E1A	$16^h 55^m$	$16^h 53^m 10^s$	$0^m.73$ (ΔV)	$0^m.530 \pm 0^m.03$ (ΔV)
					$0^m.525 \pm 0^m.03$ (ΔB)
					$0^m.450 \pm 0^m.03$ (ΔU)

Table 2. Relevant observed data on the 1979 mutual events of Jupiter satellites (Blanco 1996).

Date	Event	$\Delta\alpha\cos\delta$	$\Delta\delta$	Duration (sec)	Observed Light loss
OCT 01	1E2P	+0.072	+0.182	$445 \pm 0^s.3$	$1^m.80 \pm 0^m.05$ (ΔV)
					$1^m.83 \pm 0^m.05$ (ΔB)
					$1^m.65 \pm 0^m.05$ (ΔU)
NOV 02	1E2P	–0.160	–0.390	260	$0^m.80 \pm 0^m.03$ (ΔV)
					$0^m.80 \pm 0^m.03$ (ΔB)
					$0^m.75 \pm 0^m.03$ (ΔU)
NOV 02	3E2P	+0.094	+0.321	140	$1^m.30 \pm 0^m.03$ (ΔV)
					$1^m.10 \pm 0^m.03$ (ΔB)
					$1^m.12 \pm 0^m.03$ (ΔU)
NOV 09	1E2P	–0.262	–0.635	235	$0^m.20 \pm 0^m.03$ (ΔV)
					$0^m.17 \pm 0^m.03$ (ΔB)
					$0^m.17 \pm 0^m.03$ (ΔU)

Relevant observed data on the 1979 mutual events of Jupiters satellites are reported in Table 2 (Blanco 1996). Due to Catania geographic coordinates, the 1979 apparition was a poor one, requiring most observations to be made through large air masses, thus resulting in uncertain sky corrections and approximate lightcurves. The observations of mutual events of this campaign were best used for astrometric purposes.

A total of 37 mutual events were observed in 1985 and 22 of these were of such good quality to improve the computations of satellites' orbits. These events are reported in Table 3 (Blanco 1988; Franklin *et al.* 1991). Predicted values of the light minimum (UT) here shown were computed by Arlot (1984) by using the ephemerides given by Arlot (1982). The predicted light losses were by Aksnes & Franklin (1984). The observed light losses and values of the light minimum do not show relevant regular deviations from the predicted ones.

During the 1991 campaign we observed 29 mutual events. In Table 4 we report relevant data for some of them (Arlot *et al.* 1997).

As part of the PHEMU97 International Campaign, 28 lightcurves were obtained at M. G. Fracastoro Station. The mutual events that present little noise and allowed us to compare themself with predictions are 10 and they are reported in Table 5 (Blanco 1999; Blanco *et al.* 2001; Arlot *et al.*, 2006). A global analysis of these data shows that the times of occurrence of the observed minima normally come early compared to the predicted ones. Moreover, the observed light losses in seven cases are greater than the predicted ones while they are comparable in the remaining three cases.

Table 3. Mutual events observed in 1985 (Blanco 1988; Franklin *et al.* 1991). Predicted values of the light minimum (UT) here shown were computed by Arlot (1984) by using the ephemerides given by Arlot (1982). The predicted light losses were by Aksnes & Franklin (1984).

Date	Event	Predicted minimum	Observed minimum	Predicted Light loss (ΔV)	Observed Light loss (ΔV)
JUN 17	3O4P	$01^h44^m.91$	$01^h45^m.24 \pm 0^m.005$	$0^m.09$	$0^m.06 \pm 0^m.03$
JUL 08	3O2P	$22^h17^m.21$	$22^h18^m.20$	$0^m.31$	$0^m.36 \pm 0^m.01$
JUL 12	4O3P	$00^h39^m.86$	$00^h39^m.90$	$0^m.45$	$0^m.32 \pm 0^m.01$
JUL 16	3O2P	$01^h14^m.31$	$01^h14^m.19$	$0^m.27$	$0^m.26 \pm 0^m.08$
AUG 05	3O4P	$23^h26^m.96$	$23^h28^m.85$	$0^m.11$	$0^m.04 \pm 0^m.01$
AUG 27	3O2P	$19^h36^m.95$	$19^h36^m.60$	$0^m.18$	$0^m.25 \pm 0^m.01$
AUG 29	4O1P	$20^h18^m.57$		$0^m.02$	$0^m.03 \pm 0^m.01$
AUG 30	1E2P	$20^h03^m.90$	$20^h02^m.11$	$0^m.40$	$0^m.20 \pm 0^m.01$
SEP 03	3O2P	$22^h03^m.47$	$22^h03^m.72$	$0^m.15$	$0^m.17 \pm 0^m.01$
SEP 04	3E2T	$21^h46^m.91$	$21^h54^m.36$	$1^m.00$	$0^m.25 \pm 0^m.01$
SEP 06	1O2P	$21^h49^m.45$	$21^h51^m.88$	$0^m.04$	$0^m.03 \pm 0^m.01$
SEP 07	1E2P	$00^h29^m.73$	$00^h29^m.67$	$0^m.59$	$0^m.25 \pm 0^m.02$
SEP 07	1O3P	$20^h33^m.75$	$20^h34^m.07$	$0^m.05$	$0^m.03 \pm 0^m.01$
SEP 07	1E3P	$23^h22^m.31$	$23^h22^m.02$	$0^m.31$	$0^m.26 \pm 0^m.01$
OCT 01	4E2A	$19^h06^m.66$	$19^h07^m.20$	$0^m.68$	$0^m.34 \pm 0^m.01$
OCT 01	1E2A	$21^h52^m.16$	$21^h51^m.08$	$0^m.94$	$0^m.92 \pm 0^m.02$
OCT 02	4E1A	$19^h33^m.11$	$19^h35^m.00$	$0^m.81$	$0^m.85 \pm 0^m.01$
OCT 31	3E1P	$19^h17^m.06$	$19^h16^m.29$	$0^m.47$	$0^m.46 \pm 0^m.01$
NOV 07	2E1A	$16^h52^m.53$	$16^h52^m.09$	$0^m.55$	$0^m.65 \pm 0^m.01$
NOV 14	2E1A	$19^h06^m.56$	$19^h06^m.36$	$0^m.58$	$0^m.68 \pm 0^m.02$
DEC 04	1O2P	$17^h18^m.31$	$17^h18^m.23$	$0^m.10$	$0^m.08 \pm 0^m.01$
DEC 14	3E2P	$17^h34^m.66$		$0^m.98$	$0^m.77 \pm 0^m.03$

Table 4. Relevant data of some mutual events observed during the 1991 campaign (Arlot *et al.* 1997). Predicted values of the light minimum (UT) and the light loss were by Arlot (1997).

Date	Event	Predicted minimum	Observed minimum	Predicted Light loss (ΔV)	Observed Light loss (ΔV)
JAN 02	2O3A	$01^h 40^m 07^s$	$01^h 39^m 07^s \pm 0^s.3$	$0^m.479$	$0^m.299 \pm 0^m.02$
JAN 05	2O1P	$00^h 19^m 26^s$	$00^h 22^m 01^s$	$0^m.136$	$0^m.180 \pm 0^m.01$
JAN 09	2E3P	$03^h 03^m 00^s$	$03^h 03^m 06^s$	$0^m.232$	$0^m.253 \pm 0^m.01$
JAN 12	2O1P	$01^h 53^m 10^s$	$01^h 54^m 45^s$	$0^m.749$	$0^m.533 \pm 0^m.01$
FEB 13	2E1A	$02^h 04^m 44^s$	$02^h 05^m 21^s$	$0^m.712$	$1^m.761 \pm 0^m.02$
FEB 23	2E1A	$17^h 42^m 40^s$	$17^h 43^m 50^s$	$0^m.714$	$0^m.628 \pm 0^m.01$
MAR 09	2E1A	$20^h 58^m 13^s$	$20^h 58^m 28^s$	$0^m.221$	$0^m.228 \pm 0^m.01$
APR 29	2E4A	$22^h 18^m 12^s$	$22^h 18^m 28^s$	$0^m.434$	$0^m.291 \pm 0^m.01$
MAY 07	3E1P	$20^h 29^m 11^s$	$20^h 29^m 06^s$	$0^m.446$	$0^m.484 \pm 0^m.01$
JUL 04	3E1P	$19^h 42^m 26^s$	$19^h 43^m 03^s$	$0^m.942$	$1^m.351 \pm 0^m.01$

Table 5. Relevant data of some mutual events observed during the 1997 campaign (Blanco 1999; Blanco *et al.* 2001; Arlot *et al.* 2006). Predicted values of the light minimum (UT) and the light loss of each event are by Arlot 1999.

Date	Event	Predicted minimum	Observed minimum	Predicted Light loss (ΔV)	Observed Light loss (ΔV)
JUN 18	2E1A	$01^h 05^m 46^s$	$01^h 04^m 42^s \pm 0^s.3$	$0^m.647$	$1^m.25 \pm 0^m.01$
JUL 26	3O1P	$23^h 24^m 24^s$	$23^h 23^m 19^s$	$0^m.296$	$0^m.50$
JUL 29	1O3P	$23^h 50^m 43^s$	$23^h 49^m 30^s$	$0^m.206$	$0^m.23$
AUG 01	4E3A	$00^h 21^m 53^s$	$00^h 20^m 52^s$	$0^m.601$	$0^m.95$
AUG 02	3O2T	$02^h 54^m 07^s$	$02^h 53^m 30^s$	$0^m.259$	$0^m.57$
AUG 28	1E3P	$00^h 37^m 25^s$	$00^h 37^m 25^s$	$0^m.199$	$0^m.20$
SEP 29	3O2P	$18^h 43^m 22^s$	$18^h 42^m 21^s$	$0^m.196$	$1^m.22$
OCT 06	3O2P	$22^h 13^m 33^s$	$22^h 12^m 58^s$	$0^m.217$	$0^m.38$
NOV 17	3O1P	$16^h 48^m 30^s$	$16^h 47^m 35^s$	$0^m.278$	$0^m.52$
NOV 18	3E1	$19^h 18^m 48^s$	$19^h 18^m 48^s$	$0^m.007$	$0^m.01$

3. PHEMU09

During the 2009 observational campaign, the 91-cm Cassegrain telescope equipped with the new KODAK KAF 1001E CCD camera (1024x1024 pixels; 24-micron pixel-size) and, for the first time, the 80-cm Cassegrain telescope (APT2) equipped with the KODAK KAF 09000 CCD camera (3056x3056; 12-micron pixel-size) were used. The observations have been carried out in the *B* band and, depending on the nightly weather conditions and the telescope/CCD set-up used, with exposure time varying from 0.3 to 5 sec. Table 6 shows the mutual events observed during the 2009 campaign. Predicted values of the light minimum (UT) and the light loss, for each event, were by http://ftp.imcce.fr/pub/ephem/satel/phemu09/visibility. The acquired data are currently under reduction.

Table 6. Mutual events observed during the 2009 campaign. Predicted values of the light minimum (UT) and the light loss were by http://ftp.imcce.fr/pub/ephem/satel/phemu09/visibility.

Date	Event	Predicted minimum	Predicted Light loss (ΔV)
JUL 23	4E2	$23^h 12^m 35^s$	$0^m.139$
JUL 24	1E2P	$00^h 16^m 46^s$	$0^m.367$
JUL 24	1O2P	$01^h 15^m 09^s$	$0^m.214$
SEP 01	1O2P	$20^h 03^m 22^s$	$0^m.192$
SEP 01	1O2P	$21^h 05^m 16^s$	$0^m.349$
OCT 03	1O2P	$18^h 31^m 24^s$	$0^m.161$
OCT 03	1E2P	$20^h 23^m 36^s$	$0^m.625$

Acknowledgements

We are kindly grateful to Dr. G. Bonanno, Dr. G. Leto, P. Bruno, E. Martinetti and, to the technical staff of M. G. Fracastoro Station for their precious assistance during the observations.

References

Aksnes, K. 1974, *Icarus*, 21, 100

Aksnes, K. & Franklin, F. 1984, *Icarus*, 60, 180

Aksnes, K., Franklin, F., Millis, R., Birch, P., Blanco, C., Catalano, S., & Piironen, J. 1984, *AJ*, 89, 280

Arlot, J.-E. 1982, *A&A*, 107, 305

Arlot, J.-E. 1984, *A&A*, 138, 113

Arlot, J.-E., Ruatti, C., Thuillot, W., *et al.* 1997, *A&A SS*, 125, 399

Arlot, J.-E., Thuillot, W., Ruatti, C., *et al.* 2006, *A&A*, 451, 733

Blanco, C. 1988, *A&A*, 205, 297

Blanco, C. 1996, *Annales de Physique*, 21, C1-7

Blanco, C. 1999, *Proceeding of the Planetary Science II Italian Meeting*, 157, Bormio (IT), January 26-31, 1998, ed. by A. Manara and E. Dotto

Blanco, C. & Catalano, S. 1974, *A&A*, 33, 303

Blanco, C., Riccioli, D., & Cigna, M. 2001, *P&SS*, 49, 31

Franklin, F. A., Galilean Satellite Observers, Africano, J. *et al.* 1991, *AJ*, 102, 806

Galileo's Medicean Moons: their impact on 400 years of discovery
Proceedings IAU Symposium No. 269, 2010
C. Barbieri, S. Chakrabarti, M. Coradini & M. Lazzarin, eds.

doi:10.1017/S1743921310007453

Io, the closest Galileo's Medicean Moon: Changes in its Sodium Cloud Caused by Jupiter Eclipse

Cesare Grava[1a], **Nicholas M. Schneider**[2] **and Cesare Barbieri**[1b]

[1]Department of Astronomy, Padova University,
Vicolo dell' Osservatorio, 3, 35122 Padova, Italy
[a]email: cesare.grava@unipd.it
[b]email: cesare.barbieri@unipd.it

[2] LASP, University of Colorado
Campus Box 392, Boulder, Colorado 80309-0392515, USA
email: nick.schneider@lasp.colorado.edu

Abstract. We report results of a study of true temporal variations in Io's sodium cloud before and after eclipse by Jupiter. The eclipse geometry is important because there is a hypothesis that the atmosphere partially condenses when the satellite enters the Jupiter's shadow, preventing sodium from being released to the cloud in the hours immediately after the reappearance. The challenge lies in disentangling true variations in sodium content from the changing strength of resonant scattering due Io's changing Doppler shift in the solar sodium absorption line. We undertook some observing runs at Telescopio Nazionale Galileo (TNG) at La Palma Canary Island with the high resolution spectrograph SARG in order to observe Io entering into Jupiter's shadow and coming out from it. The particular configuration chosen for the observations allowed us to observe Io far enough from Jupiter and to disentangle line-of-sight effects looking perpendicularly at the sodium cloud. We will present results which took advantage of a very careful reduction strategy. We remove the dependence from γ-factor, which is the fraction of solar light available for resonant scattering, in order to remove the dependence on the radial velocity of Io with respect to the Sun.

This work has been supported by NSF's Planetary Astronomy Program, INAF/TNG and the Department of Astronomy and Cisas of University of Padova, through a contract by the Italian Space Agency ASI.

Keywords. Planets and satellites: Io

---

1. Io's Neutral Cloud

There are three major features in the Neutral Sodium Cloud around Io: a banana shaped cloud of slowly escaping neutral atoms, a fast "jet", produced by pickup ion neutralization in Io's atmosphere and a molecular ion "stream" resulting from ionization and pickup of sodium-bearing molecules directly from Io's atmosphere. The "banana cloud" (Brown 1974) contains slow atoms escaping from the surface, and its shape is controlled by celestial mechanics and by ionization (Burger *et al.* 1999). The "stream" is composed of fast sodium atoms which are ejected from the torus by dissociation or dissociative recombination of unidentified molecular pickup ions containing sodium (NaX^+). The "jet" and "stream" showed that Io's ionosphere and Jupiter's magnetosphere interact with each other (Schneider & Trauger 1995). The Jupiter-Io system is reviewed in Schneider & Bagenal 2007.

2. Resonant Scattering

The Sodium is only a trace component in the Io's exosphere (only few %'s). But it is by far the most visible from Earth thanks to the very high cross section for resonant scattering mechanism. A solar photon excites an electron to the first excited state, and when it returns to its ground state (after nearly 10^{-7} sec) it emits a photon. The first excited state is in reality double due to fine structure, so this process can actually emit in two possible wavelengths (5890 and 5896 Angstrom for the D2 and D1 line respectively).

The γ-factor is the ratio between the intensity of the absorption line at a given wavelength and the intensity at the continuum. It is an indicator of how much solar light a Sodium electron "sees" at a given velocity with respect to the Sun. When Io is in conjunction, its relative velocity with respect to the Sun is zero, so the electrons see a very small amount of solar light, corresponding to the bottom of the Fraunhofer line. In this case the γ-factor is $\sim$ 0.05, meaning that the Sodium atoms receive $\sim$ 5% of the solar continuum and the Sodium Cloud will be faint (right side of the Figure 1). On the contrary, when Io is at elongation, its velocity with respect to the Sun is at maximum, the atoms have more light available for scattering (the γ-factor increases up to 0.70) and the Sodium Cloud will be brighter (left side of the Figure 1).

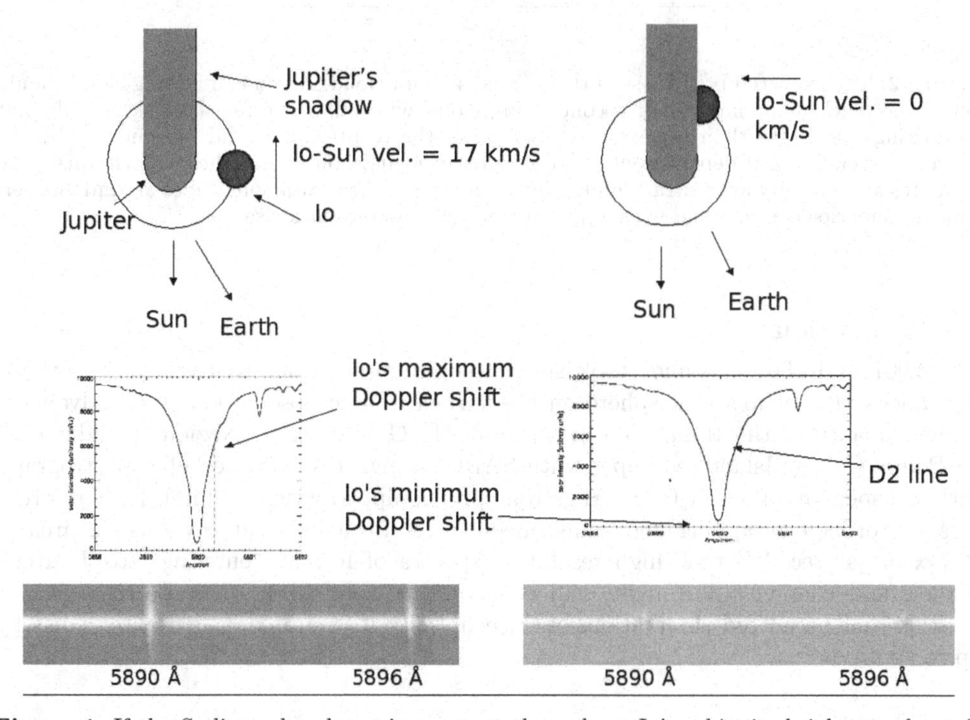

Figure 1. If the Sodium abundance is constant throughout Io's orbit, its brightness depends only on the orbital longitude. Left part: Io is at western elongation; right part: Io is at superior conjunction. Together with the geometrical configurations (upper part) there are the plots of solar radiance around the Sodium D2 line (middle) to show the dependence of the γ-factor from the orbital longitude, and the spectra of Io (bottom), showing the Sodium emission being fainter while Io-Sun relative velocity approaches zero.

3. Scientific Goals

As the γ-factor increases after the occultation, Sodium atoms see more solar photons, and the intensity of the cloud increases. Nevertheless, some past observations showed some evidence of post-eclipse brightening, i.e. the sodium cloud immediately after the eclipse is less bright than it should be, and reaches the expected values of luminosity hours after the reappearance. One possible explanation is condensation of sulfur dioxide during the eclipse, which would prevent NaCl (principal source of Na atoms) to be released. This is the fascinating hypothesis we want to test (see Figure 2).

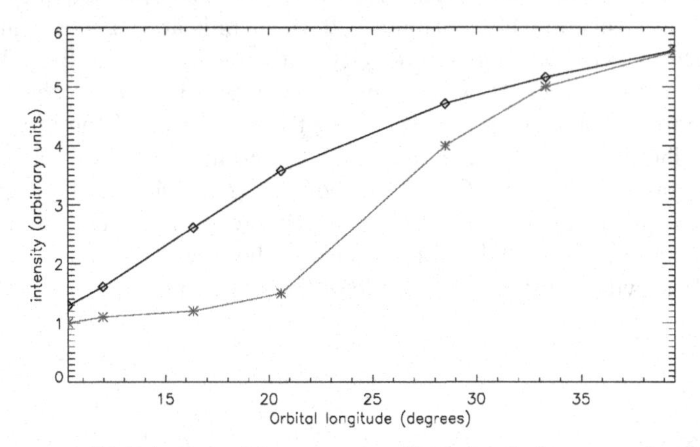

Figure 2. The expected brightness of the neutral sodium cloud. A non-condensing cloud should follow the black diamonds trend; a condensing cloud will follow the red asterisks trend, that is: starting less bright than the expected soon after the reappearance and reaching the purely geometric trend, which depends only by the radial velocity, hours after the reappearance. The ordinates are intensity at arbitrary units; the abscissae are the orbital longitudes around Jupiter: 0 means superior conjunction, with angles increasing counter-clockwise.

4. Observations

In 2007 and 2009, we undertook spectroscopic observations to test the hypothesis of condensation of Io's atmosphere on the surface. These observations took advantage of high quality of the Italian 3.6 m telescope TNG (Telescopio Nazionale Galileo) at La Palma Canary Island, equipped with SARG, a high resolution échelle spectrograph with a dispersion of about 0.022 Angstrom/pix, equipped with a Sodium filter in order to avoid order overlapping and simultaneously to use a long slit covering an area of 26.7 x 0.4 arcsec. We took high resolution spectra of Io while entering into Jupiter's shadow and coming out from it, and observations were taken when Earth was at a position that permitted observations of the eclipse itself separated from Jupiter's disk by up to 10 arcsec.

5. Reduction Steps

As we want to study the Na emission from the cloud, we have to remove the sunlight reflected from Io's surface. So, for the purposes of this research, besides the "standard" reduction steps (as bias and flat fielding), the telluric lines removal and the reflected sunlight subtraction must be performed. The conversion between ADUs and kiloRayleighs ($1R = 10^6/4\pi\ photons\ cm^{-2}\ sec^{-1}\ sr^{-1}$) has been done using the well known Jupiter's intensity of 5.4 MegaRayleighs/Angstrom. Knowing the g-factor from the γ-factor, under the assumption of optically thin cloud, it is also possible to get the column density N (expressed in *atoms* cm^{-2}) and therefore the amount of sodium present. From Brown & Yung (1976):

$$g = \left[\gamma\pi F_{Sun}(.59)\,\frac{\lambda^2}{c}\right]\frac{\pi e^2}{mc}f \qquad (5.1)$$

where g is expressed in *photons* s^{-1} *atom*$^{-1}$; $F_{Sun}(.59)$ is the solar flux at the wavelength λ of the D lines scaled to the Io-Sun distance, expressed in *photons* cm^{-2} sec^{-1} $\mathring{A}^{-1}$, and f is the oscillator strength. The other symbols are standard notation for physical constants. Finally, the intensity is $I = g \cdot N$ and is expressed in kiloRayleighs.

6. Preliminary Results

In Figure 3 we plotted the intensity in Rayleigh as a function of the orbital longitude. The black asterisks are the observed data (referring to the left ordinates), the red diamonds are the γ-factor (referring to the right ordinates), which show the expected trend in the case of a non-condensing exosphere. Observations were taken prior to eclipse in April and May, and after eclipse in June and July.

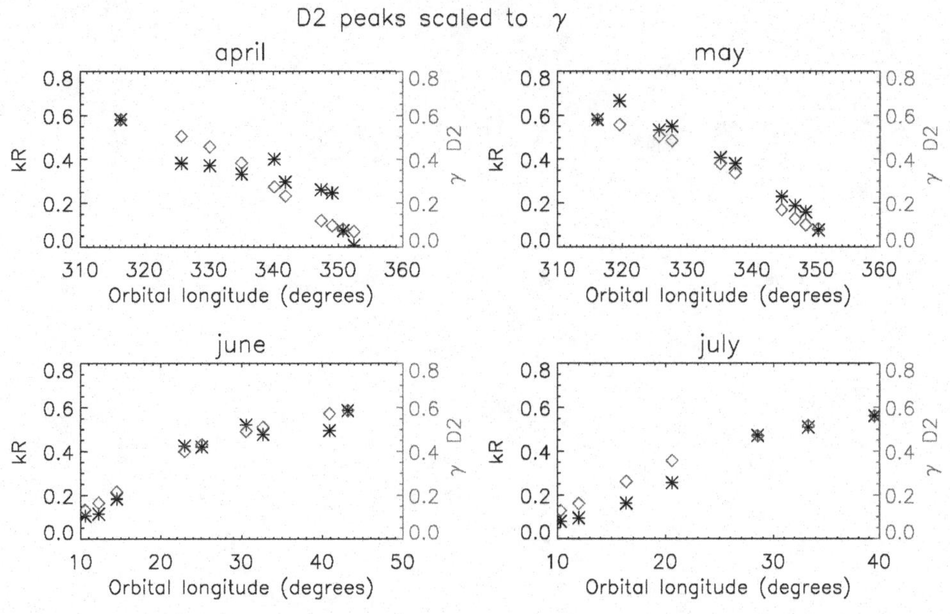

Figure 3. Preliminary results from the 2007 data. Cloud brightness is plotted versus orbital longitude. Red diamonds are the γ-factor in the non-condensing case, black asterisks represent the observed data in kiloRayleighs scaled to the D2 values to show them in the same ordinates range.

The data before eclipse follow the expected trend (in particular May), while data after the eclipse (in particular July) seem to support the condensation hypothesis, with brightness increasing slower than the expected. Further steps are required to understand the outliers (especially in April and June).

References

Brown, R. 1974, *New Scientist*, 64, 484
Brown, R. A. & Yung, Y. L. 1976, in *Jupiter*, p. 1102-1145
Schneider, N. M. & Trauger, J. T. 1995, *ApJ*, 450, 450
Burger, M. H., Schneider, N. M., & Wilson, J. K. 1999, *Geophys. Res. Lett.*, 26, 3333
Schneider, N. M. & Bagenal, F. 2007, in *Io After Galileo*, R. Lopes, ed. Springer/Praxis

Galileo's Medicean Moons: their impact on 400 years of discovery
Proceedings IAU Symposium No. 269, 2010
Barbieri C., Chakrabarti S., Coradini M. & Lazzarin M.

doi:10.1017/S1743921310007465

Exploration of the Galilean Moons using Electrodynamic Tethers for Propellantless Maneuvers and Self-Powering

Lorenzini E. C.[1], Curreli D.[2] and Zanutto D.[2]

[1]Departement of Mechanical Engineering, University of Padua,
e-mail: enrico.lorenzini@unipd.it

[2]Centre of Studies and Activities for Space, CISAS - "G. Colombo", University of Padua,
e-mail: davide.curreli@unipd.it, denis.zanutto@studenti.unipd.it

Abstract. Recent studies have demonstrated the benefits of using electrodynamic tethers (*EDT*) for the exploration of the inner region of the Jovian system. Intense planetary magnetic field and reasonable environmental plasma density make the electrodynamic interaction of the conductive tether with the plasmasphere strong. The interaction is responsible for a Lorentz force that can be conveniently used for propellantless maneuvers and extraction of electrical power for on board use. Jupiter and the four Galilean Moons represent an exceptional gravitational environment for the study of the orbital dynamics of an *EDT*. The dynamics of such a system was analyzed using a 3-body model, consisting of the planet plus one of its moons (Io in this work) and the *EDT* itself. New and interesting features appear, like for example the possibility to place the tether in equilibrium with respect to a frame co-rotating with the moon at points that do not coincide with the classical Lagrangian points for non-null electrodynamic forces.

Keywords. Jovian Environment, Plasma Torus, Space Plasma, Space Tethers, Electrodynamic Propulsion, Radiation Belts

---

1. Introduction

Jupiter's fast rotation and its strong magnetic field create a magnetosphere with unique features in the Solar System for its immense size. Jupiter's huge magnetosphere differs from most other planetary magnetospheres because it derives much of its plasma internally from the volcanic activity of the moon Io. The nature of Jovian environment (high magnetic field and moderate electron density in several regions) offers very attractive features for an *EDT* to operate. The motional electric field due to the relative motion of the *EDT* with respect to the magnetic lines drives the collection of electrons from the plasmasphere onto the anodic portion of the conducting cable. As a consequence, an electric current flows along the tether driven by the potential difference between anode and cathode (where an hollow cathode emits electrons). The Lorentz force acting on the system due to the interaction of the current with the local magnetic field can be conveniently used for propellantless maneuvers within the Jovian system. Adopting an Orbital Motion Limited (*OML*) model the averaged current collected can be calculated using the *OML* formula (Sanmartin, 1993). When the electric current on the conductive wire is closed on a load, the *EDT* allows the extraction of electric power at the expense of the energy of the corotating plasma, which is thermodynamically cooled and slowed down.

2. Jupiter's Environment

The Jovian magnetic field is very intense (4.2 Gauss at a Jovian radius) and its structure is complex because of the interactions with solar wind. Pioneer and Vojager missions and the radio observations provided some indications about the intensity of this magnetic field demonstrating that is inverted with respect to the Earth field, inclined and offset. Though in the literature fairly accurate models of magnetic field exist (for example the Octupole Model, named O4), the relatively simple Offset Tilted Dipole Model, named D4, is already a good approximation for the aims of this work.
Jupiter's magnetic field rotates at the same rotational speed of the planet, with a period of 9 h 55 m, so the plasma entraped inside the magnetosphere corotates, with the planet, at very high velocities.
Tidal deformations inside Io produce extreme tectonics and volcanism: neutral gas continuously ejected by the moon is ionized and accelerated by the fast-flowing Jovian magnetosphere. Because of this activity, Io acts as a source of neutral material that is then ionized and taken to the state of a cold plasma, forming the plasma torus that surrounds Io's orbit.
The Divine-Garrett model gives a simplified description of the plasma environment by providing the densities of protons, electrons and six positive ion species as a function of the position vector. Plasma environment has a fundamental importance in the operations of the electrodynamic tether because the tether current is directly proportional to the plasma electron density.

3. Perturbed Circular Restricted Three Body Problem

The system formed by Jupiter, Io and *EDT* satellite represents an interesting example of three body problem. The attitude and orbital motion of the *EDT* satellite center of mass is governed by the classical circular restricted three body problem perturbed by the electrodynamic force F_{el} and its moments.

$$\ddot{\vec{R}} + 2\vec{\Omega} \times \dot{\vec{R}} + \vec{\Omega} \times \vec{\Omega} \times \vec{R} = \vec{f}_{gr,1} + \vec{f}_{gr,2} + \vec{f}_{el} \tag{3.1}$$

$$[\dot{I}_c]\vec{\omega} + [I_c]\dot{\vec{\omega}} + \vec{\omega} \times [I_c]\vec{\omega} = \vec{M}_{gr,1} + \vec{M}_{gr,2} + \vec{M}_{el} \tag{3.2}$$

The tether is approximated by means of the Extended Dumbbell Model with two lumped masses at the tips of a rigid and massless cable of length L. The dynamics of the tether attitude angles, θ and φ, with respect to inertial frame, is as follows:

$$\ddot{\theta} + 2\frac{\dot{L}}{L}\dot{\theta} - 2\dot{\theta}\dot{\varphi}\tan\varphi = \frac{1}{\cos\varphi}\left[\frac{3\nu_1}{\rho_1^5}(\vec{\rho}_1 \cdot \hat{u})(\vec{\rho}_1 \cdot \hat{v}) - \frac{3\nu_2}{\rho_2^5}(\vec{\rho}_2 \cdot \hat{u})(\vec{\rho}_2 \cdot \hat{v}) + \frac{M_{\theta,el}}{I_c}\right] \tag{3.3}$$

$$\ddot{\varphi} + 2\frac{\dot{L}}{L}\dot{\varphi} + \dot{\theta}^2\cos\varphi\sin\varphi = -\frac{3\nu_1}{\rho_1^5}(\vec{\rho}_1 \cdot \hat{u})(\vec{\rho}_1 \cdot \hat{w}) - \frac{3\nu_2}{\rho_2^5}(\vec{\rho}_2 \cdot \hat{u})(\vec{\rho}_2 \cdot \hat{w}) + \frac{M_{\varphi,el}}{I_c} \tag{3.4}$$

The orbital motion of the electrodynamic tethered satellite in the Io Torus was investigated and new triangular equilibrium positions were found which are displaced with respect to the classic Lagrangian points and rotated along the orbit (see Figure 1) by an angle in the direction opposite to the *EDT* force. The value of the rotation angle is a function of the electrodynamic interaction and, consequently, of size parameters like tether length, tether width and S/C mass. With the increase of tether length the superior point (near L_4) is stable and tends asymptotically to a location very close to Io, while the inferior point (near L_5) disappears gradually, losing its stable nature. These equilibrium positions of the perturbed system were obtained numerically by using an iterative algorithm. Small

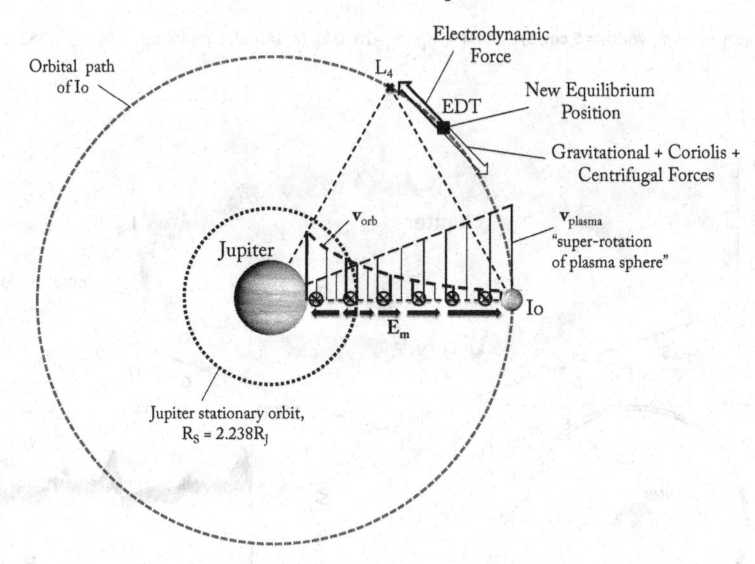

Figure 1. The *EDT* can stay in new equilibrium positions w.r.t. the synodic frame

motions around them were also studied with a linear variational analysis, by computing the eigenvalues of the linearized system. They consist of complex numbers with the same real part R_e and frequencies f_1 and f_2.

$$s_{1,2} = R_e \pm i f_1 \tag{3.5}$$

$$s_{3,4} = -R_e \pm i f_2 \tag{3.6}$$

Indeed, the magnetic field, electron density and relative velocity with respect to the plasma are variable along the orbit and consequently they modulate the electrodynamic force. The solutions of the characteristic polynomial are time and space-dependent: real parts of all eigenvalues change along the trajectory because of the modulation of B, N_e and v_{rel}. Three kinds of trajectories are possible as a function of the integral average of real parts of all eigenvalues: convergent, divergent and quasi-periodic orbits.

3.1. *Motion around the new equilibrium points*

The dynamics around the modified triangular Lagrangian points was investigated to identify mission profiles capable of exploring the inner zone of the Jovian system at the distance of Io that scans the whole Plasma Torus. The relevant trajectories exhibit large amplitude oscillations around the modified equilibrium positions. Several tether configurations of moderate sizes were studied to allow trajectories with useful levels of power extracted from the plasma environment. By adding an initial out-of-plane component the S/C develops an harmonic oscillation in z-direction that lets the system explore and scan the Torus (see Figure 2).

3.2. *Rotating tether*

Rotating tethers are necessary to maintain the cable taut at all times by means of centrifugal forces, also when the natural gravity-gradient force is not able to exert a sufficient tension. It is possible in this way to keep the lateral bowing due to Lorentz force small. The Lorentz force acts as a lateral distributed load all along the cable.
The total tension is:

$$T_{tot} = T_{gr} + T_{spin} = 3\omega_{orb}^2 m_1 L \left(\frac{m_2}{m_1 + m_2}\right) + \omega_{spin} \frac{L}{2} \left(\frac{m_1 + m_2}{2}\right) \tag{3.7}$$

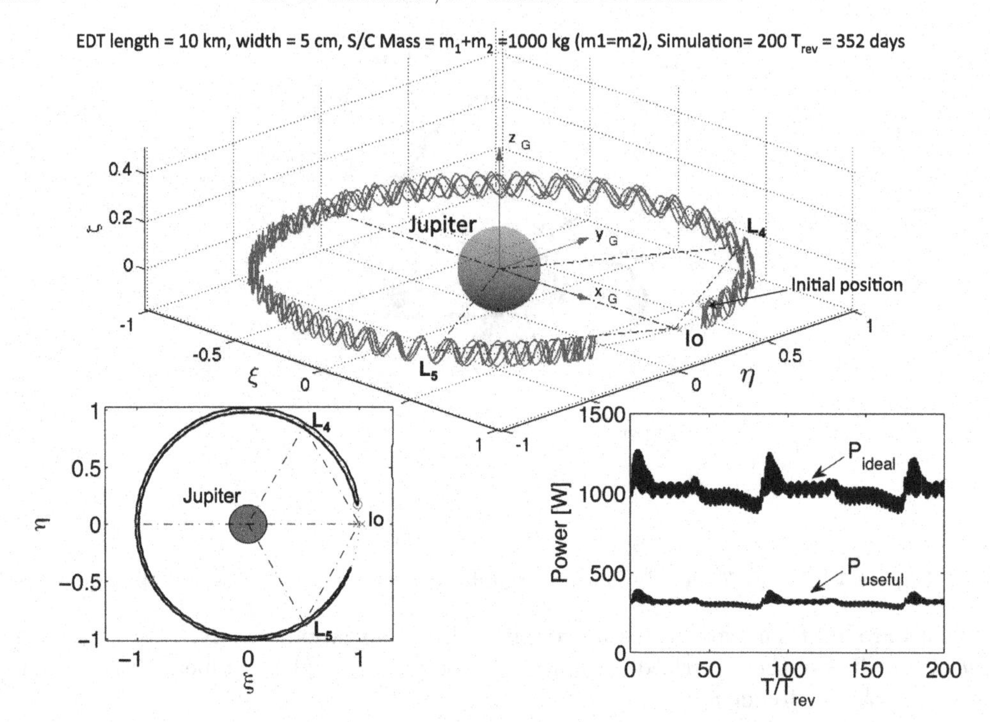

Figure 2. Horse-shoe orbit along the Io plasma torus

3.3. *Useful electrical power*

When the conductive wire is closed on an electrical load, the EDT generates useful on-board power. The power extraction occurs in conjunction with the electrodynamic force acting on the system, and it is generated at the expense of the energy of corotating plasma. Ideal power corresponds to zero load impedance and negligible ohmic losses, so useful power at the load is a fraction of P_{ideal}, typically $\eta \approx 0.3$.

4. Moons Tour With Propellantless Maneuvers

At Io's orbit the plasma density is relatively strong ($\approx 2 \times 10^3 cm^{-3}$): electrodynamic interaction between the plasmasphere and tether can be exploited to change the orbital profile by means of propellanless maneuvers. Moreover, the Plasma Torus is above the stationary orbit of Jupiter, so two orbital maneuvers are possible with an EDT:

- orbit raising
- orbit lowering

An initial orbit with perijove r_p at Io's distance and apojove r_a at Europa's orbit is considered. For orbit raising, the electrodynamic force is exploited to increase the total energy of the dynamical system. In this case the satellite orbit is prograde (counter-clockwise) with the velocity vector $\vec{v}$ aligned in the same direction of $\vec{f_{el}}$. At every perijove passage, where the plasma distribution is denser, the S/C energy increases. In this way, its possible to pump energy into the system by removing it from the plasmasphere. The greatest effect is on the apojove: in fact, while the r_p remains more or less at the same altitude, r_a tends to increase significantly.
For orbit lowering the satellite must be on a retrograde orbit and $\vec{v}$ is in the opposite direction of $\vec{f_{el}}$. In this case a drag force brakes the S/C and the mechanical energy of

the system decreases. At every perijove passage the orbit tends to change its eccentricity by lowering the apojove.

5. Radiation Dose

The magnetic field of Jupiter is very strong and its magnetosphere is the seat of intense radiation, that is seriously dangerous to the satellite. The Divine-Garretts model divides the charged particles in three main components as a function of their energetic level: the radiation belts include high energetic electrons and protons up to *1000MeV*. The GIRE (Galileo Interim Radiation Electron) model was used to quantify the fluences [$particles/cm^2$] of highly energy electrons and protons, trapped in the radiation belts, on the satellite along the trajectory. As a result, the Io's Torus, situated at 5.5-10R_J, is a region of the magnetosphere with high density of charged particles, and if not appropriately shielded a satellite, orbiting inside the Io torus, will be exposed to high radiation doses: several hundreds *MRad* in a year mission.

6. Conclusion

The dynamics of a dumbell tether placed in the proximity of triangular Lagrangian points of the Jupiter-Io system was studied. New equilibrium positions were discoverd in the corotating frame. Their location depends on the system sizes (S/C mass, tether length and width). Moreover, the stability of these new points was investigated revealing that at high values of the Lorentz force only the superior equilibrium point remains stable and it moves more and more toward Io as the Lorentz force increases. The strong Jovian magnetic field, high electron density of the Plasma Torus and relative velocity of the satellite with respect to the plasmasphere make the *EDT* a convenient propellantless system for producing thrust (and consequently orbital maneuvres) and on board power. In fact, kWs of electrical power can be extracted from the Plasma Torus: about 1kW of useful power with a rotating tether 20km long. At the same time, by exploiting the electrodynamic interaction between an *EDT* and the plasma of Io's Torus it is possible to change the orbital parameters and make propellantless maneuvers to navigate among the Galilean Moons.

References

Bombardelli C., Lorenzini E. C., Curreli D., Sanjurjo-Rivo M., Lucas F. R., Peláez J. Scheeres D. J., & Lara M. (2008), *AIAA/AAS Astrodynamics Specialist Conference*, Honolulu, Hawaii, USA

Curreli D., Lorenzini E. C., Bombardelli C., Sanjurjo-Rivo M., Lucas F. R., Peláez J., Scheeres D. J., & Lara M. (2009), *AAS 09-240, 19th AAS/AIAA Space Flight Mechanics Meeting*, Savannah, Georgia, USA

Divine N. & Garrett H. B. (1983), *Journal of Geophysical Research*, Vol. 88, No. A9, pp. 6889-6903.

Garrett H. B., Jun I., Ratliff J. M., Evans R. W., Clough G. A., & McEntire R. W. (2003), *National Aeronautics and Space Administration*, JPL Publications 03-006

Peláez J., Sanjurjo-Rivo M., Lucas F. R., Lara M., Lorenzini E. C., Curreli D., & Sheeres D. J. (2008), *Final Report Ariadna Study 07/4201*

Sanmartín J. R., Martínez-Sanchez M., & Ahedo E. (1993), *Journal of Propulsion and Power*, Vol. 9, No.3, pp. 353-360

Galileo's Medicean Moons: their impact on 400 years of discovery
Proceedings IAU Symposium No. 269, 2010
C. Barbieri, S. Chakrabarti, M. Coradini & M. Lazzarin, eds.

doi:10.1017/S1743921310007477

Near-Earth Objects 400 Years after Galileo: Physical Properties and Internal Structure

Dmitrij Lupishko[1] **and Zhanna Pozhalova**[2]

[1]Institute of Astronomy of Kharkiv V. N. Karazin National University,
Sumska str., 35, Kharkiv 61022, Ukraine
email: lupishko@astron.kharkov.ua

[2]Research Institute "Nikolaev Astronomical Observatory",
Observatornaya str. 1, Nikolaev 54030, Ukraine
email: zhanna@mao.nikolaev.ua

Abstract. The review contains the most recent data on near-Earth objects such as their sizes and densities, rotation and shapes, taxonomy and mineralogy, optical properties and structure of their surfaces, binary systems among the NEOs and internal structure of asteroids and comets constituted the NEO population.

Keywords. Asteroids, near-Earth objects, physical properties, internal structure.

1. Introduction

Near-Earth objects (NEOs) are defined as asteroids and comets having orbits with perihelion distances of 1.3 *AU* or less. About 30% of the entire NEO population may reside in orbits having a Jovian Tisserand parameter <3, and among them roughly half are observed to have comet-like physical properties such as albedos and spectra. Thus, about 10-15% of the NEO population may be comprised by extinct or dormant comets (Lupishko & Lupishko 2001; Binzel & Lupishko 2006; Michel & Bottke 2009). The rest are the near-Earth asteroids, which are traditionally divided into three groups (the relative abundances are estimated by Bottke *et al.* (2002)):

Amor	a $\geqslant$ 1.0 AU	1.017 < q $\leqslant$ 1.3 AU	(32 ± 1 %)
Apollo	a $\geqslant$ 1.0 AU	q < 1.017 AU	(62 ± 1 %)
Aten	a < 1.0 AU	Q > 0.983 AU	(6 ± 1 %)

Besides, there is an additional group of rather dangerous asteroids whose orbits reside entirely inside of the Earth's one (Q < 0.983 AU). According to (Michel *et al.* 2000) objects of this inner-Earth asteroid group and Aten group together can constitute about 20% of the km-sized Earth-crossing population. About 6730 NEOs are discovered by the beginning of 2010. They are the objects of a special interest from the point of view not only of the basic science but of the applied science as well (the problem of asteroid and comet hazard, the NEAs as the potential sources of raw materials in the nearest to the Earth space, etc.).

2. Sizes, densities and axis rotation

The size distribution of NEOs can be approximated as N(>D km) = k D^{-b} with an exponent b = 1.95 and k = 1090 (Stuart 2003). This expression indicates that there are 1090 NEOs with D$\geqslant$1 km. Including uncertainties, Stuart and Binzel (2004) give this result as 1090±180 objects that are 1 km or lager within the NEO population. Below (Table 1) the sizes of some individual objects are presented which display the whole range of sizes of cataloged NEOs overlapping four orders of magnitude.

Table 1.

Largest NEOs	**D[km]**	**Smallest discovered NEOs**	**D[m]**
1036 Ganymed	38.5	...	...
433 Eros	16.5	2000 WL107	38
3552 Don Quixote	12÷15	2003 QB30	17
1866 Sisyphus	8.9	2003 SQ222	10
...	...	2008 TC3*	4

* discovered on 6 Oct. 2008, came into collision with the Earth on 7 Oct. 2008 and disintegrated in atmosphere over northern Sudan.

The most reliable estimates of bulk densities (g/cm^3) for S, Q and C-type NEOs are summarized in Table 2. Discovery of binary NEOs gives a good opportunity to determine their bulk densities, however those estimates are usually not accurate enough due to an uncertainty of binary system parameters.

Table 2.

433 Eros"	2.67±0.03	S
6489 Golevka*	2.7(+0.4,-0.6)	Q
25143 Itokawa"	1.95±0.14	S,Q
1999 KW4*	1.97±0.24	S
2100 Ra-Shalom*	1.1-3.3	C
1996 FG3	1.4±0.3	C
2000 DP107	1.6(+1.2,-0.9)	?
2000 UG11	1.5(+0.6,-1.3)	?

Notes:
" space mission data; * radar data

Comparing bulk densities of these NEOs with those of their meteorite analogues (ordinary or carbon chondrites) we have to suppose about 30÷50% of the NEO porosity. It means that at least some of NEOs are not monolithic bodies but "rubble-pile" structures, which have no coherent tensile strength and weakly held together by their own mutual gravity.

The distribution of the rotation rates of NEOs (Figure 1) is quite different in comparison with that for small main-belt asteroids (MBAs) and it shows the prominent excesses of slow and fast rotators (Lupishko *et al.* 2007). Among the reasons for that can be the difference in asteroid diameter distributions within these two populations, influence the radiation pressure torques (YORP-effect), the influence of the rotational parameters of binaries and may be some selection effects. The whole interval of NEO rotation periods ranges over four orders of magnitudes from 500-600 hrs (96590 1998 XB and 1997 AE12) to 1.3 min (2000 DO8). It is clear that such small (tens meters in sizes) and super-fast rotating bodies are beyond the rotational breakup limit for aggregates like "rubble piles" and they are monolithic fragments.

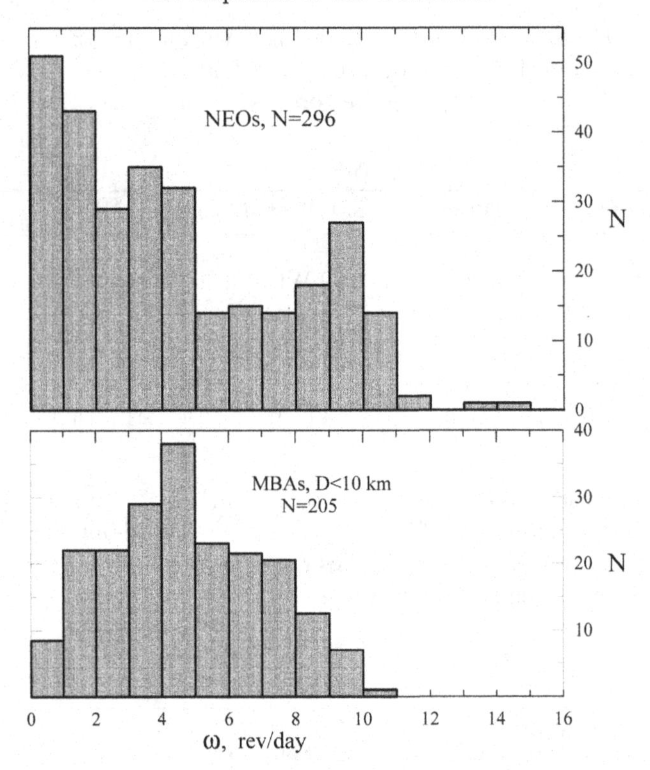

Figure 1. Distribution of the rotation rates of NEOs and small (D⩽10 km) MBAs.

3. Taxonomy and mineralogy

As a first step toward estimating the nature of any NEO is determination of its taxonomic class, that is, the object total mineralogy. Practically all taxonomic classes identified among main-belt asteroids have been also found in NEO population, including the C, P and D classes that are typical of outer main belt. Binzel *et al.* (2004) from their spectroscopic survey of 252 NEAs and Mars-crossers noted that 25 of 26 Bus' taxonomic classes of MBAs are represented in the NEO-population. The most common taxonomic classes among them are however S and Q (silicate) types. Recent spectroscopic investigation of 150 NEAs (Lazzarin *et al.* 2008) have summarized that 62% of them belong to S-complex, 20% to X-complex, 12% to C-complex and 6% to other classes of Bus' taxonomy. Stuart and Binzel (2004) modeled the bias-corrected distribution of taxonomic classes and obtained that C and other low-albedo classes consist of 27% and S+Q classes 36% of all NEOs.

Observing smaller and smaller S-objects Binzel *et al.* (2001) showed a continuous range of NEO spectra from those of S-types to ordinary chondrites. That is, there is a continuous transition from spectra of S-types to those of Q-types. At the same time Q-objects are smaller in sizes and brighter than S-objects, that is, their surfaces are "younger, fresher". Therefore, this continuum is interpreted as a result of space weathering process, that is, the process of alteration of the young surface of Q-asteroid to look more and more redder like S-type surface (Binzel *et al.* 2004). Lazzarin *et al.* (2008) found that only the 17% of NEOs and 6% of MBAs are compatible with ordinary chondrite spectra but other objects are much redder. They also found the statistically valid linear increase of spectral slope with increase of asteroid exposure (that is, amount of Sun's radiation that a body receives along its orbit) what support the idea of space weathering. Fevig and Fink (2007) reported the results of spectrophotometry of 55 NEOs which revealed the evidence for orbit-dependent trends in their data: while observed S-types reside in orbits which do not cross the asteroid main belt, the majority of objects with spectra of ordinary chondrites (Q-types, fresh and relatively unweathered NEOs) are in highly eccentric Apollo orbits which enter the asteroid main belt. Very likely that these objects have recently been injected into such orbits after a collision in the main belt.

4. Optical properties and surface structure

The analysis of available data clearly demonstrates that the surfaces of NEOs display in general the same optical properties as the surfaces of MBAs (Binzel *et al.* 2002; Lupishko & Di Martino 1998; Lupishko *et al.* 2007). The whole range of NEO albedos (0.05÷0.50) is basically the same as that of MBAs and it corresponds to the same in general mineralogy within these two populations. But the strict similarity of the other photometric and polarimetrical parameters (such as phase coefficient, polarization slope and others, which are related to surface structure) gives evidence of the similar surface structures at submicron scale.

The polarimetric, radiometric data and direct imaging of Eros and Itokawa give evidence that most of NEOs are covered with regolith (fine granulated rocks and dust). Despite their low gravities, even the smallest NEOs appear capable of retaining some regolith coating. As it was estimated, a minimum 2.3±0.4 m thick layer of regolith exists in the lowlands of Itokawa, which, if spread evenly across the entire asteroid, corresponds to a 42±1 cm layer. The recent studies of NEO thermal IR emission showed that the average thermal inertia of km-size NEOs is 200±40 $Jm^{-2}s^{-0.5}K^{-1}$, that is about four times that of the Moon (Delbo *et al.* 2007). The authors identify also a trend of increasing thermal inertia with decreasing asteroid diameter. Radar observations showed that even the relatively small NEOs 4179 Toutatis and 1999 JM8 (D$\sim$3 km both) are cratered at about the same extent as MBAs 951 Gaspra and 243 Ida. The radar data also evidence that NEO surfaces are rougher than surfaces of large MBAs at the scale length of decimeters and meters. Recently the radar observations have also revealed a link between NEO composition and centimeter-to-decimeters surface roughness (see http://echo.jpl.nasa.gov). The most rough are the high-albedo objects of E and V-types, the meteorite analogs of which are enstatite chondrites and HED-meteorites (basalts), and they are more rough most probably due to higher strength of their material.

5. Binary and triple systems among the NEOs

By the beginning of 2010 37 binary near-Earth asteroids (two of them with two satellites) have been discovered. They show the similarity of their parameters, for example, rotation periods of primaries are within the interval of 2.3÷3.6 hrs and orbital periods of secondaries are in the range of 0.5÷1.8 days (what may be due to observational selection effects). A fraction of binary systems among the NEAs is estimated to be 15-17% (Merline *et al.* 2002), though among the Aten-asteroids the fraction can be significantly higher (Polishook & Brosch 2008).

The NEA 2001 SN263 has been revealed as the first near-Earth triple asteroid ever found. It was discovered by Mitchal Nolan and his colleagues using the Arecibo radar. The central body is spherical of D$\approx$2 km across, while the larger of the two moons is about half that size. The smallest object is about the size of the Arecibo telescope. Pravec and Harris (2007) suggest that binaries formed from parent bodies spinning at the critical rate by some sort of fission or mass shedding, and the YORP-effect is a candidate to be the dominant cause of spin-up to instability. This suggestion is in a good agreement with results obtained by Walsh and Richardson (2008) that tidal disruption due to close planetary encounters should account for about 1-2% of NEAs being binaries and that there are other formation mechanisms that contribute significantly to this population. Discovery and study the binary or triple systems allow us to determine the density of the NEOs and type of their material.

6. On the internal structure of NEOs

There are only indirect data on the internal structure of NEOs such as bulk densities and porosities of them, their rotational rates, the events of comet nuclei disintegration, existence of large craters, crater chains and grooves on asteroids and satellites and the recent data on asteroid Itokawa. Campo Bagatin (2008) analyzed these evidences in order to extract information on the internal structure of NEOs. Taking into account the results of his analysis and other available data one can summarize that the NEO population presents at least three very different types of body internal structures. They are: a) monolithic objects (the fragments of larger parent main-belt asteroids) including the metal ones with a tensile strength of about 10^9 dyne/cm^2, b) the structures of "rubble-piles" type or gravitational agregates and there are data that fraction of such bodies can be rather big (25143 Itokawa is considered as the most striking example of such structure, taking into account its density and macroporosity, an availability of large blocks on its surface and other evidences of a catastrophic disruption scenario for its formation) and c) about 10-15% of extinct or dormant comet nuclei with a tensile strength of about 10^2 -10^3 dyne/cm^2.

One can expect that forthcoming ESA space mission ISHTAR (Internal Structure High-resolution Tomography by Asteroid Rendezvous), which foresees the investigation of two Apollo-objects of different types, will supply us with new and valuable information on the internal structure of NEOs.

References

Binzel, R. P., Harris, A. W., Bus, S. J., & Burbine, Th. H. 2001, *Icarus*, 151, 139

Binzel, R. P. & Lupishko, D. F 2006, *Asteroids, Comets, and Meteors*, Eds. D. Lazzaro *et al.* Proc. of IAUS 229 "ACM 2005" Aug. 7-12, 2005. Buzios, Rio de Janeiro, Brazil. Cambridge Univ. Press, 207

Binzel, R. P., Lupishko, D. F., Di Martino, M., Whiteley, R. J., & Hahn, G. J. 2002, *Asteroids III* , (W. Bottke *et al.*, eds.) Univ. Arizona Press, Tucson, 255

Binzel, R. P., Rivkin, A. S., Scott, J. S., Harris, A. W., Bus, S. J., & Burbine, Th. H. 2004, *Icarus*, 170, 259

Bottke, W. F. Jr., Morbidelli, A., Jedicke, R., Petit, J-M., Levison, H. F., Michel, P., & Metcalfe, T. S. 2002, *Icarus*, 156, 399

Campo Bagatin, A. 2008, *Mem. S.A.It. Suppl.*, 12, 150

Delbo, M., Dell'Oro, A., Harris, A. W., Mottola, S., & Mueller, M. 2007, *Icarus*, 190, 236

Feving, R. A. & Fink, U. 2007, *Icarus*, 188, 175

Harris, A. W. & Pravec, P. 2006, *Asteroids, Comets, and Meteors.*, (Lazzaro D. *et al.*, eds.) Proc. of IAUS 229, Aug. 7-12, 2005. Buzios, Rio de Janeiro, Brazil. Cambridge Univ. Press, 439

Lazzarin, M., Magrin, S., & Marchi, S. 2008, *Mem. SAI Suppl.*, 12, 20

Lupishko, D. F. & Di Martino, M. 1998, *Planet. Space Sci.*, 46, No.1, 47

Lupishko, D. F., Di Martino, M., & Binzel, R. P. 2007, *Near Earth Objects, our Celestial Neighbors: Opportunity and Risk*, Proc. IAUS236, 2006 (A. Milani *et al.*, eds.), 251

Lupishko, D. F. & Lupishko, T. A. 2001, *Solar System Research*, 35, No. 3, 227

Merline, W. J., Weidenschilling, S. J., Durda, D. D., Margot, J-L., Pravec, P., & Storrs, A. D. 2002, *Asteroids III*, (W. F. Bottke *et al.*, eds.), Univ. of Arizona Press, Tucson (USA), 289

Michel, P. & Bottke, W. F. 2009, *Book of Abstracts of Intern. Conf. "Asteroid-Comet Hazard - 2009*, Russia, St. Petersburg, Sept. 21-25, 14

Michel, P., Zappala, V., Cellino, A., & Tanga, P. 2000, *Icarus*, 143, 421

Polishook, D. & Brosch, N. 2008, *Icarus*, 194, 111

Pravec, P. & Harris, A. W. 2007, *Icarus*, 190, 250

Stuart, J. S. 2003, *Ph. D. Thesis. Massach. Inst. of Technology, Cambridge, Mass.*

Stuart, J. S. & Binzel, R. P. 2004, *Icarus*, 170, 295

Walsh, K. J. & Richardson, D. C. 2008, *Icarus*, 193, 553

Galileo's Medicean Moons: their impact on 400 years of discovery
Proceedings IAU Symposium No. 269, 2010
C. Barbieri, S. Chakrabarti, M. Coradini & M. Lazzarin, ed.

doi:10.1017/S1743921310007489

Theory of the rotation of the Galilean satellites

Benoît Noyelles

University of Namur – Dept of Mathematics
Rempart de la Vierge 8 – B-5000 Namur – Belgium **and**
IMCCE (Paris Observatory, USTL, UPMC) – CNRS UMR 8028
77 avenue Denfert-Rochereau – 75014 Paris – France
email: noyelles@imcce.fr

Abstract. As most of the natural satellites of the Solar System, the Galilean moons are since a long time assumed to be tidally locked in a spin-orbit synchronous resonance. Thanks to the mission Galileo, we now dispose of enough gravity data to perform 3-dimensional theories of the rotation of these satellites, in particular to model the departure from the exact synchronous rotation. We here present such theories depending on the interior model we consider, in highlighting some observable output data. Inverting them will give us information on the internal structure of these bodies.

Keywords. planets and satellites: general

1. Context

As most of the natural satellites in the Solar System, the Galilean satellites of Jupiter are assumed to be in a synchronous rotation. This corresponds to a stable dynamical equilibrium, that makes the satellites always present the same face to a fictitious Jovian observer. This equilibrium exists because of the asphericity of these bodies, that are in fact ellipsoids.

The gravitational data of these satellites given by the Galileo spacecraft allow us to give first 3-dimensional studies of their rotations. We here present first our analytical formulation of these problems, then our results, before introducing some ways to improve the modelisation.

2. Analytical study

The analytical study starts from the following Hamiltonian:

$$\mathcal{H} = \underbrace{\frac{nP^2}{2} + \frac{n}{8}\left[4P - \xi_q^2 - \eta_q^2\right]\left[\frac{\gamma_1+\gamma_2}{1-\gamma_1-\gamma_2}\xi_q^2 + \frac{\gamma_1-\gamma_2}{1-\gamma_1+\gamma_2}\eta_q^2\right]}_{\text{Kinetic energy}}$$
$$+ \underbrace{n\left(\frac{d_0}{d}\right)^3\left(1+\delta_s\left(\frac{d_0}{d}\right)^2\right)\left[\delta_1(x^2+y^2)+\delta_2(x^2-y^2)\right]}_{\text{Jovian perturbation}} \tag{2.1}$$

with the following canonical variables:

$$p = l + g + h \qquad P = \frac{G}{nC}$$
$$r = -h \qquad R = \frac{G-H}{nC} = P(1 - \cos K) = 2P\sin^2\frac{K}{2}$$
$$\xi_q = \sqrt{\frac{2Q}{nC}}\sin q \qquad \eta_q = \sqrt{\frac{2Q}{nC}}\cos q$$

where n is the satellite's mean orbital motion, $q = -l$, and $Q = G - L = G(1 - \cos J) = 2G\sin^2\frac{J}{2}$. The coefficients of the Hamiltonian are defined as follows:

$$\gamma_1 = J_2\frac{MR^2}{C} \qquad \delta_1 = -\frac{3}{2}\left(\frac{n^*}{n}\right)^2\gamma_1$$
$$\gamma_2 = 2C_{22}\frac{MR^2}{C} \qquad \delta_2 = -\frac{3}{2}\left(\frac{n^*}{n}\right)^2\gamma_2$$
$$\delta_s = \frac{5}{2}J_{2♃}\left(\frac{R_♃}{d_0}\right)^2$$

and the angles can be seen on the figure above, reproduced from Henrard (2005). They use 2 sets of Euler angles, the first one (h, K, g) locates the position of the angular momentum in the first frame $(\vec{e_1}, \vec{e_2}, \vec{e_3})$, while the second (g, J, l) locates the body frame $(\vec{f_1}, \vec{f_2}, \vec{f_3})$ in the second frame tied to the angular momentum.

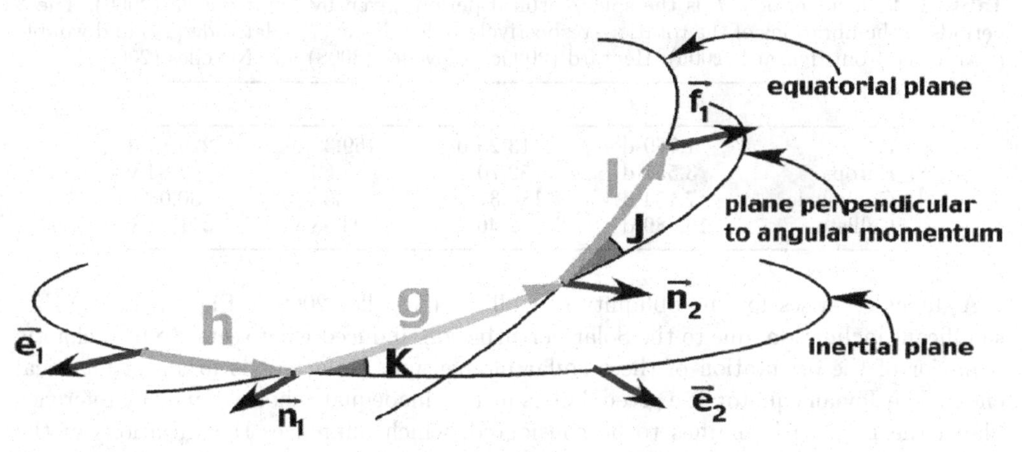

Figure 1. The angles (reproduced from Henrard (2005)).

x and y are the first two coordinates of the center of the perturber (here Jupiter) in the frame $(\vec{f_1}, \vec{f_2}, \vec{f_3})$ bound to the satellite. Then we use the model of Henrard & Schwanen (2004) to obtain the equilibrium (K^*, P^*) and the fundamental periods of the 3 proper librations around it: T_u, T_v and T_w. At the strict Cassini state, $\sigma = p - \lambda + \pi = 0$, $\rho = r + \Omega = 0$, $\xi_q = 0$ and $\eta_q = 0$ (i.e. the wobble angle J is null), where λ and Ω are respectively the body's mean longitude and ascending node in an inertial frame.

After some canonical transformations introducing the slow angles, we find the equilibrium of the system, confirming the synchronous rotation and giving the obliquity of the body. Then, a centering of the Hamiltonian on the equilibrium and a conversion into polar coordinates yields:

$$\mathcal{H}(u, v, w, U, V, W) = \underbrace{\omega_u U + \omega_v V + \omega_w W}_{\text{3-d oscillator}} + \underbrace{\mathcal{P}(u, v, w, U, V, W)}_{\text{Perturbation}}, \tag{2.2}$$

where (u, v, w) are angles, (U, V, W) the actions associated, and $(\omega_u, \omega_v, \omega_w)$ the 3 frequencies of the small oscillations about the equilibrium. These oscillations are expected to be damped, but knowing them gives information on the response of the system to the perturbation. The perturbation comes from the non-spherical motion of the studied body, and is given by L1 ephemerides (Lainey *et al.* (2006)). The influence of the perturbations can be used either analytically by way of Lie transforms (cf. Deprit (1969)), or numerically.

3. Results

We have shown that the equilibrium obliquity of the body with respect to the normal of the orbital plane is:

$$\epsilon \approx \frac{\dot{\Omega} I}{n(\delta_1 + \delta_2 + \dot{\Omega}/n)}, \tag{3.1}$$

where $\dot{\Omega}$ is the precessional rate of the ascending node of the body, I its orbital inclination, and n its mean frequency, i.e. its spin rate in the case of the synchronous rotation. The proper periods are given in Tab. 1.

Table 1. Proper periods. T is the spin / orbital period, given by Lainey *et al.* (2006). The 3 periods of the librations of the rotation, respectively in longitude (T_u), latitude (T_v) and wobble (T_w), come from Henrard (2005), Henrard (2005c), Noyelles (2008) and Noyelles (2009).

	T	T_u	T_v	T_w
Io	1.769 d	13.25 d	159.39 d	229.85 d
Europa	3.551 d	52.70 d	3.60 y	4.84 y
Ganymede	7.154 d	186.37 d	23.38 y	30.08 y
Callisto	16.689 d	2.46 y	203.58 y	317.11 y

A difficulty arises for the obliquity of Callisto (Noyelles 2009 & Fig. 2), because its significant inclination, due to the Solar perturbation, induced what seems to be a chaotic behavior of the orientation of the angular momentum. In fact, it is just a geometrical effect. The Jovian equator at a given date is just an inadequate choice of inertial reference plane, the Laplace Plane has to be considered, which minimizes the variations of the orbital inclination. For most of the natural satellites, it is close enough to the planet's equator, while for Callisto its inclination is $\approx 0.2°$.

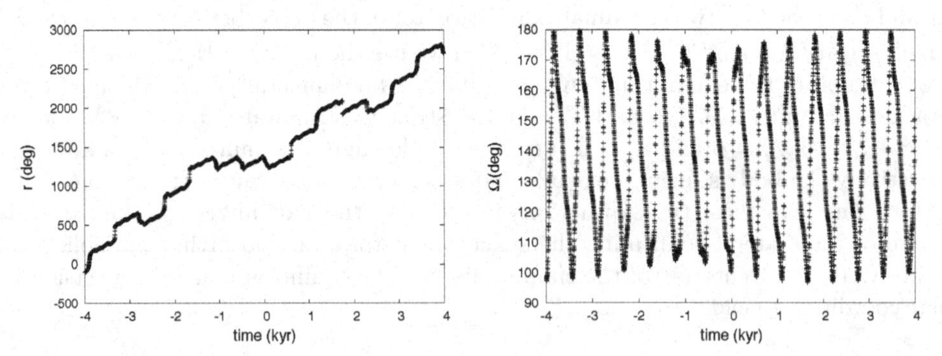

Figure 2. The problem with the inertial reference plane of Callisto. The rotational node of Callisto (left) and the orbital one (right) should have the same mean precession rate. This is in fact a geometrical effect, due to an improper choice of reference plane.

An interesting output variable to look at is the longitudinal motion, that can be seen in longitudinal libration of the body about the synchronous rotation or about the planet-satellite direction, or in the L.O.D. (length of day) (see Fig. 3). This departure from the exact synchronous rotation is due to the orbital eccentricity, i.e. to the variations of the planet-satellite distance.

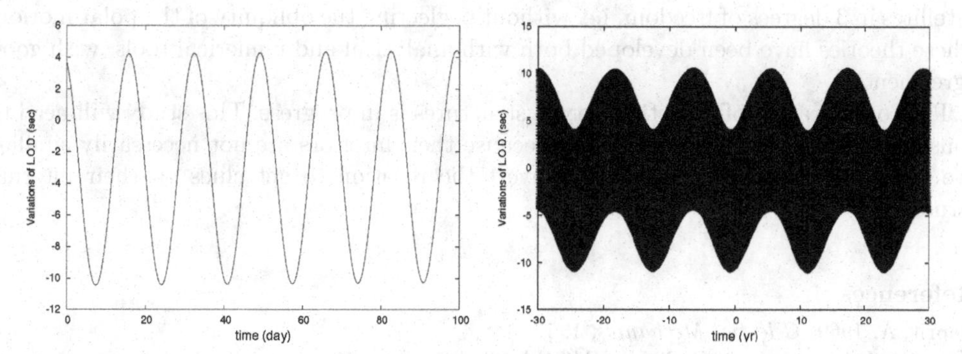

Figure 3. Variations of the Length-Of-Day of Callisto, the mean value being 16.686 days. We can see variations in 16.686 days, i.e. the orbital period of Callisto. Over a larger timescale, we can see a 11.86-yr modulation, due to the Solar perturbation.

4. Introducing a fluid core

Considering these satellites as rigid bodies is a first approximation. They are in fact differenciated, multi-layered bodies (perhaps with the exception of Callisto), that have a liquid core (like Io) or an internal ocean (like Europa). So, the next step is to introduce a multi-layered internal structure.

The easiest way to introduce a fluid core is to consider it as spherical. This way, we have no core-mantle interaction and we can just remove the core in the equation, considering the body as an empty body.

Henrard (2008) proposed to consider Io as a body composed of a rigid mantle and a ellipsoidal cavity filled with an inviscid, incompressible and non viscuous fluid. This way, we get core-mantle interactions that should be included in the kinetic energy, i.e.:

$$\begin{aligned}
\mathcal{T} = & \frac{n}{2(1-\delta)}\left[P^2 + 2P - 2(1+P-Q)(\delta + P_c - Q_c) + (P_c^2 + 2P_c)/\delta\right. \\
& \left. + 2\sqrt{QQ_c(2+2P-Q)(2\delta+2P_c-Q_c)}\cos(q+q_c)\right] \\
& + \frac{n\epsilon_1}{2(1-\delta)}\left[(2+2P-Q)Q + (2\delta+2P_c-Q_c)Q_c/\delta\right. \\
& \left. + 2\sqrt{QQ_c(2+2P-Q)(2\delta+2P_c-Q_c)}\cos(q+q_c)\right] \\
& - \frac{n\epsilon_2}{2(1-\delta)}\left[(2+2P-Q)Q\cos 2q + (2\delta+2P_c-Q_c)(Q_c/\delta)\cos q_c\right. \\
& \left. + 2\sqrt{QQ_c(2+2P-Q)(2\delta+2P_c-Q_c)}\cos(q-q_c)\right],
\end{aligned} \tag{4.1}$$

where the subscript c refers to the core, δ to its size, and ϵ_1, ϵ_2 to its shape. Here, a 4th degree of freedom is added, related to the motion of the liquid filling the cavity. The proper frequency associated is close to the spin frequency.

The most spectacular observational consequence is in the amplitude of the forced longitudinal librations. It can be shown that it depends on the size of the core δ. Hence, an observation of these librations can be inverted to measure the size of the core, as has been done for Mercury (Margot *et al.* (2007)).

5. Conclusion

This contribution presents the elaboration of models of the rotation of the Galilean satellites in 3 degrees of freedom, i.e. without neglecting the obliquity of the polar motion. These theories have been developed both with analytical and numerical tools, with good agreements.

The consideration of realistic internal structures is in progress. This study will need to consider each satellite as individuals, because their interiors are not necessarily similar. We hope that it will be possible to invert the rotation to get clues on their internal structure.

References

Deprit, A. 1969, *Celestial Mechanics*, 1, 12
Henrard, J. & Schwanen, G. 2004, *CM&DA*, 89, 181
Henrard, J. 2005, *Icarus*, 178, 144
Henrard, J. 2005b, *CM&DA*, 91, 131
Henrard, J. 2005c, *CM&DA*, 93, 101
Henrard, J. 2008, *CM&DA*, 101, 1
Lainey, V., Duriez, L., & Vienne, A. 2006, *A&A*, 456, 783
Margot, J.-L., Peale, S. J., Jurgens, R. F. *et al.* 2007, *Science*, 316, 710
Noyelles, B. 2008, *CM&DA*, 101, 13
Noyelles, B. 2009, *Icarus*, 202, 225

Galileo's Medicean Moons: their impact on 400 years of discovery
Proceedings IAU Symposium No. 269, 2010
C. Barbieri, S. Chakrabarti, M. Coradini & M. Lazzarin, eds.

doi:10.1017/S1743921310007490

Study of mutual occultation phenomena of the Galilean satellites at radio wavelengths

S. Pluchino[1], E. Salerno[1], G. Pupillo[1,2], F. Schillirò[1], A. Kraus[3], K.-H. Mack[1]

[1] INAF - Istituto di Radioastronomia, via Gobetti 101, I-40129 Bologna, Italy
[2] INAF - Osservatorio Astronomico di Torino, Strada Osservatorio 20, I-10025 Torino, Italy
[3] Max-Planck-Institut für Radioastronomie, Auf dem Hgel 69, D-53121 Bonn, Germany

Abstract. We present preliminary results for our study of mutual phenomena of the Galilean satellites performed at radio wavelengths with the Medicina and Noto antennas of the Istituto di Radioastronomia - INAF, and with the Effelsberg 100-m radio telescope of the Max-Planck-Institute for Radioastronomy, Bonn. Measurements of the radio flux density variation during the mutual occultations of Io by Europa and Ganymede were carried out during the PHEMU09 campaign at 22 GHz and 43 GHz. Flux density variations observed at radio wavelengths are consistent with the typical optical patterns measured when partial occultations occur.

Keywords. Occultations, Planets and satellites: individual (Europa, Ganymede, Io, Jupiter), Radio continuum: solar system

---

1. Introduction

Twice every 11.8 years Jupiter transits the nodes of its orbit. Since the inclination of the orbital planes of Galilean satellites with respect to the planet's equatorial plane is very small, for a few months during these passages, the satellites either occult or eclipse each other, depending on whether they are collinear with the Earth or with the Sun, respectively. These events are referred to as mutual phenomena or PHEMU (Figure 1). Their observation can be used, for instance, to derive corrections of the orbital parameters or to study the surface properties of the involved Jovian satellites. A typical example is given by the activity of Io's volcanoes. The Galileo mission has revealed 74 active volcanic centers on Io's surface (73 hot spots plus the Ra Patera plume), and 23 additional sites that were identified as probable active volcanic centers. There are two types of hot spot activity in terms of duration: persistent (active for periods longer than one year) and sporadic (events that persist up to 3 months) (Lopes-Gautier *et al.* 1999). The persistent hot spots are particularly important for the study of Io activity because they most likely represent the major pathways of magma to the satellite surface. Imaging of Io at shorter IR wavelengths (from 1 to 5 micrometers) has revealed a higher number of hot spots whose temperatures are typically included within the range of 650 K and 750 K reaching sometimes even higher values (Spencer *et al.* 1994, McEwen *et al.* 1997). Variations of the hot spot activity in terms of power output have been measured with ground-based optical and IR observations. Occultation phenomena were used to study the temperature of the Loki volcano - the most powerful volcano in the solar system - that is known to undergo periods of brightening and to be the site of giant outbursts. Optical and infrared data obtained during an occultation at a wavelength of 4.8 microns showed a large drop/jump of the flux density simultaneously with the disappearance/reappearance of Loki behind the limb of the occulting satellite (Howell 1998, Arlot *et al.* 2006). So far mutual phenomena, and in particular mutual satellite occultations, have been rarely studied at radio wavelengths. The major problem in observing the Galilean satellites with single dish radiotelescopes is in fact the strong Jovian radio-emission that may fall within the primary beam pattern of the antenna. The Jupiter flux density observed with the Very Large Array (VLA) radio interferometer is about 10 Jy at 6 cm and 35 Jy at 2 cm (de Pater *et al.* 1984). Observations of occultations by Jupiter do not provide the high angular resolution possible in mutual events whereas they can be obtained when Io is in eclipse and therefore radio flux density measurements are more sensitive to faint hot spots. Since no thick atmosphere surrounds the Galilean satellites, the observations of these phenomena are extremely accurate for astrometric purposes. In the case of Io, flux density variability depends on the wavelength, the time and the location on the surface of the hot spots. Ground-based observations at radio wavelengths are therefore extremely important in order to collect further information on the characterization, localization and time evolution of Io's volcanoes.

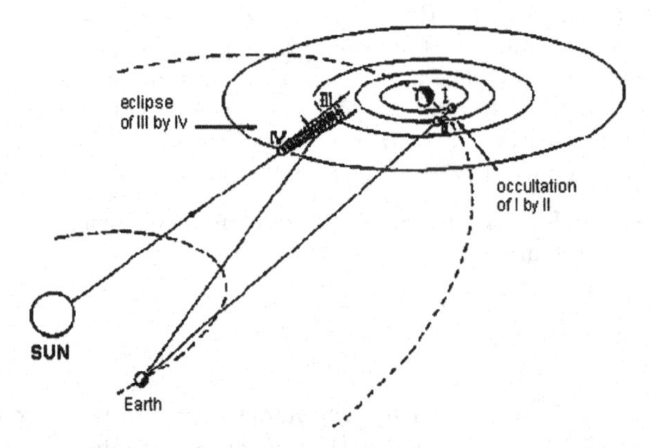

Figure 1. Mutual events (occultation and eclipses) between Jupiter satellites.

2. Observations of PHEMU at radio wavelengths

The total flux density of the eclipsed satellite measured by a ground-based radiotelescope is defined as the integral over all points of its surface, each considered as a source. Taking into account the fraction of brightness due to each point (including hot-spots), the resulting integral is calculated as sum over finite satellite surface elements. By partitioning the surface of the eclipsed satellite into a number of finite elements and assuming that radio flux density is isotropically irradiated by the satellite surface in all directions, the flux density sent by a surface element toward a ground-based observer will then be proportional to the area of this element, its brightness temperature, and the cosine of the angle between the normal to the surface of the current element and the direction of the ground-based observer. For each occultation event, a numerical model was computed and displayed by a customized software which was able to simulate the flux density variation by setting as input parameters the satellites radii, the mean flux densities, the impact parameter and the time duration. Figure 2 shows the simulations performed for the 2OCC1 PHEMU events occurred on June 10th (on the left) and December 18th (on the right) 2009. At the top of the figure the astrometric maps of the PHEMU events are shown. The maps were derived from JPL ephemeris values of the Jovian satellites. Blue and orange circles indicate side and central positions respectively of Io during the occultation. Red circles indicate the Europa position during the maximum of the occultation.

In the middle of the figure the numerical model of the occultation curves computed by the software is shown.

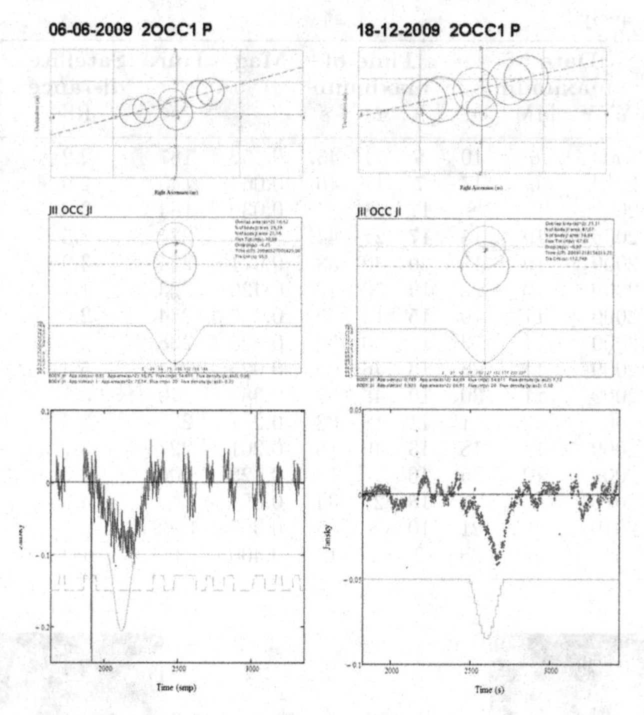

Figure 2. Top: astrometric maps of the 2OCC1 PHEMU events occurred on June 10th (on the left) and December 18th (on the right) 2009. Middle: numerical model of the occultation curves (see text for details).

A summary of the occultation events observed during the PHEMU09 campaigns is presented in Table 1. Observations were performed with the Medicina and Noto 32-m antennas of the Italian Istituto di Radioastronomia and with the 100-m radiotelescope of the Max-Planck-Institute for Radioastronomy, Bonn. Observational frequencies were 43 GHz at Noto, 22 GHz at Medicina and 10 or 32 GHz (according to the weather conditions) at Effelsberg.

At Medicina and Noto data were recorded with the Mark IV backend used for the Very Long Baseline Interferometer (VLBI) observations. At Effelsberg data were taken with the Digital Backend used for continuum observations. Data acquired by the Italian antennas were post-processed with the Advanced Software Tools for Radio Astronomy (ASTRA) (Pluchino 2008). Data collected at Effelsberg were analyzed with the CONT2 subpackage of the Toolbox package for the analysis of single-dish radio observations (von Kap-herr 1977).

Radio occultation measurements require to point the occulted satellite (Io in this case) and to track it while it is occulted by the other one (Europa or Ganymede). Assuming an event-length nodding cycle with a block integration time of 1 s per point, we sampled the occultation curves with a number of ON-source total power measurements equal to the expected duration of the event in seconds. Absolute flux density calibrations were performed by using the radio source J2131-1207 characterized by flux densities of 1.2, 0.55 and 0.49 Jy at 5, 22 and 43 GHz respectively. The averaged difference of the calibrated ON-source and OFF-source gave as a flux density of 0.5 ± 0.1 Jy for Io, 0.6 ± 0.3 for

Table 1. List of observed events. In column one, n. 1 indicates the Io satellite, n. 2 and 3 indicate the Europa and Ganymede satellites, respectively. P and A indicates partial and annular occultation respectively.

Phenomena	Date of maximum			Time of maximum			Mag.	Dur.	Satellite distance	Elevation at		
										Md	Nt	Eff
	YYYY	MM	DD	h	m	s		s	RS	deg	deg	deg
2 OCC 1 P	2009	6	10	5	11	43.	0.152	187	2.2	+29	+35	+24
2 OCC 1 P	2009	6	17	7	17	46.	0.064	145	2.0	+12	+14	+11
3 OCC 1 P	2009	10	8	17	51	6.	0.030	184	3.3	+25	+33	+18
2 OCC 1 A	2009	10	15	17	21	44.	0.426	221	1.5	+24	+33	+18
3 OCC 1 P	2009	10	15	20	30	33.	0.071	254	3.9	+25	+30	+20
2 OCC 1 P	2009	10	22	19	30	19.	0.426	223	1.7	+27	+33	+22
2 OCC 1 A	2009	11	16	15	14	29.	0.426	234	2.6	+25	+33	+18
2 OCC 1 A	2009	11	23	17	30	54.	0.426	238	2.8	+28	+34	+23
3 OCC 1 P	2009	11	28	13	46	42.	0.025	254	5.4	+20	+28	+14
2 OCC 1 P	2009	11	30	19	48	53.	0.388	240	3.1	+12	+14	+10
2 OCC 1 P	2009	12	11	11	18	38.	0.281	237	3.5	+07	+13	+01
2 OCC 1 P	2009	12	18	13	40	16.	0.201	227	3.7	+27	+35	+20
2 OCC 1 P	2009	12	25	16	3	14.	0.123	206	4.0	+28	+35	+24
2 OCC 1 P	2010	1	1	18	27	34.	0.054	169	4.2	+10	+11	+09
2 OCC 1 P	2010	3	21	10	8	35.	0.397	1882	1.4	+38	+46	+32
2 OCC 1 P	2010	3	28	9	1	15.	0.400	1980	1.3	+32	+31	+30

Figure 3. From the left: the 32-m Medicina, 32-m Noto and 100-m Effelsberg radiotelescopes.

Europa and 0.6 ± 0.3 Jy for Ganymede. The distance between the flux calibrator and Io ranged from 2.6 deg to 40 arc min. Different types of OFF-source measurements were also performed during occultations in order to subtract background noise and to monitor the stability of the receivers gain. The OFF-source measurements (obtained with a RA offset of 1 deg) were subtracted from the ON-source to also remove possible contributions of the Jupiter radio emission in the side lobes of the antenna's beam. Since the beam width at 43 GHz is approximately 1 arcmin we can exclude at this frequency the strong contribution of Jupiter inside the antenna's main lobe, whereas any contribution into the secondary side lobes would attenuate the signal by a factor of 100. Due to the proper motion of the satellites, we performed an accurate tracking by constantly pointing the antenna according to the coordinates provided by the JPL Horizon Solar System Dynamics facilities. The 22 GHz receiver at Medicina allowed us to integrate with a bandwidth of 2 GHz, dual polarization, obtaining a sensitivity of approximately 12 mJy with an integration time of 0.68 s achieved with an original sampling interval of 40 ms. Recent observational tests performed with the 43 GHz receiver at Noto on the Jupiter satellites showed the feasibility of the high-frequency front-end for this kind of measurements.

3. Results

We have studied PHEMU of the Galilean satellites at radio wavelengths by using the radio occultation technique. Figure 4 shows two examples of the flux density variations measured during occultations. Signals are total power, already ON-OFF corrected and calibrated. Blue curves indicate the flux density measured at 22 GHz by the Medicina antenna during the 2OCC1 PHEMU events occurred on June 10th (on the left) and December 18th (on the right) 2009. Signal integrations were performed over the entire available bandwidth. Green curves indicate the expected flux density variation according to the numerical model.

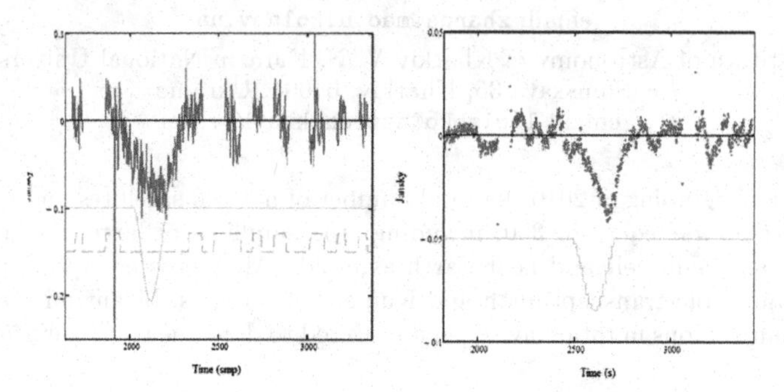

Figure 4. Radio flux density measurements of the 2OCC1 PHEMU events occurred on June 10th (on the left) and December 18th (on the right) 2009. In green: expected flux density variation according to the numerical model. In blue: flux density measured at 22 GHz by the Medicina antenna.

As can be seen in the graph, during the observed occultations, a trough of the flux density was measured in perfect agreement with the expected variation. Besides, the minimum value of the flux density resulted time-shifted by few tens of seconds in accordance with O-C (Observed-Calculated) variations (~1 min) also observed at optical wavelengths. The observational campaign is still ongoing and a numerical model, which is able to simulate the diffusion of the electromagnetic radiation near the satellite limb, is under development in order to better explain structures and possible anomalies in the measured flux density curves.

References

Arlot, J.-E. *et al.* 2006 *A&A*, 451, 733

de Pater, I., Brown, R. A., Dickel, J. R. 1984 *Icarus*, 57, 93

Howell, R. R. 1998 *Proceedings of the 29th Annual Lunar and Planetary Science Conference, Houston, Texas, March 16-20, 1998.*, 57, 93

Lopes-Gautier, R., McEwen, A. S., Smythe, W. D., Geissler, P. E., Kamp, L., Davies, A. G., Spencer, J. R., Keszthelyi, L., Carlson, R., Leader, F. E., Mehlman, R., Soderblom, L. 1999 *Icarus*, 140, 243

McEwen, A. S., Simonelli, D. P., Senske, D. R., Klaasen, K. P., Keszthelyi, L., Johnson, T. V., Geissler, P. E., Carr, M. H., Belton, M. J. S. *Geophysical Research Letters*, 24, 2443

Pluchino, S. 2008 *IRA Internal Report 421/08*

Spencer, J. R., Clark, B. E., Toomey, D., Woodney, L. M., Sinton, W. M. 1994 *Icarus*, 107, 195

von Kap-herr A. 1977 *Technischer Bericht Nr. 40, MPIfR Bonn*

Galileo's Medicean Moons: their impact on 400 years of discovery
Proceedings IAU Symposium No. 269, 2010
C. Barbieri, S. Chakrabarti, M. Coradini & M. Lazzarin, eds.

doi:10.1017/S1743921310007507

How many satellites have been discovered in the Solar System after Galileo?

Zhanna Pozhalova[1] **and Dmitrij Lupishko**[2]

[1]Research Institute "Nikolaev Astronomical Observatory",
Observatornaya 1, Nikolaev, 54030, Ukraine
email: zhanna@mao.nikolaev.ua

[2]Institute of Astronomy of Kharkiv V. N. Karazin National University,
Sumskaya 35, Kharkiv, 61001, Ukraine
email: lupishko@astron.kharkov.ua

Abstract. By the beginning of 2010 the total number of natural satellites and multiple systems in the Solar System was equal to 350, including: 168 satellites of large planets, 119 multiple asteroids (including main-belt and near-Earth asteroids, Mars-crossers and Jupiter Trojan asteroids) and 63 multiple transneptunian and Kuiper-belt objects. Meanwhile, we cannot count precisely how many moons in total have been discovered to date due to the deficiency of accepted definitions.

Keywords. Planets and satellites, binary asteroids, discoveries.

1. Discoveries

The Moon as a natural satellite of the Earth was "discovered" by Nicolaus Copernicus (1473-1543) in his heliocentric system and it was the only known satellite in the Solar System before Galileo Galilei (1564-1642). At the end of 1609 Galileo had made the new adjustment to his instrument and turned his attention to Jupiter. Early in 1610 he discovered four satellites orbiting Jupiter, that are the satellites of other planet than the Earth. They are well known as Galilean moons (or satellites) Io, Europa, Ganymede and Callisto. His book, Sidereus Nuncius, in which his discovery was described, came off the press in Venice in the middle of March 1610 and made Galileo famous.

It was not until 45 years later (13 years after Galileo's death) that the next satellite in the Solar System, namely the biggest Saturn's moon Titan, was discovered by Christian Huygens (1629-1695). He first "published" his discovery as an anagram, sent out on June 13, 1655; later he published in pamphlet form as De Saturni luna Observatio Nova and in full in Systema Saturnium (July 1659). After that the discoveries of satellites happened more frequently and by the end of the 17th century the total number of discovered moons was ten. Giovanni Domenico Cassini (1625-1712) was the first to discover the four of Saturn's moons (Iapetus in 1671; Rhea in 1672; Tethys and Dione in 1684).
From 1684 till 1787 (during more than one century!) no satellites were discovered and the next 18th century added only four satellites (two of Uranus and two of Saturn). They were discovered by William Herschel (1738-1822), the most famous astronomer of the 18th century, who constructed more than four hundred telescopes and using his own telescopes, discovered Titania and Oberon in 1787 as well as Enceladus and Mimas in 1789. Herschel also discovered the 7th planet Uranus in 1781. It should be stressed that during that century there were no more discoveries of either planets or satellites. Thus, we can conclude that while Galileo Galilei was the greatest discoverer of the 17th century, W. Herschel was the greatest discoverer of the 18th century.

Table 1. Timeline of planet satellite's discoveries.

Planet	Number of moons	17th century	18th century	19th century	20th century	21th century	By space mission
Earth	1						
Mars	2			2			
Jupiter	63	4		1	13	45	3
Saturn	62	5	2	2	21	32	12
Uranus	27		2	2	17	6	11
Neptune	13			1	7	5	6
Total	168	9	4	8	58	88	32

Table 2. Sizes and masses of the biggest satellites.

	Ganymede	Titan	Callisto	Io	Moon	Europa	Triton
R (km)	2631	2575	2403	1815	1737,4	1569	1350
Mass (kg)	$1.48 \cdot 10^{23}$	$1.35 \cdot 10^{23}$	$1.08 \cdot 10^{23}$	$8.94 \cdot 10^{22}$	$7.35 \cdot 10^{22}$	$4.8 \cdot 10^{22}$	$2.14 \cdot 10^{22}$

In the 19th century the 8th planet Neptune and its moon Triton, two Uranus' and two Saturn's moons, one of Jupiter's and both of Mars' moons (Phobos and Deimos) were discovered. The available theoretical considerations guided Asaph Hall (1829-1907) to conclusion that any Martian satellite must revolve very close to the planet. He undertook systematic search for possible Mars' satellites and found them successfully in 1877.

The next 20th century and the beginning of 21st one turned out to be much more abundant in satellite discoveries due to both ground-based observations and space missions (Voyager-1, Voyager-2, Cassini-Huygens, Galileo). The Voyagers, which were launched in 1977 and flew by the Jupiter, Saturn, Uranus and Neptune, found 27 new satellites: 3 of Jupiter, 7 of Saturn, 11 of Uranus and 6 of Neptune. Six new moons of Saturn were discovered by Cassini-Huygens mission in the last five years. In 1993 the first asteroid satellite (Dactyl) orbiting the main-belt asteroid 243 Ida was discovered by the space mission Galileo. It was the beginning of abundant discoveries of asteroid satellites by photometric and radar observations and then of satellites of Kuiper-belt and transneptunian objects. The first minor planet with two satellites, 87 Sylvia, was identified by adaptive optics telescope direct imaging in 2005. Pluto's first satellite Charon was discovered in 1978. Other binary/multiple TNOs have been discovered since 2000. Since 15 May 2005 Pluto is known to be the only quadruple system.

2. Modern world of the satellites and binaries

Satellites of the planets. Six planets of the Solar System have satellites, total amount of which up-to-date is 168. Timeline of their discoveries is presented in Table 1. It is interesting to note, that the amount of satellites increased more than 2.5 times during the 20th century and was increased again 1.5 times during the first ten years of the 21st century. Meanwhile, the space mission discoveries contribute about 20 percents. Table 2 contains the principal parameters (radius and mass) of seven biggest satellites in the Solar System.

Multiple systems among asteroids and transneptunian objects. By the end of February 2010 there were discovered 182 objects with companions: 172 binaries, 9 triple systems, and 1 quadruple system (Pluto). They count 193 companions total (see website "Asteroids with satellites" by Wm. R. Johnston www.johnstonsarchive.net/astro/asteroidmoons.html).

These systems include the following:

- 37 near-Earth asteroids (2 NEAs with two satellites each);
- 9 Mars crossing asteroids (MCAs);
- 69 main-belt asteroids (5 MBAs with two satellites each);
- 4 Jupiter Trojan asteroids (JTAs);
- 63 transneptunian objects (2 of them 136108 Haumea and 47171 1999 TC36 with two and 134340 Pluto with three satellites - Charon, Nix and Hydra). Among these TNOs 3 objects (134340 Pluto, 136108 Haumea and 136199 Eris) now are recognized as dwarf planets under the current IAU nomenclature.

The near-Earth binary asteroids consist of 6 Atens, 21 Apollos, and 8 Amors, plus the triple Amor asteroid (153591) 2001 SN263 and the triple Apollo asteroid (136617) 1994 CC. A fraction of binary systems among the NEOs is estimated to be 15-17% (Merline *et al.* 2002).

The first confirmed binary Trojan asteroid 617 Patroclus has very similar in size components. The 2nd identified binary Trojan asteroid 624 Hektor has a primary which is a contact binary accompanied by a smaller secondary.

The first companion discovered among more than 1300 known TNOs and Centaurs was that of 1998 WW31, found in December 2000. Now the total number of multiple systems constitutes 60 binary TNOs, 2 triple TNO system, and the quadruple system of Pluto. Thus, as well as in case of asteroids, the multiple systems among TNOs seem to be relatively common.

One might note that only one satellite Dactyl was discovered due to the direct imaging by NASA space probe Galileo. The contribution of different methods to discoveries of companions is as following:

- photometric lightcurve method 41% (48 MBAs, 14 NEAs, 9 MCAs, 2 JTAs and 1 TNO);
- HST-imaging 30% (3 MBAs and 51 TNOs);
- ground-based imaging 16.5% (17 MBAs, 2 JTA and 11 TNOs);
- radar observations 12.5% (only 23 NEAs).

3. Discussion and Conclusion

In spite of the fact that all discoveries of satellites and binary systems among asteroids and transneptunian bodies are registered and cataloged, we cannot precisely count how many moons in total are discovered to date. The reason is that there is no established definition of what should be considered as a moon and it concerns first of all asteroid and transneptunian body populations. Indeed, there is no rule to discriminate between a double asteroid and asteroid with a satellite, both of which are binary systems. What binary system do we have to consider with certainty as a primary body and its satellite? What should be the upper limit of their mass ratio or at what minimum distance from primary the barycentre of binary system should be located? Some people assume that it should be located below the surface of the larger body, though this postulate is somewhat arbitrary and not generally accepted. Due to this uncertainty, for example, some authors consider the Pluto-Charon system to be a double (dwarf) planet.

The analysis of binary system parameters (which in general are not determined accurately enough) shows that the fraction of systems with nearly the same in sizes components ($R_{prm}/R_{scd} \leqslant 2$) is ~14% among the NEAs, ~20% among the MBAs, ~60%

among the JTAs and ∼90% among the TNOs. It means that there is a strong observation selection effect in discovery of satellites: due to a weakness of asteroid's and especially of TNO's brightness it is much more complicated to detect a small companion near a bright primary than to detect two nearly equal in brightness components.

In any case, the population of discovered natural satellites, orbiting the large planets, numbered in 168. Besides, roughly about 25% of all discovered binary asteroids and transneptunian bodies can be considered as having moons. Thus, one can conclude that about two hundred satellites are discovered in the Solar System to date.

They have a surprising variety of their orbits and physical properties. Nevertheless, it is impossible to overestimate the discovery of the first four Jupiter's moons Io, Europa, Ganymede and Callisto made by Galileo Galilei 400 years ago. Galileo's discovery of Jupiter moons had a major impact on cosmology of that time. In 1610 the traditional Aristotelian cosmology recognized only one center of motion, the center of the universe which was the place of the Earth. But according to the Copernican theory, the Earth went around the Sun while the Moon went around the Earth. Galileo discovered one more center of motion Jupiter, a new system of planetary bodies in miniature, and it was a strong argument in support of the new Copernican System.

Nowadays the discoveries of binary asteroids and TNOs objects give a good opportunity to determine their masses and bulk densities, and hence a type of their material and internal structure. However those estimates usually are still not accurate enough due to uncertainty of binary system parameters.

Acknowledgements

The authors are grateful to SOC and LOC of the IAU Symposium 269 for the IAU grants, which gave them the opportunity to participate in this unforgettable meeting in the Galileo's homeland.

References

Merline, W. J., Weidenschilling, S. J., Durda, D. D., Margot, J-L., Pravec, P., & Storrs, A. D. 2002, *Asteroids III. (W. F. Bottke et al., eds.)*, Univ. of Arizona Press, Tucson (USA), 289

Galileo's Medicean Moons: their impact on 400 years of discovery
Proceedings IAU Symposium No. 269, 2010
C. Barbieri, S. Chakrabarti, M. Coradini & M. Lazzarin, eds.
doi:10.1017/S1743921310007519

A virtual tour of the Galilean Satellites

Paul Schenk
Lunar and Planetary Institute, Houston
USA
email: schenk@lpi.usra.edu

1. Introduction: the four galilean satellites

Galileo's imagination was quick to comprehend the importance of the 4 starry objects he observed near Jupiter in January 1610, not only for himself as a scientist but for our common understanding of the place of the Earth and our species in the cosmos. Even he, however, could not have imagined what those four objects would actually look like once humans got their first good look. Some 369 years the fast traveling Voyager 1 and 2 spacecraft provided that first good look during 1979, followed by an even closer look from the Galileo Orbiter beginning in 1996 through 2001. The following mosaics represent some of the best of those views. They include views of impact craters young and ancient, icy terrains that have been intensely faulted, eroded or disrupted, mountains towering 10 or more kilometers high, and volcanic eruptions hotter than those on Earth. Each of the four Galilean satellites is geologically distinct, betraying very diverse global histories and evolutions. Images and other observations of these 4 objects revealed the importance of tidal heating and subsurface water oceans in planetary evolution, but mapping is very incomplete. New missions to explore these planetary bodies are being planned and the images and observations of the missions that went before will lay the groundwork for these new explorations as we begin the 5th Galilean century.

2. The four galilean satellites

This montage (Figure 1) shows each of the four Galilean satellites at different scales. From left to right the satellites are: Io, Europa, Ganymede, and Callisto. The top two rows show global views, one with cut-away views showing our best concepts of what the interiors likely look like. Although each has a rocky deep core, ice-rich crusts and mantles of differing depths dominate the outer three satellites. The bottom views show how the surface changes as we zoom in on selected features at progressively higher resolution. Each row down represents an increase in resolution of roughly a factor of ten: the first two rows are at 10 kilometers, increasing to ~1 kilometer, ~100 meters, and finally ~10 meters (bottom). What stands out is the diversity of features at different scales and on the different satellites.

3. Callisto

CALLISTO GLOBAL (LEADING HEMISPHERE)
Most of the large multiring impact basins that characterize Callisto formed on this hemisphere (Figure 2). Linear crater chains were formed by disrupted comets, which strike Callisto while outbound after a close passage to Jupiter. Also visible are innumerable bright spots, the marks of countless impact craters on this ancient surface. Orthographic Projection: Center 0^oN, 90^oW.

Figure 1.

CALLISTO: ASGARD MULTIRING BASIN

This spectacular radial swath (Figure 3) shows the transition from Asgard center (including floor and rim deposits) to ancient cratered plains. The prominent 115-kilometer-wide penedome impact crater Doh obscures the center of Asgard at the top of the mosaic. A large intensely fractured central dome 25-kilometers-wide dominates Doh. 1500-meters-high arcuate ridges in the upper half of the mosaic are inner rings of the Asgard structure. Darker areas to the south are heavily cratered plains, crossed by two of the basin's outer gràben rings. Small landslides can be also seen within some of these craters. This terrain is generally degraded and partly eroded. The mosaic is ~800 kilometers long.
ENCOUNTER: Galileo C10 RESOLUTION: 90 meters/pixel.

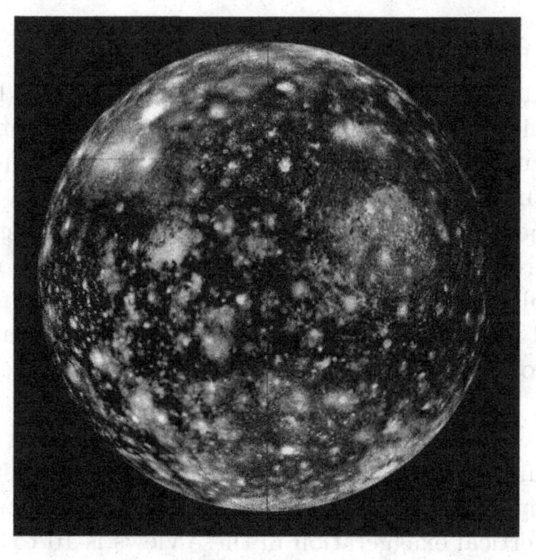

Figure 2.

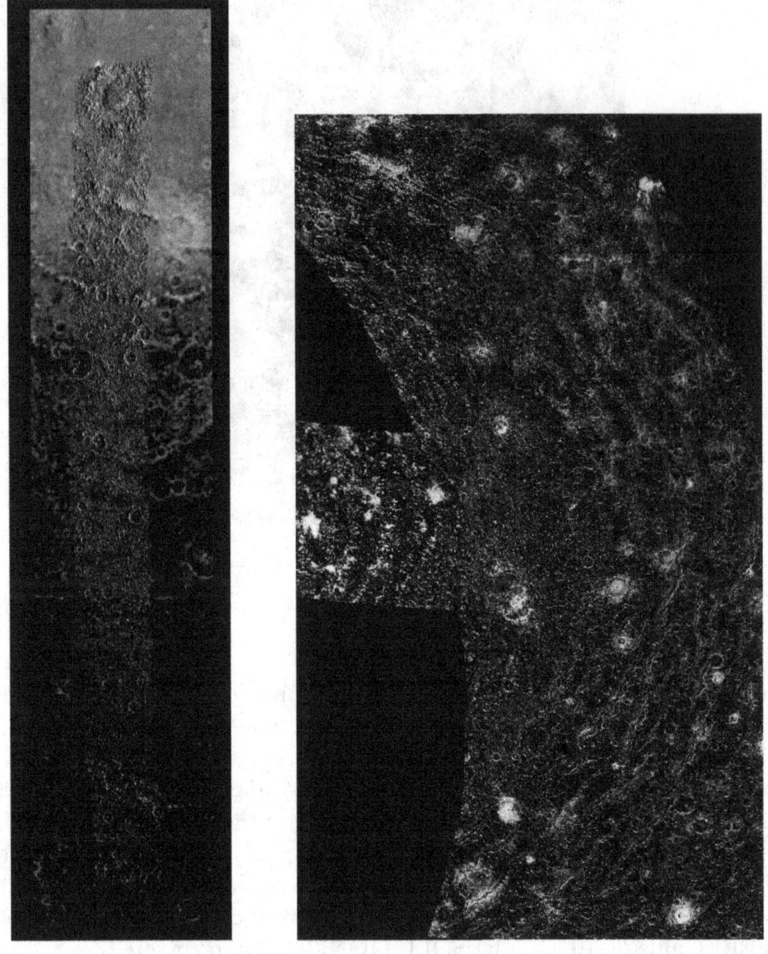

Figure 3. Figure 4.

CALLISTO: VALHALLA MULTIRING BASIN

Galileo acquired this multiframe mosaic (Figure 4) of the eastern half of Valhalla to further investigate how this global-scale event formed. The central zone at left is a relatively bright but rugged circular patch, probably formed by refrozen impact deposits on the floor of the impact. Unlike lunar basins there is no deep central depression here. Outward up to 2000 kilometers from the center extends a zone numerous concentric ridges, many of which retain relief of 2 to 3 kilometers, transitioning to graben and scarps superposed on cratered plains. Although the original crater rim can no longer be identified, the size of the original impact crater is estimated to be approximately 1000 kilometers.
ENCOUNTER: Galileo C9 RESOLUTION: 410 meters/pixel

CALLISTO: PERSPECTIVE VIEWS

These four views (Figure 5) show the relief of Callisto from different perspectives. These renderings are made digitally by combining the original image mosaics with topographic maps of the surface. Vertical exaggeration in these views is 10 to 30 times normal.

Figure 5.

Upper left: Cratered Plains - Very High Resolution.
This is one of the highest resolution views we have of Callisto. Most of the icy surface is gone, replaced by a dark non-icy layer that is now heavily cratered and has a rugged topography. The larger older craters are partly eroded, leaving sharp-peaked knobs several hundred meters high.
ENCOUNTER: Galileo C21 RESOLUTION: 15 meters/pixel.

Upper right: Valhalla - Outer Ring Scarp.
Our best observation of one of the Valhalla basins many rings shows a prominent fault scarp over 1700 meters high. Several minor fault scarps and aligned knobs are also visible on the floor of the graben itself (although we do not see the opposing wall scarp), indicating that these terrains also experienced faulting.
ENCOUNTER: Galileo C3 RESOLUTION: 46 meters/pixel

Lower right: Hár and Tindr.
This combined pair of images compares craters of different ages. Tindr, the younger large central pit crater to the right is 70 kilometers wide and ~1 kilometer deep. Radial textures outside the rim are due to ejecta blasted out during impact that scoured nearby surfaces. The ancient impact scar Hár is the concentric set of degraded topographic features. Surrounding the 25-kilometer-wide smooth rounded central dome is a 15-kilometer-wide ring depression and an annular plateau. The dome and ring depression are part of the central uplift complex and the annular plateau is the remains of the crater floor impact deposit.
Tindr:
ENCOUNTER: Galileo C10 RESOLUTION: 400 meters/pixel
Hár:
ENCOUNTER: Galileo C9 RESOLUTION: 145 meters/pixel

Lower left: Asgard graben.
This view of the outermost ring graben at the Asgard impact basin shows the erosion that has partly destroyed older topography, including the graben walls that curve across the center of the view, as well as older crater rims. Total relief across the graben is roughly 1.5 kilometers.
ENCOUNTER: Galileo C10 RESOLUTION: 90 meters/pixel

Figure 6.

4. Ganymede

GANYMEDE GLOBAL (GALILEO REGIO, POLAR CAP)
The huge dark feature in this view (Figure 6), Galileo Regio, is the largest contiguous geologic feature on Ganymede. Astronomers may have seen Galileo Regio and Pioneer saw but did not resolve it. Galileo Regio is a large block of older dark terrain, predating the bands of icy volcanic and tectonic bright terrain surrounding it. Remnants of dark terrain now cover only 33% of Ganymede. Arcuate furrows formed during a very large ancient impact event. The bright polar frost cap is believed due to sputtering and redistribution of water ice controlled by Ganymede's magnetic field.
Orthographic Projection: Center 42°N, 130°W.

GANYMEDE: NIPPUR SULCUS This three-frame Galileo mosaic (Figure 7) shows several crosscutting bands of bright terrain that have been deformed to different degrees. The smooth narrow diagonal band Nippur Sulcus was the last formed and least deformed. Smoothness is consistent with volcanic resurfacing by water lavas. To the north lies a wide zone of intensely grooved and fractured bright terrain. Dark terrain to the south is itself heavily deformed by numerous narrow grooves and fissures. Impact craters have been relaxed or distorted, indicative of higher heat flow in the distant past.
ENCOUNTER: Galileo G2 RESOLUTION: 100 meters/pixel

GANYMEDE: NICHOLSON REGIO AND ARBELA SULCUS
This 9-frame mosaic (Figure 8) illustrates the complex geologic history of these ancient terrains. Belts of narrow fractures can be seen, especially towards the south and east. Most (though not all) large craters are degraded, either by fracturing and disruption, landform degradation, viscous relaxation, or a combination of these. Relatively bright circular patches are probably ancient impact scars. Two bands of bright terrain cross Nicholson Regio here, including the smooth bright band Arbela Sulcus across scene center. Most lanes of bright terrain formed by flooding by icy lavas and tectonic fracturing. The scene is ~600 kilometers across.
ENCOUNTER: Galileo G7 RESOLUTION: 180 meters/pixel
ENCOUNTER: Galileo G28 RESOLUTION: 132 meters/pixel

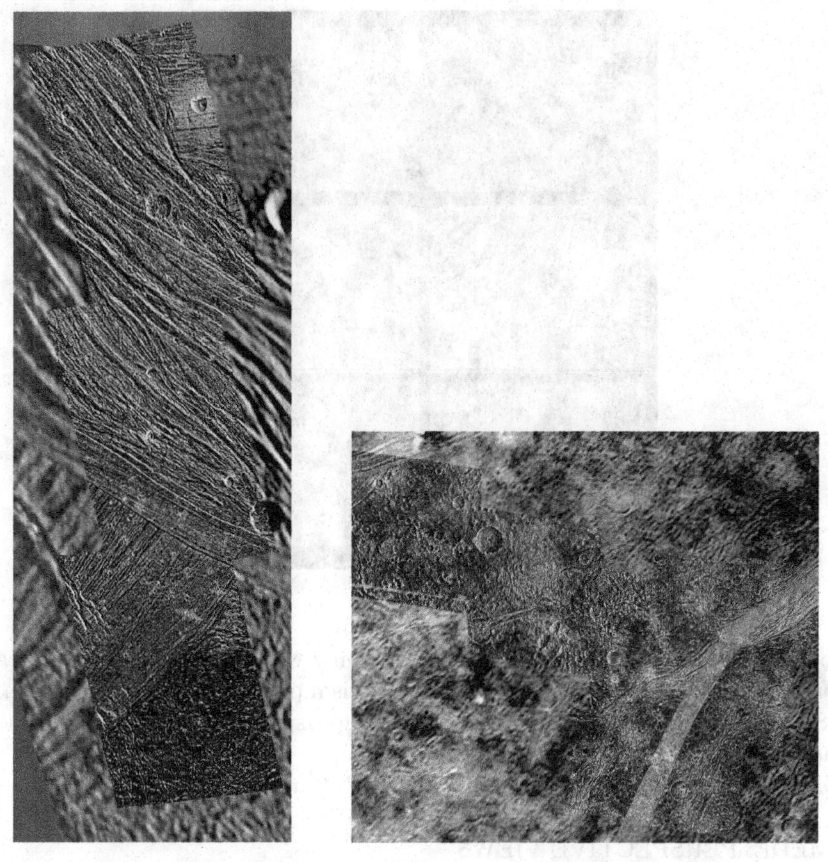

Figure 7. **Figure 8.**

GANYMEDE: CALDERAS This six-frame mosaic of bright terrain features 6 irregularly shaped rimmed depressions interpreted to be volcanic calderas (Figure 9). Most are cut by the narrow band of smooth bright terrain crossing the scene east to west. The largest caldera has a rugged surface characteristic of thick viscous lavas. The smooth

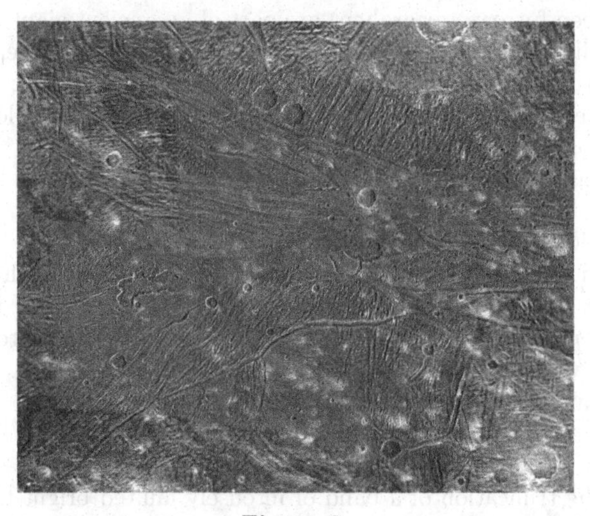

Figure 9.

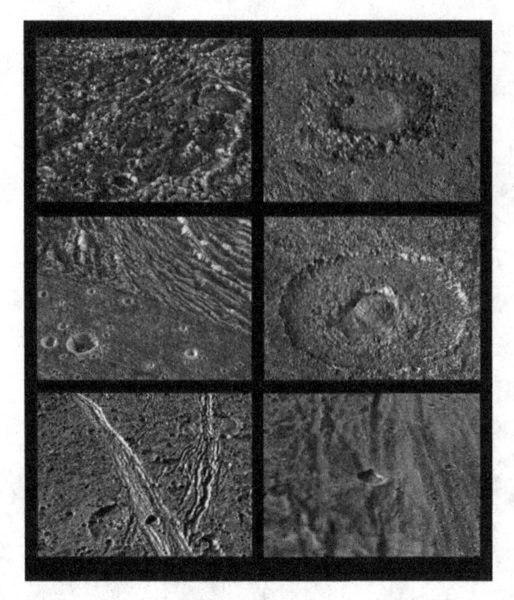

Figure 10.

surface of the band, in contrast, is indicative of runny watery lavas. At least 40 calderas have been identified globally, indicating that volcanism played an important role in bright terrain. Secondary craters from Osiris form the bright splotches. The scene is ~600 kilometers across.
ENCOUNTER: Galileo G8 RESOLUTION: 180 meters/pixel

GANYMEDE: PERSPECTIVE VIEWS
These six views show the relief of Ganymede from different perspectives (Figure 10). These renderings are made digitally by combining the original image mosaics with topographic maps of the surface. Vertical exaggeration in these views is 10 to 30 times normal.

Upper left: Dark terrain - Galileo Regio.
This observation of dark terrain is centered near the intersection of several furrows. The 1-to-2 kilometer deep furrows (or graben) are part of an ancient giant impact. Most features were originally deeper and have been modified by viscous creep of ice at higher temperatures than are prevalent now. Bright and dark materials are also partly segregated by downslope creep into bright slopes and peaks and darker valleys and low areas.
ENCOUNTER: Galileo G1 RESOLUTION: 77 meters/pixel
ENCOUNTER: Galileo G2 RESOLUTION: 90 meters/pixel

Upper right: Penedome crater Neith.
A 1-kilometer-high, 45-kilometer-wide central dome dominates Neith and surrounding ring of massifs but the crater rim normally present is mostly absent. Penedome craters like Neith no longer form today. The unusual morphology is related to higher heat flow in the ancient past of Ganymede and Callisto.
ENCOUNTER: Galileo G7 RESOLUTION: 140 meters/pixel

Center left: Bright terrain - Arbela Sulcus.
This view shows the truncation of a band of intensely faulted bright terrain (at top) by the smooth band of Arbela Sulcus in the foreground. Ridges are typically 200 meters

high, the tallest almost 500 meters. The largest crater is ~500 meters deep and 2.5 kilometers across.
ENCOUNTER: Galileo G7 RESOLUTION: 180 meters/pixel
ENCOUNTER: Galileo G28 RESOLUTION: 132 meters/pixel

Center right: Central dome crater - Melkart.
Melkart, a 100-kilometer-wide central dome crater, is relatively young and well preserved. The central dome is ~30 kilometers wide and ~1 kilometer high. Domes like this, seen only on Ganymede and Callisto, may be material uplifted from several kilometers deep during impact, exposing Ganymede's interior.
ENCOUNTER: Galileo G8 RESOLUTION: 180 meters/pixel

Lower left: Dark Terrain and Anshar Sulcus- Marius Regio.
Several generations of fracture sets cut across Marius Regio in this area, with the most prominent of these leading to the formation of the narrow lane of grooved or fractured bright terrain. Dark terrain features several craters 20 to 30 kilometers across that have been fractured and topographically relaxed.
ENCOUNTER: Galileo G8 RESOLUTION: 145 meters/pixel

Lower right: Bright and dark Terrain - Harpagia Sulcus.
This view shows the boundary between dark terrain (left) and bright terrain (right). Here the boundary has been faulted in several places as the icy crust has been stretched, forming relatively steep cliff walls several hundred meters high within both dark terrain and younger bright terrain.
ENCOUNTER: Galileo G28 RESOLUTION: 20 meters/pixel

5. Europa

EUROPA GLOBAL (ICE RIFTING) This global view of Europa (Figure 11) highlights a vast zone of dark dilational bands known as Argadnel Regio forming the largest, most coherent zone of rifting evident on Europa today. These short stubby dark bands

Figure 11.

(formerly known as wedge-shaped bands) are typically 25 kilometers across and up to 300 kilometers long and formed when the icy shell cracked and was pulled apart in a north-south direction. The bright band Agenor Linea at south center runs along the southern border of this rift zone. Long dark arcuate bands (formerly known as triple bands) dominate the regions north of the dilational bands. Many of these features may have formed together, possibly during an episode of polar wander on Europa.
Orthographic Projection: Center 25°S, 215°W.

EUROPA - POLE-TO-POLE
This pole-to-pole mosaic (Figure 12), was obtained during four orbits: E11 to the west, E15 to the north, E19 to the far north, and E17 to the south. Here we see a transition from ridged plains up north featuring long narrow bands and small dark spots (probably the surface expressions of diapirs) to disrupted terrains near the equator, as well as a number of narrow dark dilational bands. Ridged plains lie to the south, as do the large dark spots Thrace and Thera Macula.
ENCOUNTER: Galileo E11 RESOLUTION: 220 meters/pixel
ENCOUNTER: Galileo E15 RESOLUTION: 230 meters/pixel
ENCOUNTER: Galileo E17 RESOLUTION: 225 meters/pixel
ENCOUNTER: Galileo E19 RESOLUTION: 200 meters/pixel

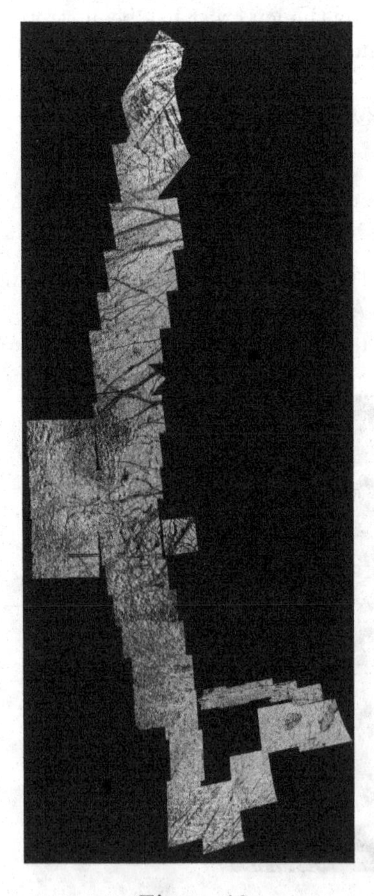

Figure 12.

Figure 14.

Figure 13.

EUROPA: CONAMARA CHAOS

Galileo acquired a series of progressively higher resolution mosaics of Conamara Chaos (Figure 13), the disrupted zone nestled under the crossing of the two large dark bands crossing the top scene. Conamara Chaos itself consists of numerous broken blocks of ridged plains intermingled with intensely disrupted matrix material. Erosional debris has accumulated at the base of the sheer cliffs on the edges of larger intact blocks, which stand up to 100 meters high. Small secondary craters can be resolved within the bright splotches, parts of a bright ray from the recent crater Pwyll. Chaos such as Conamara formed either by melt-through of the icy shell or more likely from diapiric blobs of warm ice that penetrated to and disrupted the surface.
TOP:
ENCOUNTER: Galileo E6 RESOLUTION: 175 meters/pixel
MIDDLE:
ENCOUNTER: Galileo E6 RESOLUTION: 55 meters/pixel
BOTTOM:
ENCOUNTER: Galileo E12 RESOLUTION: 9 meters/pixel

EUROPA: RIDGED PLAINS

The prominent double ridge, Androgeos Linea, rises 350 meters high (Figure 14). It also sits in a shallow 100-meter deep depression flanked by small cracks, formed by the weight of the ridge itself. Shear heating along faults is currently thought to be important, but the origin of the topographic ridge is unclear. Strike-slip faulting is also common on Europa, including this region, breaking the ice shell into numerous plates. Tidal deformation and internal convection move these plates around.
ENCOUNTER: Galileo E6 RESOLUTION: 20 meters/pixel

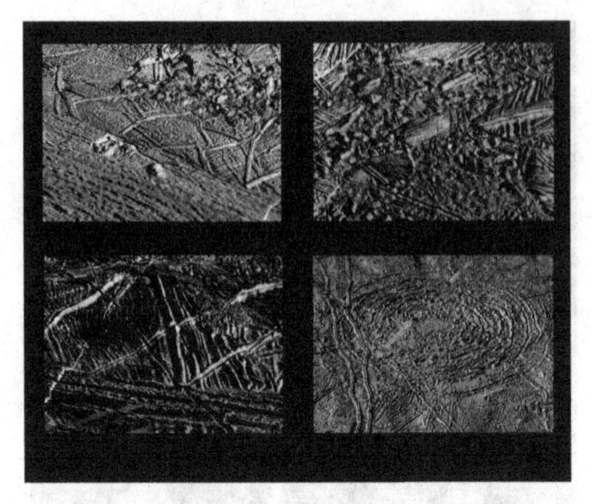

Figure 15.

EUROPA: PERSPECTIVE VIEWS
These four views show the relief of Europa in different perspectives (Figure 15). These renderings are made digitally by combining the original image mosaics with topographic maps of the surface. Vertical exaggeration in these views is 10 to 30 times normal.

Upper left: Agenor Linea.
Agenor Linea is over 1000 km long and has two major components: a darker southern lane and a brighter northern component. Deformation is complex, involving both compression and lateral shear, not unlike some strike-slip faults on Earth. Several small walled depressions and chaos units have subsequently altered the surface. Relief across this 40-kilometer-wide band seems to be limited to a few hundred meters.
ENCOUNTER: Galileo E17 RESOLUTION: 42 meters/pixel

Upper right: Conamara Chaos.
This perspective view shows several of the tilted fault blocks remaining within this large patch of chaos. These blocks can be up to 100 meters high. The rugged matrix material between the blocks is probably highly disrupted diapiric material and has relief of tens of meters.
ENCOUNTER: Galileo E6 RESOLUTION: 55 meters/pixel

Lower left: Ridged plains.
This mosaic show the complexity of ridged plains in some areas. Double ridges and broader ridge complexes of varying complexity are present. Some of the bands formed as new material intruded from below when the ice shell pulled apart. Several patches of ridged plains are relatively smooth and may have escaped deformation. The small smooth feature a few kilometers across appears to be liquid water that ponded here and has now frozen over.
ENCOUNTER: Galileo E4 RESOLUTION: 25 meters/pixel

Lower right: Tyre multiring impact basin This multi-ring impact structure stretches ~150 kilometers across and resembles much larger ancient impact structures on Ganymede and Callisto, such as Asgard. Tyre features concentric inner ridges and outer graben fractures, surrounded by hundreds of small secondary craters. The fractures form only if the icy lithosphere is very thin compared to the size of the impact cavity.
ENCOUNTER: Galileo E14 RESOLUTION: 170 meters/pixel

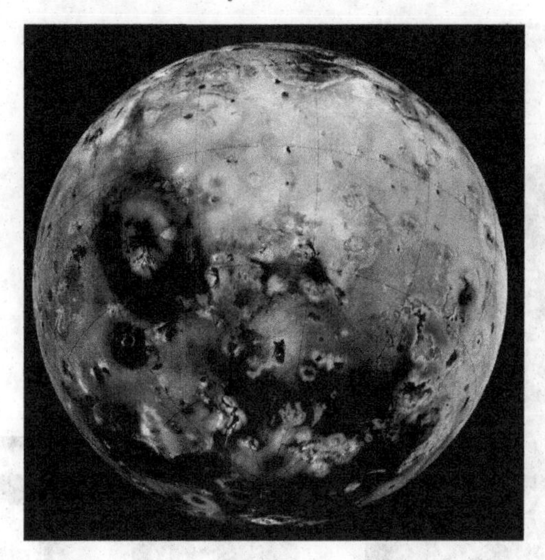

Figure 16.

6. Io

IO GLOBAL (SOUTHERN HEMISPHERE)

The gigantic reddish Pele and dark Babbar Patera eruption sites dominate this hemisphere (Figure 16). These huge deposits form as volcanic plumes of gas and dust fall back to the surface. Smaller reddish fan-shaped deposits betray the locations of recent or ongoing eruptions at Marduk and Culann Paterae. The different colors likely represent differences in sulfur composition. The dark deposits at Babbar probably include magnesium-rich orthopyroxenes. Reds in the Pele plume are the signature of short-chain sulfur molecules, but at polar latitudes may be due to radiation damage to short-chain sulfur molecules. Whitish materials near the South Pole are probably sulfur dioxide frosts outgassed from mountain slopes or at scarp failures. Orthographic Projection: Center 35^oS, 240^oW

IO - NORTH-TO-SOUTH

The southern of these two mosaics (Figure 17) includes rugged Tohil Mons, reddish Culann Patera, the large deep caldera Michabo Patera, a bright and dark radial flow complex and small shield volcano, and a variety of pits, lineations and scarps within the intervening volcanic plains. Michabo Patera is 1.4 km deep and surrounded by a low mesa ~100 meters thick. Tohil Mons towers 8 kilometers high, but no relief is evident at Culann Patera. Two 2-kilometer-high shield volcanoes, including Zamama, can be seen in the northern mosaic, as well as the newly formed and very hot Thor volcanic flow complex (which was producing a 500-kilometer high plume) at the northeastern end of the mosaic.

ENCOUNTER: Galileo I32 RESOLUTION: 340 meters/pixel

IO: PROMETHEUS

First observed by Voyager in 1979 as a ~100-kilometer high plume of gas and dust (Figure 18, upper left), Prometheus is one of the most consistently active volcanoes on Io, being observed in all subsequent Galileo encounters (top). The large dark lobate feature seen at higher resolution is a complex flow field composed of hundreds of overlapping flows. The bright plume itself (composed of SO_2) forms when hot dark flows move across the surface, volatilizing the sulfur dioxide into a gas, which jets into space and back onto the surface.

Figure 17.

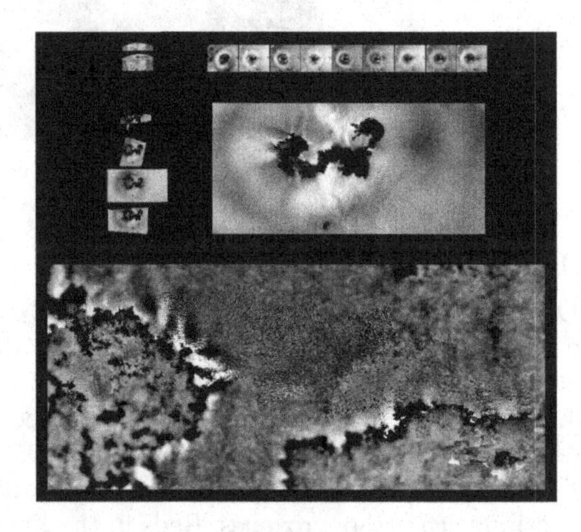

Figure 18.

ENCOUNTER: Galileo I27 RESOLUTION: 170 meters/pixel
ENCOUNTER: Galileo I27 RESOLUTION: 12 meters/pixel

IO: TOHIL MONS

Towering 9.5 kilometers above the plains, Tohil Mons is comprised of several distinct parts (Figure 19). A rectangular eastern plateau rises gently toward the west, culminating in the craggy peaks seen here. The lobate outlines of this plateau suggest it has undergone

Figure 19.

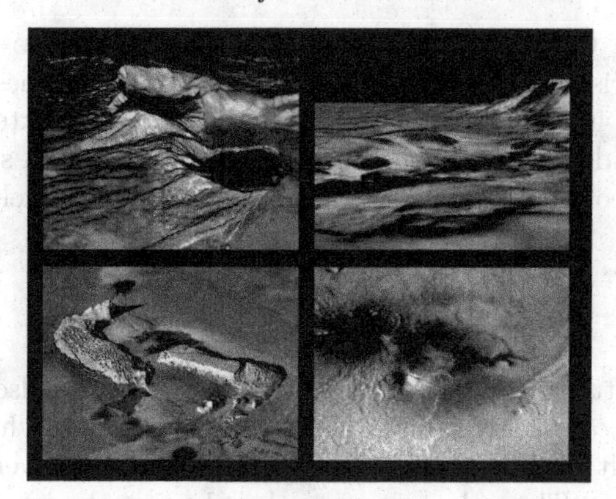

Figure 20.

partial collapse. An intensely fractured northern plateau to the north rises ~3.5 kilometers. Tohil Patera lies immediately north of Tohil Mons and includes dark and bright deposits of various colors. The smaller, dark-floored Radagast Patera, a hotspot, lies between the northern mountain and the southeastern plateau. Along the crest of Tohil Mons is a large quasi-circular amphitheatre, mostly in shadow, whose floor sits roughly 2 kilometers above the plains.
ENCOUNTER: Galileo I32 RESOLUTION: 52 meters/pixel

IO: PERSPECTIVE VIEWS
These four views show the relief of Io from different perspectives (Figure 20). These renderings are made digitally by combining the original image mosaics with topographic maps of the surface. Vertical exaggeration in these views is 10 to 30 times normal.

Upper left: Tohil Mons.
Tohil Mons is a complex mountain structure rising 8 to 9 kilometers high. This view shows the prominent axial peaks and the deep central amphitheatre. A scarp surrounds dark-floored Radagast Patera caldera in foreground 300-500 meters high. Striations on the flanks are of unknown origin.
ENCOUNTER: Galileo I32 RESOLUTION: 52 meters/pixel

Upper right: Telegonus Mensae.
This tabular mesa-like mountain stands 1.5 kilometers high. Several landslides and slumps have eroded into the edge of the cliff in the distance. Several low-lying volcanic fissures and flows formed in the foreground.
ENCOUNTER: Galileo I32 RESOLUTION: 10 meters/pixel

Lower left: Hi'iaka Montes.
Hi'iaka Montes is split into two parts. The southeastern section is a flat-topped mesa 2.5 kilometers high. Small mounds of debris can be seen along the edge, indicating the edge is slowly crumbling. The northwestern section is a complex structure featuring a ridged plateau 3 to 5 kilometers high, with a prominent set of peaks at the end rising 9 kilometers. Most mountains on Io are thrust blocks forced upward.
ENCOUNTER: Galileo I25 RESOLUTION: 270 meters/pixel

Lower right: Zamama Of Io's 500 or more volcanoes, only these two (and one other) exhibit relief characteristic of classic shield volcanoes on Earth. These two are ~2 kilometers high and roughly 40 kilometers wide. The active dark flow extending from Zamama is likely mafic lava, and produces a plume sometimes visible from spacecraft.
ENCOUNTER: Galileo I32 RESOLUTION: 340 meters/pixel

7. The movies

A series of videos (http://www.youtube/com/galsat400) have also been produced from the *Voyager* and *Galileo* spacecraft data. These videos simulate what we might see if we had a mobile spacecraft capable of maneuvering at low altitudes over the surfaces of these icy and volcanic moons. The videos were produced using the image mosaics combined with topographic data produced by P. Schenk from stereo and low-Sun images, allowing us to view the relief of these moons from any perspective.

8. The Atlas

All images were produced by Paul Schenk, Lunar and Planetary Institute, Houston, based on Voyager and Galileo imaging data. Map images are excerpted from the *Atlas of the Galilean Satellites* (P. Schenk, Cambridge Univ. Press, Cambridge, 2010). The *Atlas* includes global maps of all four moons at 1-kilometer resolutions and all high-resolution mosaics from *Voyager* and *Galileo* spacecraft. A description of the satellites, their geologic and observational histories and the Atlas format, as well as detailed Appendices are also featured.

Author Index

Subject Index